Advances in
MICROBIAL ECOLOGY

Volume 2

ADVANCES IN MICROBIAL ECOLOGY

Sponsored by International Commission on Microbial Ecology,
a unit of International Association of Microbiological Societies
and the Division of Environmental Biology of the
International Union of Biological Societies

A Continuation Order Plan is available for this series. A continuation order will bring
delivery of each new volume immediately upon publication. Volumes are billed only upon
actual shipment. For further information please contact the publisher.

Advances in
MICROBIAL ECOLOGY

Volume 2

Edited by
M. Alexander
Cornell University
Ithaca, New York

PLENUM PRESS · NEW YORK AND LONDON

The Library of Congress cataloged the first volume of this title as follows:

Advances in microbial ecology. v. 1–
  New York, Plenum Press c1977–
    v. ill. 24 cm.
  Key title: Advances in microbial ecology, ISSN 0147-4863

  1. Microbial ecology—Collected works.
QR100.A36   576′.15   77-649698

Library of Congress Catalog Card Number 77-649698
ISBN 0-306-38162-1

© 1978 Plenum Press, New York
A Division of Plenum Publishing Corporation
227 West 17th Street, New York, N.Y. 10011

Printed in the United States of America

Contributors

L. W. Belser, Department of Soil Science, University of Minnesota, St. Paul, Minnesota, U.S.A.

John M. Bremner, Department of Agronomy, Iowa State University, Ames, Iowa, U.S.A.

Y. R. Dommergues, Centre Nationale de la Recherche Scientifique, ORSTOM, Dakar, Senegal

E. Kvillner, Department of Plant Ecology, University of Lund, Ostra Vallgatan 14, S-223 61 Lund, Sweden

O. M. Neijssel, Laboratorium voor Microbiologie, Universiteit van Amsterdam, Plantage Muidergracht 14, Amsterdam, The Netherlands

W. C. Noble, Department of Bacteriology, Institute of Dermatology, Homerton Grove, London E9 6BX, U.K.

A. N. Nozhevnikova, Institute of Microbiology, U.S.S.R. Academy of Sciences, Moscow, U.S.S.R.

D. G. Pitcher, Department of Bacteriology, Institute of Dermatology, Homerton Grove, London E9 6BX, U.K.

T. Rosswall, Department of Microbiology, Swedish University of Agricultural Sciences, S-750 07 Uppsala 7, Sweden

E. L. Schmidt, Department of Soil Science, University of Minnesota, St. Paul, Minnesota, U.S.A.

Charlene G. Steele, Department of Agronomy, Iowa State University, Ames, Iowa, U.S.A.

D. W. Tempest, Laboratorium voor Microbiologie, Universiteit van Amsterdam, Plantage Muidergracht 14, Amsterdam, The Netherlands

L. N. Yurganov, Institute of Atmospheric Physics, U.S.S.R. Academy of Sciences, Moscow, U.S.S.R.

Preface

The substantial and impressive changes in microbial ecology can scarcely be chronicled in a meaningful fashion, and a review series such as *Advances in Microbial Ecology* can thus not do justice to the numerous studies that have been published in recent years. On the other hand, the mere existence of this series bears testimony to the many and diverse activities.

The growing concern with microbial communities and processes in natural ecosystems is not restricted to scientists in one region and is not limited to particular groups of organisms or to individual theoretical or applied problems. The recent and successful international symposium on microbial ecology held in New Zealand—sponsored in part by the International Commission on Microbial Ecology, as is the *Advances*—and the general microbiology and ecology conferences and congresses have included reports from investigators from all corners of the globe and have explored both new and traditional areas, agricultural and public health problems, individual species and complex communities, and heterotrophs and autotrophs as well as ecosystem models relying on mathematical concepts and environmental processes needing sophisticated chemistry for their definition.

The reviews in the present volume thus can offer only a minute sampling of the multitude of topics being actively explored at the present time. Two of the reviews focus attention on biogeochemical cycles regulated by microorganisms, in particular the way these organisms contribute to or control the levels and identities of chemical substances in the atmosphere. The chapter by Y. Dommergues, L. W. Belser, and E. L. Schmidt deals with specific factors limiting growth and biochemical transformations in terrestrial communities, a subject of broad interest to many ecologists investigating natural ecosystems. Nutrient limitation as viewed from a physiological vantage point, by contrast, is the approach of D. W. Tempest and O. M. Neijssel, and their contribution provides new insights on the application of physiological principles to environmental phenomena. W. C. Noble and D. G. Pitcher explore our knowledge of a fascinating ecosystem inhabited by a specialized microflora, an ecosystem whose residents have a profound impact on the macroscopic bearer of this microflora. Although microbial populations have been characterized and described in many ways, the review by T. Rosswall and E. Kvillner opens new doors and discloses new means for describing microbial components of a variety of habitats.

We wish to stress that *Advances in Microbial Ecology* is sponsored by international agencies and seeks to satisfy an international audience. We have been gratified by the breadth of the response to this series and hope that suggestions for future directions and reviews will continue to be submitted. By responding to the needs of microbial ecologists and by providing a wide audience for reviews of current importance, this series can serve a critical and useful role.

M. Alexander, Editor
T. Rosswall
M. Shilo
H. Veldkamp

Contents

Chapter 1
**Principal-Components and Factor Analysis for the
Description of Microbial Populations**
T. Rosswall and E. Kvillner

Chapter 2
Limiting Factors for Microbial Growth and Activity in Soil
Y. R. Dommergues, L. W. Belser, and E. L. Schmidt

Chapter 3

Eco-Physiological Aspects of Microbial Growth in Aerobic Nutrient-Limited Environments

D. W. Tempest and O. M. Neijssel

Chapter 4

Role of Microorganisms in the Atmospheric Sulfur Cycle
John M. Bremner and Charlene G. Steele

Chapter 5

**Microbiological Aspects of Regulating the Carbon Monoxide
Content in the Earth's Atmosphere**
A. N. Nozhevnikova and L. N. Yurganov

Chapter 6
Microbial Ecology of the Human Skin
W. C. Noble and D. G. Pitcher

Principal-Components and Factor Analysis for the Description of Microbial Populations

T. ROSSWALL AND E. KVILLNER

1. Introduction

The ecosystem ecologist and the population ecologist often set out to describe the structure and function of an ecosystem or of a population. The biotic structure is given by a description of the species present and their abundance, while the function of the biotic component of the ecosystem calls for a fairly detailed analysis of the role of individual populations (species or functional groups) in, for example, energy flow or nutrient cycling.

There is often a fundamental difference between botanists and zoologists on the one hand and microbiologists on the other in their approach to studying their respective groups of organisms in an ecosystem. Whereas it is natural for a plant ecologist to start an investigation with a structural description of the plant community, the microbial ecologist often tries to avoid this approach. The nature of the organisms to be studied and the techniques available limit the microbiologist, and autecological studies are rare although certain methods, such as immunofluorescence, have during the past 10 years greatly promoted studies of single microbial species or genera in natural environments. Microbiological studies are instead often process oriented, and measurements are

T. ROSSWALL • Department of Microbiology, Swedish University of Agricultural Sciences, S-750 07 Uppsala 7, Sweden.   E. KVILLNER • Department of Plant Ecology, University of Lund, Ostra Vallgatan 14, S-223 61 Lund, Sweden.

made of, for example, respiration, litter decomposition, or nitrogen fixation. Studies of processes are usually more important to the microbiologist than those of organisms, but a knowledge of the organisms responsible is necessary for an understanding of how processes are regulated.

The difference in approach between botanists/zoologists and microbiologists has in some ways hampered the development of multidisciplinary studies since the difficulties seemingly caused by semantic problems have often been left untackled. Perhaps the process-oriented microbiological studies in the ecosystem projects, developed during the International Biological Program (IBP), have brought us to an endpoint with regard to process studies, and we must now turn to microorganism-oriented studies for obtaining a further understanding of the structure and function of ecosystems, communities, and populations.

To the microbiologist, a population description is always a hazardous undertaking. Robert Koch introduced the plate-count technique, which has been used ever since, especially for bacteria, despite its severe limitations (Jensen, 1968; Schmidt, 1973), and a host of papers describing variations in plate-count numbers in different environments and the influence of various external factors on plate counts have appeared.

Pochon and Tardieux (1962) developed the use of the most-probable-number (MPN) technique for determinations of "total numbers" of soil bacteria as well as for various physiological groups, and the technique was later modified by Darbyshire *et al.* (1974) and Rowe *et al.* (1977) by the use of automatic diluters and microtiter plates. Skerman (1969) compiled data on methods available for the selective cultivation of various taxonomic and physiological groups of bacteria.

Comparison between direct microscopic measurements and plate counts have shown that only a minor proportion of the soil bacteria are able to grow on any one isolation medium (see, e.g., Nikitin, 1973). Only a small part of the fungal mycelium seen in an ordinary stained preparation is metabolically active (Söderström, 1977), and the same limitations with regard to selectivity of isolation media apply to fungi. It is thus impossible to isolate a truly representative fraction of the bacteria and fungi living in natural environments for taxonomic purposes. All these factors, together with taxonomic difficulties, make a conventional description of the number and species composition of microorganisms from natural environments difficult, if not impossible.

These inherent difficulties have forced microbiologists to look for nonconventional ways of analyzing populations of microorganisms and their functional relationships. One such approach is the use of principal-components analysis and factor analysis in describing complex populations of microorganisms from natural environments in functional terms. This chapter describes the use of factor analysis in microbial ecology. It is not intended to be a technical paper, and the reader is referred to references given in each section for details of mathematical and computational procedures.

It is hoped that the presentation will show that these techniques may afford a powerful tool in analyzing the complex interactions between organisms and processes and the possible influence of external factors on them. It is not written for the specialist in the use of numerical methods but for the general microbial ecologist who wishes to obtain a glimpse of the possibilities that factor analysis can offer for the analysis of specific problems.

The chapter focuses mainly on bacteria. Bacteria are more suitable for this type of approach inasmuch as an isolate is more easily identified in relation to its occurrence in nature; microfungi isolated from spores or pieces of mycelia can only with difficulty be related to their functional occurrence in nature.

2. Adansonian Classification and Microbiology

Numerical and multivariate methods have found wide application in such diverse fields as the social sciences (Alker, 1969), psychology (Cattell, 1966), medicine (Baron and Fraser, 1968), archaeology (Clarke, 1968), anthropology (Driver, 1965), atmospheric-pollution studies (Gaarenstroom et al., 1977), and ecology (e.g., Webb et al., 1970). The field of numerical taxonomy has witnessed a rapid evolution, resulting in a second, greatly expanded edition of Sokal and Sneath's (1963) classic textbook on numerical taxonomy after only 10 years (Sneath and Sokal, 1973). A publication solely on the methodology of the use of numerical taxonomy in microbiology has also appeared (Lockhart and Liston, 1970).

The use of numerical methods for classification purposes is based on the assumption that an organism, a population, and a community can be expressed in numerical terms describing the characteristic features. The method was first developed by a French naturalist, M. Adanson, who, during his travels in Senegal, laid the foundation stone of numerical taxonomy (also called Adansonian taxonomy). The principle of Adanson's approach was to take all measurable characters into consideration and to give them all equal weight. The relationship between taxa was thus based on overall similarity, resulting in a phenetic classification. Adanson in this way described the molluscs (Adanson, 1757) and plants (Adanson, 1763) from Senegal, and although his work was unrecognized for nearly two centuries, he is today regarded as the father of numerical taxonomy (for a further discussion, see, e.g., Sokal and Sneath, 1963).

Sneath and Sokal (1973) point out that the use of numerical methods for taxonomic purposes has generally been more readily acceptable and has provoked far less criticism when applied to bacteria than when used for plants and animals. One reason for this is probably that microbiologists, faced with seemingly insurmountable problems of taxonomy and classification, will grasp at any straw. Microorganisms, especially bacteria, are also suitable as test organisms for the development of suitable methods in numerical studies, since the possibilities for repetition, multiplication, and control of tests are greater than with most, if not all, other organisms (Bonde, 1975).

Numerical taxonomy has been used extensively for the classification of bacteria (see Sneath and Sokal, 1963, 1973, for references), while only a limited number of papers have been published on its use for the classification of microfungi (e.g., Ibrahim and Threlfall, 1966). Numerical methods for classification purposes are based on the assumption that all characters have equal weight. This has been questioned (Adams, 1964) but no practicable alternatives have resulted, and dichotomous (two-state) tests are usually used, although there are exceptions (e.g., Harman and Kocková-Kratochvílová, 1976). The use of multistate tests has been discussed by Beers and Lockhart (1962), Sundman and Gyllenberg (1967), and Lockhart (1970).

In a study on the gram reaction of soil bacteria, Gyllenberg (1968) quantified his observations and ranged the results on a scale from 0 to 1. It should similarly be possible to use quantitative metabolic fingerprints as conceived by Kühn and Hedén (1976) as a basis for a factor-analytical approach, thus avoiding the necessity of only using yes/no, $+/-$, 0/1 dichotomous tests.

Rapid developments in the use of numerical taxonomy have resulted in techniques suitable for the analysis of large numbers of microbial isolates with regard to their physiological/biochemical capacity. The techniques have mainly been used for bacteria, but the use of similar methods for microfungi should also prove possible.

3. Methods for Collecting Primary Data

3.1. Introduction

Two events have been prerequisite for the development of the use of factor analysis in the study of microbial populations, viz., the introduction of automation and rapid miniaturized techniques for the collection of the primary results and the development of computers for handling the statistical/mathematical treatment of the primary data.

The new techniques recently developed for the rapid testing of microbial cultures are mainly a result of demands from the clinical microbiology sector (Rosswall, 1976). The growth in the number of publications on the subject of rapid methods and automation in microbiology has resulted in a separate bibliography on this topic (Palmer and LeQuesne, 1976); in addition, two international symposia have been devoted to it (Hedén and Illeni, 1975a,b; Johnston and Newsom, 1976), and a review article has been published (Isenberg and MacLowry, 1976).

The result of these activities has been that a large number of bacterial strains can be investigated for a multitude of physiological and biochemical characteristics in a fraction of the time it would have taken with the conventional test-tube equipment. Although the principles have not changed much, there has been a methodological revolution with regard to the hardware equipment.

3.2. Multipoint Techniques

Multipoint techniques, whereby a number of microbial strains can be transferred from a master plate containing the various cultures to be tested to appropriate test media, is a development of the replica plating technique described by Lederberg and Lederberg (1952) for the selection of bacterial mutants. This technique was based on the use of a velveteen cloth that was pressed gently onto the surface of an agar plate with bacterial colonies. The velveteen was then pressed onto the surface of a sterile agar plate, and the bacteria that could grow on this second plate formed colonies.

Many multipoint inoculation devices have been constructed (see for example Garrett, 1946; Beech et al., 1955; Quadling and Colwell, 1964; Corlett et al., 1965; Ridgway Watt et al., 1966; Seman, 1967; Lovelace and Colwell, 1968; Lighthart, 1968; Hill, 1970; Clarholm and Rosswall, 1973; Joseph et al., 1975). Wilkins et al. (1975) have described an inoculator especially adapted for use in an anaerobic glove box.

Although most of the techniques described were geared to work on bacteria, the first multipoint inoculating device described was for use with fungi (Garrett, 1946), and some later techniques were developed for the same purpose (Cooke, 1965; Fusaro, 1972; Littlewood and Munkres, 1972). The further development of techniques suitable for use with microfungi is urgently needed not only for the inoculation step but also for all subsequent steps in the testing procedure.

A multipoint inoculator should have the following features (Hill, 1970):

(a) The risk of airborne contamination must be no greater than with conventional techniques.
(b) Cross contamination between the bacterial cultures during transfer must be non-existent.
(c) The technique shall be significantly faster and less costly than conventional inoculation methods.
(d) The inoculating loops or needles must transfer a relatively constant volume of bacterial culture to the test medium. This necessitates the elimination of variation experienced in manually controlled devices due to differences in the speed of operation.
(e) The ability to re-sterilize the loops or needles shall be incorporated in the operating cycle. This reduces the risk of aerial contamination, but in particular enables rapid changeover of the series of cultures being inoculated and prevents carryover of nutrients.

The first inoculating devices were constructed for the transfer of strains to agar plates. Although the procedure is simple, the use of conventional agar plates has a number of disadvantages:

(a) Rapidly growing and spreading bacteria tend to grow over the entire plate, even if the latter has been dried in advance.
(b) Slow-growing bacteria cannot efficiently compete for available space, energy, and nutrients.
(c) Metabolites diffuse between colonies.
(d) Extracellular enzymes may diffuse over large areas and produce clear zones from proteolytic or amylolytic colonies covering nonactive colonies on the same plate.

These drawbacks have been circumvented by the use of individual agar disks in Petri dishes (Harris, 1963), perspex sheets with wells to hold liquid media (Smith, 1961), separate test tubes (Quadling and Colwell, 1964; Sundman, 1970; Clarholm and Rosswall, 1973), Petri dishes with separate compartments (Sneath and Stevens, 1967; Hill, 1970) or other types of presterilized plastic trays with individual compartments (Joseph *et al.*, 1975), autoclavable multiwell dishes (Clarholm and Rosswall, 1973), small glass vials held in position by a template (Goodfellow and Gray, 1966), or, more recently, various types of commercially available microtiter plates (Fung and Hartman, 1972; Jayne-Williams, 1975). The use of microtiter plates has also been described for growth of anaerobic bacteria (Wilkins and Walker, 1975).

Suitable techniques for the growth of fungal strains do not seem to have been developed, although D.Y.C. Fung (personal communication) claims that two microtiter plates, one inverted on top of the other, can be used. Taylor (1974) examined basidiomycetes with biochemical tests but did not use a multipoint approach. It may also be possible to use the conventional microtiter plates and, instead of loose plastic lids, adhesive tape permeable to O_2 and CO_2 (Rosswall, 1976).

The most common technique now in use is the microtiter system originally developed for tissue cultures and phage titrations. Various types of commercial plates are available as well as equipment for disp nsing the media and manual or automatic inoculators. A typical plate with dispenser and manual inoculator is shown in Fig. 1. The plates consist of 96 wells, each holding approximately 0.35 ml of medium. The use of these not only saves time but also cuts costs, since much smaller amounts of growth media are required.

Various types of automatic dispensers based on the models described by Middlebrook *et al.* (1970) and Wilkins *et al.* (1975) are commercially available.

In many instances, it is necessary to incubate the test plates for periods of weeks. Under such circumstances, the microtiter plates must be protected against dehydration. The simplest way would be to use the types of clear adhesive sheets that can be pressed onto the surface of the plastic microtiter plates. The commercially available types do not, however, seem to be sufficiently permeable to O_2 and CO_2 to permit their

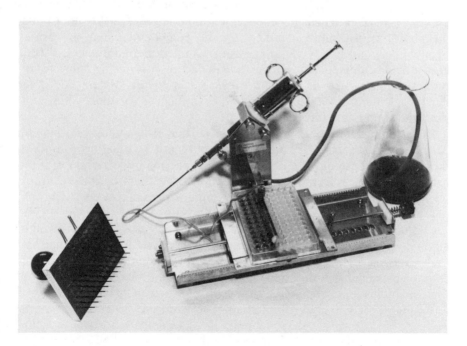

Figure 1. Multiwell dispenser, microtiter plate with 96 wells and a multipoint inoculator.

use for this purpose. Storage in plastic bags, e.g., Mylar (Green and Iljas, 1969), with moist pads seems to be sufficient to keep the plates in aseptic condition without dehydration for weeks.

3.3. Choice of Tests

If conventional techniques are used for the characterization of microbial isolates, a variety of methods is available for investigating the individual strains. Compilations of such methods have been published (e.g., Skerman, 1969; Holding and Collee, 1971). The prerequisites for the various types of tests were discussed by Sneath (1972).

The rapid development of miniaturized techniques offers a great spectrum of tests, and these have been reviewed by Hartman (1968), Fung and Hartman (1975), and Fung (1976). Thus, procedures seem to be available by which conventional testing procedures can be performed by the use of microtiter plates.

The tests used for describing bacterial populations from natural environments such as soil and water do not differ substantially from those used in identification procedures and taxonomic work. The tests selected should, however, be of a functional nature and should relate to processes. Thus, tests like resistance to crystal violet, growth at 48°C, and morphology for all practical purposes could be omitted. On the other hand, the ability to use more complex organic compounds, such as various phenolic substances—while at present of limited taxonomic value—is of great ecological importance in relation to decomposition and humification processes. The use of selective dye media, while of limited direct ecological importance, could, however, through taxonomic differentiation, give functional information.

It is important that the meaning of a certain test result always be kept in mind, as the basic principles behind many routine tests used in diagnostic bacteriology are often forgotten, and many descriptions of the original procedures lie hidden in publications from the turn of the century, when they were described by, for example, Voges, Proskauer, Ehrlich, and Kovacs (Blazevic and Ederer, 1975).

It should also be realized that tests usually do not give absolute results, but state only whether certain characters can be distinguished under a given set of conditions. Thus, it may be better to give a test result as Kovacs negative rather than oxidase negative (Sneath, 1972).

Examples of tests which could be used when making functional descriptions of bacterial populations from soil and water are found in Table I. Taylor (1974) selected 24 tests for use in characterizing mycelial cultures of basidiomycetes. The use of selective dye media for taxonomic purposes was described by Fung and Miller (1973) and Hagedorn and Holt

Table I. Examples of Tests Used for Describing Bacterial Populations from Soil and Water[a]

	A	B	C	D	E	F
1. Physiology						
a. Temperature for growth, °C						
0		X				
2	X		X			X
4		X		X	X	
5	X					
7		X				
8	X					
10		X				
15		X				
20		X				
24 (25)	X	X	X	X	X	X
30		X				
37	X	X	X		X	X
40		X				
44–48	X	X		X		
b. pH for growth						
3.5			X			
4.0	X	X	X		X	X
4.5		X	X	X		X
5.0	X	X	X			
5.5	X	X				
6.0	X	X				
6.5		X				
7.0	X	X	X	X	X	X
7.5		X				
8.0		X				
8.5		X				
9.0	X	X		X	X	
2. Biochemistry						
a. Carbohydrates (acid production)						
Glucose	X	X	X	X	X	X
Glucose, gas production	X	X				
Glucose (anaerobically)	X	X	X	X	X	X
Glucose, gas production (anaerobically)	X	X				
Arabinose	X		X	X	X	
Cellobiose	X		X	X		X
Dulcitol	X					
Fucose				X		
Fructose				X	X	
Galactose	X	X		X		
Lactose	X	X		X		
Maltose	X		X	X	X	X
Mannose	X		X	X		X
Mannitol		X		X	X	

(Continued)

Table I (*Cont.*)

	A	B	C	D	E	F
Melibiose				x		
Melizitose	x					
Raffinose				x		
Rhamnose	x		x			x
Ribose		x		x		
Sorbose				x		
Sucrose	x	x	x	x	x	x
Trehalose				x		
Xylose	x		x	x	x	x
b. Other compounds as sole carbon sources						
Acetate	x	x		x	x	
Acetylglucosamine				x		
Adonitol				x		
Aesculin				x		
Anthranilic acid				x		
Benzoate				x		
Citrate	x	x		x	x	
Dulcitol				x		
Gentisic acid				x		
Glycerol				x		
Gluconate				x		
p-Hydroxybenzoic acid	x			x		
8-Hydroxyquinoline				x		
Inositol				x		
Inulin				x		
Lactate	x					
Lignosulfonate			x			
Nicotinic acid				x		
Malonate	x					
Protocatechuic acid			x			x
Salicin				x		
Sedoheptolusan				x		
Succinate	x			x	x	
Vanillic acid			x			
Vanillin	x		x			
c. Polysaccharides (hydrolysis)						
Agar	x	x		x		
Cellulose	x		x			
Chitin	x		x			x
Starch	x	x	x	x	x	x
d. Lipids (hydrolysis)						
Egg yolk agar	x		x			x
Lecithin	x	x	x			x
Tributyrin	x					
Tween		x	x		x	x

(*Continued*)

Table I (*Cont.*)

	A	B	C	D	E	F
e. Nitrogen metabolism						
Proteolysis—casein	x	x	x	x		
—gelatin	x	x	x		x	x
Ammonification	x	x	x		x	
Urease	x		x		x	x
Nitrate reduction		x	x	x	x	x
Denitrification	x	x		x		x
Decarboxylase—lysine		x				
—ornithine		x				
—arginine		x				
Growth on "N-free" medium	x					
f. Growth factors						
Vitamins				x	x	
Soil extract				x	x	x
Amino acids		x		x	x	x
g. Other enzymes						
Acetamidase			x			x
Catalase		x			x	
Cytochrome oxidase	x				x	
Oxidase	x	x	x			x
Phosphatase	x	x	x		x	x
h. Miscellaneous						
Antibiotic sensitivity						
Aureomycin	x					
Bacitracin	x					
Chloramphenicol	x				x	
Chloromycetin		x				
Colimycin		x				
Novobiocin	x					
Penicillin	x	x	x		x	x
Polymyxin sulfate B	x					
Streptomycin		x			x	
Tetracycline		x				
H$_2$S production from cysteine		x	x		x	
CaHPO$_4$ dissolution			x		x	x
Spores	x				x	x

[a]A, Clarholm and Rosswall (1973); B, Colwell and Wiebe (1970); C, Niemelä and Sundman (1977); D, Skyring *et al.* (1971); E, Soumare *et al.* (1973); and F, Sundman (1970).

(1975b). The tests to be selected depend on the problem at hand, and while some—such as resistance to herbicides—may be extremely important in some cases, they may be of no value in others.

The number of tests to be selected depends on many factors, the most important being the manpower available. It should be kept in mind that the number of tests should be small in comparison with the number of isolates to be tested; Allais (1964; cited by Sundman and Gyllenberg, 1967) proposed as a rule of thumb that the number of isolates should be at least ten times the number of tests. One of the reasons given was that computer time is markedly influenced by the number of tests and only to a minor extent by the number of isolates. For today's fast computers, however, this may not be of importance. Debette *et al.* (1975), in classifying 165 nonfermentative, gram-negative bacteria from soil, found that the number of tests could be decreased from 51 to 21 without significant loss of information.

It is important that the selected tests should be as varied as possible, and lists of methods used for classification purposes in numerical taxonomy should only be taken as a starting point from which an ecologically relevant set of tests can be developed. As pointed out by Sneath (1957) and Sundman and Gyllenberg (1967), it is usually preferable to commence with a large number of tests which can then be reduced before the factor analysis is carried out. The elimination of tests is discussed below (Section 4.3).

After a set of relevant tests and suitable methods has been selected, consideration should be given to the reproducibility of the test results. Although it is very difficult to obtain identical results from any test when performed at different times or in different laboratories, the possibility of eliminating tests which may give variable results or which are difficult to read objectively should be borne in mind.

Sneath and Johnson (1972) and Sneath (1974) concluded that, even under conditions of careful standardization, experimental errors are usually quite appreciable. If the probability that a result is erroneous (p) is more than 10%, the error in the similarity values used in numerical taxonomy becomes unacceptable. Sneath and Johnson (1972) concluded that the value of p can usually be kept below 5% if tests are performed in the same laboratory, but p is often larger when results obtained in different laboratories are compared. For numerical taxonomy, Sneath and Johnson (1972) recommended that tests with more than 10–15% error should be rejected.

Few investigations have been made on test reproducibility, but Sneath and Collins (1974) published a comprehensive study on the reproducibility of tests used in classifying *Pseudomonas* spp. Only a few of the ten tests gave truly reproducible results. Arginine hydrolysis gave the most reproducible results, while growth at 37°C, denitrification, and urea hydrolysis showed poor consistency (Table II).

Sneath and Johnson (1972), in analyzing results obtained on strains of *Bordetella* and related genera obtained in the same laboratory but at different times, found that 19 of 54 tests were uniformly positive or

Table II. The Proportion of Errors, p_i, for Tests of 27 Pseudomonad Strains Investigated in Different Laboratories [a,b]

Test	p_i
Growth at 4°C	0
Arginine hydrolysis	3.3
Lactose oxidation	5.5
Gluconate oxidation	5.5
Growth in filaments	6.3
Coccoid morphology	7.1
Gelatin hydrolysis (stab method)	7.3
Glucose oxidation	7.8
Nitrate reduction—gas production	7.8
Fluorescein pigment	7.8
Casein hydrolysis	9.8
Sucrose oxidation	10.3
Motility	10.6
Aesculin hydrolysis	11.8
Nitrate reduction—presence of nitrite	11.9
Oxidase	12.2
Gelatin hydrolysis—plate method	13.9
Egg-yolk reaction	14.3
Morphology—pairs or chains	15.3
Growth at 42°C	16.2
Denitrification—gas production (Stanier)	19.6
Urease	22.2
Growth at 37°C	24.5
Denitrification—growth (Stanier)	33.3

[a]From Sneath and Collins (1974).
[b]For a discussion of p_i, which is related to the variance s_i^2, see Sneath and
Johnson (1972).

negative on both occasions, while the peroxidase test showed a 25%
discrepancy. Oxidase, phosphatase, urease, ammonification, H_2S production, and nitrate reduction tests were among those which showed incon-
sistent results. Gottlieb (1961), reporting on a joint investigation on
actinomycetes, obtained an overall agreement of 87% for 14 carbohy-
drate-utilization tests and 74% agreement when testing seven nitrogen
sources for growth, while H_2S production (96%) and nitrate reduction
(92–96%) showed more consistent results.

In conclusion, the reproducibility of the tests to be performed should
be kept in mind, and the investigation should strive for standardization in
every possible way. Lockhart (1967) stated that inaccuracy can result
when classifications are based on unstable properties or imperfectly
standardized techniques. However, the numerical taxonomist did not
invent these sources of error; he simply made them painfully obvious by
quantifying them.

Test results can also be analyzed for their ability to distinguish between different populations, and such results can be used for designing a test battery for factor-analytical descriptions of populations. According to the scheme of Gyllenberg (1963), the characterization figure (C), separation figure (S), and rank figure (R) are computed.

The measure of intragroup variation (characterization figure) is given by the equation

$$C = \frac{M}{N}$$

where M is the higher of separate counts of the $(+)$ and $(-)$ records for a specific test and N is the total number of isolates in the group. The significance of C is determined by the chi-square test. If M is significantly different from $N/2$, the test is considered characteristic of the group. The sum of C values for the different populations of a given test is a general measure of the value of this test in group characterization (Gyllenberg, 1963).

The separation figure (S) is computed as (Gyllenberg, 1963):

$$S = U \times V$$

where U is the number of groups or isolates characterized by a positive $(+)$ record and V the number characterized by a negative $(-)$ record.

Gyllenberg (1963) continued by calculating the rank figure (R) as $R_n = \Sigma C_n \times S_n$. This is then used to rank the tests in decreasing order of discriminating ability. This was further discussed by Niemelä and Gyllenberg (1968).

S values were also used by Rypka et al. (1967) and by Piguet and Roberge (1970) for discriminating useful tests for classification purposes.

3.4. Data Collection and Storage

Generally, only binary data are used, and the results are scored as positive or negative. Naturally, much important information is lost in this way, and procedures whereby quantitative aspects can be taken into account may prove useful in the future. The use of multistate tests as a basis for factor analysis was discussed by Sundman and Gyllenberg (1967), for example, and used by Harman and Kocková-Kratochvílová (1976). Up to now only dichotomous tests have been used as a basis for factor analysis of results from studies on mixed microbial populations.

Results can be recorded as $+/-$ or 1/0 in any way the investigator chooses, but they should preferably be transferred directly to standard forms for computer card punching where the 80 columns are used for the various tests, with one row for each isolate. Results can also be recorded directly on data cards using manually or electrically operated card punches (Cobb *et al.*, 1970; Rosswall, 1976).

It should be possible to develop automatic recorders for microtiter plates, especially if use is made of techniques like the one described by Bochner and Savageau (1977) in which triphenyl tetrazolium chloride (TTC) is incorporated into the growth medium. Colonies capable of catabolizing the test substrate produce a deep red formazan, and it should be possible to determine automatically the production of color in some microtiter plate compartments. A similar approach was previously suggested by Piguet and Roberge (1970).

Other techniques can be foreseen, and the possibility of using a fully automated system, such as the Autoline system described by Hedén (1975) and Kühn and Hedén (1976), may prove useful.

4. Factor Analysis for Describing Microbial Populations

4.1. Principal-Components Analysis and Factor Analysis

Principal-components analysis and factor analysis represent a branch of multivariate analysis that is concerned with the internal relationships of a set of variables. Initially, factor analysis was developed mainly by psychologists, with Spearman, Thomson, Thurstone, and Burt as the foremost pioneers. Later, the two methods were introduced to biological sciences, and Defayolle and Colobert (1962) and Hill *et al.* (1965) were the first to use factor analysis on microbiological data.

The primary task of principal-components analysis and of factor analysis is to express a large number of observed variables (tests and characters) in terms of a smaller number of hypothetical and "nonobserved," new variables (axes, principal components, or factors), thereby allowing the maximum variance in the original set of data to be evaluated (Kvillner, 1978). This approach is well-suited to the requirements of the microbiologist wishing to analyze a complex population of isolates, since the directions of variation in the observed combinations of characteristics are given by means of these new axes. Whether the axes are used to interpret taxonomical or ecological relationships depends naturally on the microbial data and on the purpose of the study.

The major steps of principal-components analysis are shown in Fig. 2 for a simplified case of 12 multistate tests (points in the graph) applied to two isolates (axes or dimensions). The isolates are obviously correlated with each other, as are the tests, and the test-result points fall in an elliptic

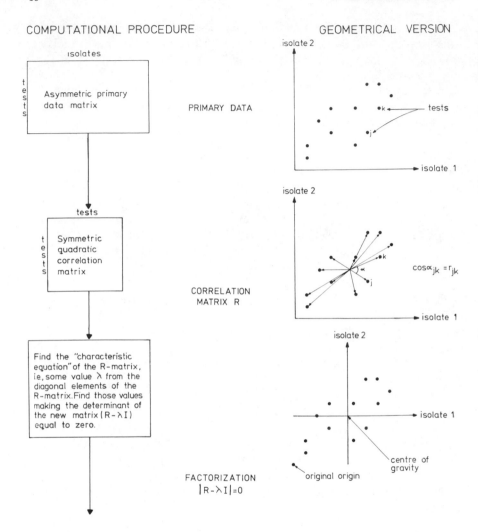

COMPUTATIONAL PROCEDURE

GEOMETRICAL VERSION

isolates

Asymmetric primary
data matrix

PRIMARY DATA

tests

Symmetric
quadratic
correlation
matrix

CORRELATION
MATRIX R

$\cos \alpha_{jk} = r_{jk}$

Find the "characteristic
equation" of the R-matrix,
i.e, some value λ from the
diagonal elements of the
R-matrix. Find those values
making the determinant of
the new matrix $(R-\lambda I)$
equal to zero.

FACTORIZATION
$|R-\lambda I|=0$

cluster in the two-dimensional plane (or in a hyperellipse in multidimen-
sional space). The points can also be illustrated by a bundle of vectors,
provided that it is possible to find a center of gravity from which they can
originate. In the beginning of the analysis, the origin given by the mean of
the two isolate axes has to be used. The angles between the vectors,
drawn from the origin to the points, give a valuable measure of the
interrelationships between the tests. By calculating the correlation coefficient (the cosines of the angles), numerical values of the interrelationships
are obtained from which a new center of gravity can be found. New axes

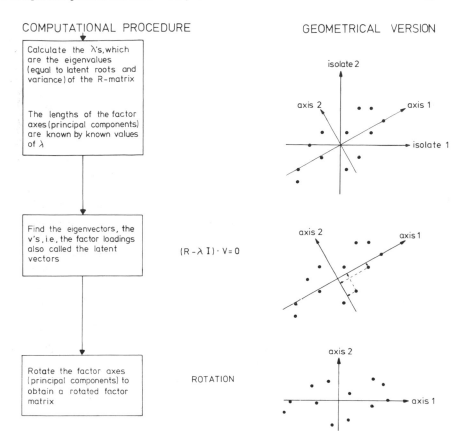

Figure 2. Major steps in the *R*-technique of principal-components analysis for 12 multistate tests applied to two isolates. The geometrical version is illustrated together with some mathematical synonyms for matrix computations.

are drawn from this center; it is important that they lie in the direction of maximum variation within the test results, i.e., pass through the point cluster in the most "efficient" way. This is done in a process called factorization. The lengths of these new axes can be calculated from the proportions of the test variance (the eigenvalues) accounted for by the axes. Naturally, the longest axis will be the most "efficient" one and is called the first principal axis. The second principal axis is orthogonal to the first. As is easily seen, the tests can be plotted or projected down to the new axes, thereby obtaining new coordinates (the factor loadings or

Table III. Variables Used in the Classification of *Torulopsis* spp.[a]

Character	Number of states[b]	Mean	s.d.[d] (n = 130)	s.d.[d] in % of mean
RQ for galactose metabolism	c[c]	0.42	1.18	280
Maltose uptake rate	4	15.11	26.20	173
Sensitivity to actidione	c	28.03	34.63	124
Assimilation of ethandiol	c	4.62	6.43	139
Autolysis	3	0.22	0.30	136
Growth at 40°C	c	2.29	3.07	134
Galactose respiration	4	30.74	39.18	127
Oxidation of acetate	5	5.00	5.99	119
Ethanol dehydrogenase activity	6	9.26	10.62	114
Sedimentation rates	5	15.45	15.42	99
Sucrose uptake rate	5	3.28	2.84	86
Phosphomolybdic acid test	5	1.14	0.97	85
Succinate dehydrogenase activity	6	5.83	4.80	82
Lactate dehydrogenase activity	6	13.17	10.15	77
Fermentation types	5	1.98	1.28	64
Growth at 5°C	4	39.13	23.10	59
Sensitivity to tartaric acid (2% w/v)	c	57.07	32.12	56
Character of pseudomycelium	5	1.30	0.72	55
Assimilation of ethanol	3	0.75	0.35	46
Assimilation of glycerol	3	0.78	0.33	42
Requirements of vitamins	5	48.20	20.45	42
Tolerance for ethanol	4	45.68	18.53	40
Sensitivity to lactic acid (2% w/v)	c	67.91	26.11	38
Formation of starchlike polysaccharides	3	1.18	0.40	33
Giant colonies	4	1.88	0.60	32
Growth on liquid media	9	2.88	0.88	30
Growth rate of giant colonies	c	1.23	0.32	26
Catalase activity	c	73.31	18.67	25
Correlation coefficients of lengths and widths of cells	c	0.67	0.17	25
Average length of cells	c	4.98	1.24	24
Average width of cells	c	3.85	0.86	22
Quotient surface/volume of cells	c	1.52	0.32	21
Osmophily	5	86.07	13.92	16
Length : width ratio of cells	c	1.29	0.17	13
Growth at 28°C	4	93.56	7.63	8

[a]From Harman and Kocková-Kratochvílová (1976). Only the variables with number of states >2 have been included.
[b]Number of alternatives.
[c]c, continuous.
[d]s.d., standard deviation.

Table IV. Percentage Positive Strains in Eight Different Bacterial Populations with Regard to Some Ecologically Important Tests and Groups of Tests[a]

	Tundra soils[b]				Pine forest soil[c]		Deciduous forest soil[d]	
	N	S	U	I	G_A	G_C	H_H	H_S
Acid from sugars ($n = 13$)	40	35	56	16	37	31	17	23
Organic salts as sole carbon source ($n = 5$)	58	9	3	20	25	23	8	8
Hydrolytic ability ($n = 5$)	14	35	50	36	55	26	32	47
Urease	22	22	22	16	0	6	17	22
Oxidase	82	70	86	57	93	29	17	14
All tests ($n = 25$)	40	31	43	23	39	27	18	25

[a]Compiled from Rosswall and Clarholm (1974).
[b]From Clarholm and Rosswall (1973): N, Norway, Hardangervidda, wet meadow ($n = 88$); S, Sweden, Stordalen, mire ($n = 60$); U, United Kingdom, Moor House, blanket bog ($n = 90$); I, Ireland, Glenamoy, blanket bog ($n = 63$).
[c]From Goodfellow (1966, 1968): G_A, United Kingdom, Freshfield, A horizon ($n = 257$); G_C, United Kingdom, Freshfield, C horizon ($n = 270$).
[d]From Hissett and Gray (1973): H_H, United Kingdom, Meathop Wood, humus ($n = 100$); H_S, United Kingdom, Meathop Wood, soil ($n = 100$).

influence of soil-management practices or other man-made influences on the microbial populations. Other ratios used earlier for this purpose are, for example, the ratios of filamentous fungi to yeasts and of actinomycetes to bacteria used by Maltby (1975) in a study of ecological indicators of changes resulting from moorland reclamation.

4.3. Correlation Matrix between Variables

A correlation matrix is computed from the test results of the individual isolates. The correlation between the individual tests, "R correlation," is computed and an $n \times n$ matrix formed (Fig. 3), while the correlation matrix between the individual strains ("Q correlation," $N \times N$ matrix; Fig. 3) is used in numerical taxonomy as a basis for describing the similarities between the strains investigated. The R and Q techniques were first described by Cattell (1952) for use in factor analysis; these concepts were later adopted by numerical taxonomists (see, for example, discussion in Sneath and Sokal, 1973). In 1963, when Sokal and Sneath (1963) published their monograph on numerical taxonomy, they pointed out that usually fewer operational taxonomic units (OTUs) than characters had been measured. Because of limitations in computer capacity, Q correlations were generally calculated. Harman (1972) considered this limitation to be valid no longer and found no reason not to use R

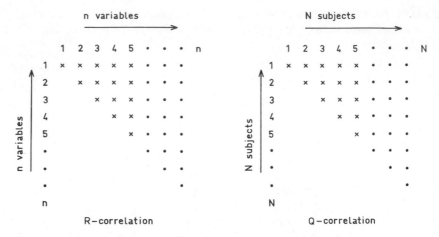

Figure 3. Correlation matrices between tests ($n \times n$; R = correlation) and isolates ($N \times N$; Q-correlations).

correlations also for taxonomic purposes. He also gave some reasons why correlations between characters are to be preferred:

  (a) avoidance of problems of deficient rank that would arise in a matrix of correlations among the strains when these exceed the number of characters.
  (b) factor analysis of the characters requires knowledge of the character for scientific interpretation, while an analysis of the OTUs requires knowledge about these elements themselves.
  (c) the principle of objectivity seems to be served better by determining the factors from the relationships among characters. (Harman, 1972; Harman and Kocková-Kratochvílová, 1976)

In numerical taxonomy, different types of similarity coefficients are employed to characterize the similarities between two OTUs, and these are mainly of four types, viz., distance coefficients, association coefficients, correlation coefficients, and probabilistic similarity coefficients (Sneath and Sokal, 1973). The Pearson product–moment correlation coefficient is the most frequently employed similarity coefficient in numerical taxonomy (Sneath and Sokal, 1973) and is used on data where most of the characters are present in more than two states. This coefficient is calculated according to the formula:

$$r_{jk} = \frac{\sum\limits_{i=1}^{n} (X_{ij} - \overline{X}_j)(X_{ik} - \overline{X}_k)}{\sqrt{\sum\limits_{i=1}^{n} (X_{ij} - \overline{X}_j)^2 \sum\limits_{i=1}^{n} (X_{ik} - \overline{X}_k)^2}}$$

where X_{ij} and X_{ik} are results obtained in test j and k, respectively, with isolate i, and X_j and X_k, are mean results for tests j and k, respectively.

The denominator in the equation above is a rough measure of the standard deviation of test j times the standard deviation of test k. This means that a standardization of the variables is performed, i.e., all the tests are treated as if they are of equal weight.

The limits of this correlation coefficient are -1 and $+1$. According to Fig. 2, r_{jk} is geometrically equal to the cosine of the angle between vectors j and k; i.e., the two tests are negatively correlated at a maximum ($r_{jk} = -1.00$) if the angle is 135°, not correlated at all ($r_{jk} = 0$) if the angle is 90°, and postively correlated at a maximum ($r_{jk} = +1.00$) if the angle is 45°.

Ranges between -1 and $+1$ are also obtained with the fourfold point correlation coefficient (ϕ):

$$\phi = \frac{(n_{JK}\, n_{jk} - n_{jK}\, n_{Jk})}{\sqrt{(n_J n_j n_K n_k)}}$$

where the following 2×2 table clarifies the four possible combinations between two-state tests (from Sokal and Sneath, 1963):

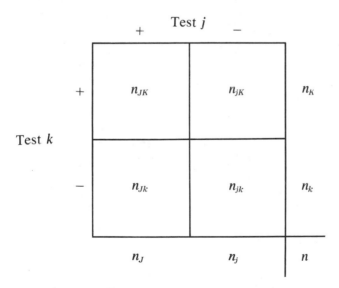

The fourfold point correlation coefficient has been recommended (Sokal and Sneath, 1963) for use in the case of "presence-absence" characters, as is the case in most instances when factor analysis or principal-components analysis is used for characterization of microbial populations.

Other coefficients have, however, also been used, and Brisbane and Rovira (1961) utilized the coefficient of Yule (Q coefficient of Yule and Kendall, 1950) for the characterization of rhizosphere populations of bacteria. Pike (1965) utilized the same coefficient for a study of *Micrococcus* spp. Sundman and Carlberg (1967) used the correlation coefficient of Ochiai (1957):

$$\mu_{jk} = \frac{n_{JK}}{\sqrt{n_J \, n_K}}$$

Gyllenberg *et al.* (1967), Gyllenberg and Eklund (1967), and Pohja and Gyllenberg (1967) used the above-mentioned correlation coefficient of Pearson.

The correlation matrix is usually not published, but a large amount of important information can be received from it. The matrix is inspected for high positive or negative correlations; tests should be neglected if they give high positive or negative correlations with other tests by definition (e.g., gram-positive and gram-negative) or by reflecting attributes which are a manifestation of the same single genetic locus, in which case little information is obtained (Quadling, 1967). Zones of proteolysis on skim milk agar plates and peptonization of litmus milk in test tubes are really tests for the same property, and the lack of an enzyme essential to glucose metabolism might influence the utilization of fructose and the ability to grow anaerobically (Sneath, 1957). Similarly, one of the two tests for dissolution of apatite and acid production should be disregarded for the same reasons (Brisbane and Rovira, 1961). In such a way, redundant tests can be eliminated, but it should be kept in mind that, even if new tests seemingly add no new discriminative information, they may serve to strengthen conclusions which could be made without them (Quadling, 1967).

As an example, the correlation matrix from a study by Sundman and Gyllenberg (1967) is shown in Table V. The numerically largest coefficient for each test has been printed in bold face. It can be seen that only about one-third of the maximal correlations are negative, a finding that is dependent on the more common occurrence of tests that refer to genetically or physiologically linked characters. The correlations between characters are a consequence of natural selection, which during the course of evolution has suppressed some of the theoretically possible character combinations (Gyllenberg, 1965). It is thus, from an evolutionary and physiological standpoint, more likely to find strong positive rather than negative correlations between tests. It should also be realized that loss of a certain ability occurs fairly frequently through mutations, while the gain of the same ability in a strain which has not earlier possessed it usually requires a more complex series of events (Krassilnikov, 1959). The results of most tests do not depend on single enzymes or genes but are the combination of a number of factors that affect the final result. Negative correlations can, however, occur as a result of the presence of suppressor genes or inhibitors of given metabolic activities (Hill *et al.*, 1961) and may thus be of a similar evolutionary origin as the positive correlations.

As was pointed out by Sundman and Gyllenberg (1967), the two numerically highest correlations in their matrix were truistic, viz., gram-positive/gram-negative ($\phi = -0.83$) and surface growth/deep growth ($\phi = -0.73$). In the correlation matrix, the five highest correlations were attributed to logical consequences of test definitions.

Character correlations can often be postulated from existing knowledge of the physiology of microorganisms. This knowledge is, however, often based on culture-collection isolates which have been kept in the laboratory for considerable lengths of time. It is probable that, after repeated subculturing, they have lost some of their original capacity. An improved knowledge of actually existing character correlations in microorganisms would result in a better understanding of the dynamics of microbial associations of various kinds (Gyllenberg, 1965). Investigations of correlation matrices of physiological characteristics of bacteria isolated from natural environments may prove to be one key to the understanding of the complex interrelationships that exist with regard to the metabolic and catabolic activities of soil and water microfloras. Character correlations can also form a useful basis for taxonomic purposes (e.g., Pike, 1965).

4.4. Eigenvalues, Variance, and Communality

The correlation matrix has several "valuable" qualities from the mathematical point of view. It is a symmetric matrix, since r_{jk} is equal to r_{kj}, and the r_{jj}'s (the diagonal elements, i.e., the correlations of the tests with themselves) are all equal to 1.00. The sum of the diagonal elements (also called the trace of the matrix) is equal to the total test variance. Thus, the absolute values of this sum depend on the number of tests included in the analysis.

Because of its qualities, new submatrices, e.g., factor matrices, can easily be formed from the *R*-matrix, measuring the importance of each new factor (principal component) and the importance of these factors for each observed test. In fact, a new suitable coordinate system is created that has a smaller number of axes or factors than the original observed system. The factors can be found by subtracting a certain value (λ) from

Table V. Correlation Matrix ($n \times n$) for a Sample

n	tests	1	2	3	4	5	6	7	8	9	10	11	12	13	14	
1	Gram-positive	1.	−83	24	−19	−14	04	−02	29	−13	11	−08	−05	13	−03	−
2	Gram-negative		1.	−23	20	17	01	01	−25	15	−14	11	07	−12	08	(
3	spores present			1.	12	11	−22	11	07	10	−22	17	08	−07	31	−
4	rod form predominates				1.	29	07	−02	−27	18	−16	02	07	01	−00	(
5	motile					1.	02	12	−00	65	−62	05	03	07	12	(
6	insoluble pigment present						1.	15	03	07	01	13	06	26	10	(
7	acid produced from glucose aerobically							1.	25	18	−15	56	24	28	19	
8	» » » » anaerobically								1.	02	−02	15	04	10	06	−
9	surface growth in semisolid stab cultures									1.	−73	07	05	10	19	
10	deep growth » » » »										1.	−06	02	−10	−08	−
11	acid produced from xylose											1.	35	04	24	
12	» » » rhamnose												1.	−09	14	
13	litmus milk peptonized													1.	01	−
14	» » reduced														1.	−
15	» » rendered slimy															
16	» » » alkaline															
17	CaHPO$_4$ dissolved															
18	phosphatase present															
19	lipolysis of Tween 80															
20	growth on NH$_4$-lactate as sole source of C and N															
21	resistante to 100 i.u. penicillin/ml. medium															
22	nitrate reduced to nitrite in peptone medium															
23	gelatine hydrolyzed															
24	growth at temperature below 5°															
25	» » 37°															
26	urease present															
27	H$_2$S produced from cysteine															
28	methylene blue reduced															
29	distinct growth on basal medium (Taylor, 1951)															
30	amino acids (casamino acids, Difco) necessary for growth															
31	yeast and/or soil extract necessary for growth															
32	oxidase present															

[a]Sundman and Gyllenberg (1967).
[b]Only the first two figures after the decimal point are shown.

of 681 Isolates of Soil Bacteria and 32 Tests [a,b]

16	17	18	19	20	21	22	23	24	25	26	27	28	29	30	31	32	n
6	26	-01	00	-03	-10	20	24	-17	20	08	09	11	02	03	01	08	1
4	**-28**	-03	05	03	06	-24	-23	14	-14	-11	-05	-07	01	-04	-04	-12	2
9	18	-18	-11	05	-22	05	**38**	-24	20	02	-08	34	25	-08	-13	-08	3
7	-01	00	05	01	03	-13	05	-06	05	-11	-04	12	01	05	-07	-16	4
1	-00	-03	20	05	11	-11	10	01	-01	-22	21	17	03	12	-19	07	5
1	03	**34**	33	04	09	-14	16	19	17	-11	15	05	19	-10	-14	-12	6
9	14	16	09	11	13	-01	32	**28**	01	-03	15	20	27	-03	-28	08	7
8	26	08	05	04	-01	29	20	04	**23**	05	18	12	16	-08	-02	**30**	8
5	04	-01	16	08	12	-12	14	04	-05	**-24**	28	22	08	08	-22	04	9
3	-13	15	-02	01	-09	11	-15	-04	-03	22	-16	-23	-01	-11	-15	-04	10
5	19	09	06	16	01	-04	18	21	11	-07	-02	13	35	-14	-26	-04	11
4	05	02	-02	13	01	-06	05	-00	-06	-02	01	18	18	-09	-14	-07	12
7	10	24	31	05	**33**	-01	**39**	26	04	02	17	14	02	07	-12	-05	13
	03	05	10	10	-18	-23	26	-01	-05	-07	12	31	30	-09	-28	-12	14
)	-01	00	02	-02	02	-09	-07	04	-07	-08	-02	-09	-06	01	04	-02	15
.	-14	-19	-17	12	12	26	-33	01	-14	17	04	-17	-16	14	01	25	16
	1.	-03	-06	10	08	22	19	20	16	-03	03	13	10	02	-11	19	17
		1.	**34**	02	15	08	12	15	14	-03	15	00	11	03	-15	03	18
			1.	13	17	-03	19	22	08	-08	20	07	13	03	-20	-10	19
				1.	23	28	03	22	-09	-11	21	21	**41**	-09	-40	15	20
					1.	14	08	28	01	07	27	15	04	10	-18	14	21
						1.	07	07	21	21	16	-01	09	05	-09	**62**	22
							1.	07	19	-05	19	**43**	28	-08	-19	01	23
								1.	-15	-06	14	01	13	01	-19	07	24
									1.	14	-03	12	13	-05	-06	02	25
										1.	00	-01	-08	08	02	14	26
											1.	**30**	05	17	-25	21	27
												1.	24	08	-37	01	28
													1.	**-64**	-37	-01	29
														1.	**-40**	14	30
															1.	-12	31
																1.	32

the diagonal elements of the R-matrix, thereby extracting the roots of the matrix (the eigenvalues or latent roots) by the determinantal equation:

$$|R - \lambda I| = 0$$

where I is the identity matrix.

Accordingly, each factor is characterized by a certain eigenvalue, the absolute value of which directly corresponds to the length of the new axis. However, the eigenvalue of a factor is more often related to the portion (percent variance) of the total test variance it accounts for. The first factors explain a major portion of the total variance of the data.

The eigenvalues of the factors are usually given in a general printout together with their contribution to the total test variance. As pointed out by Sundman and Gyllenberg (1967), the ways in which the eigenvalues decrease is a measure of the heterogeneity of the population under study. Fig. 4a is taken from their study and shows that the more homogeneous the population, the steeper the curve for the cumulated contribution of eigenvalues to the total test variance. For a given set of populations, the shape of the curve of cumulative eigenvalues versus numbers of factors

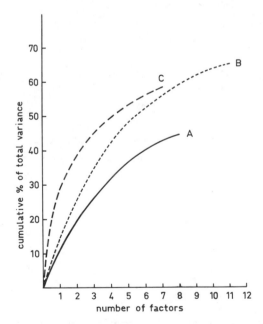

Figure 4a. Cumulated contributions of eigenvalues to total test variance for the first factors. A : $n = 29$, $N = 681$; soil isolates. B : $n = 29$, $N = 83$; soil isolates with simple nutritional demands. C : $n = 18$; $N = 192$; coliforms isolated from milk. From Sundman and Gyllenberg (1967).

gives an indication of the heterogeneity of the populations and thus a pictorial representation of the physiological diversity. As an example, Fig. 4b shows the result from a study of changes of bacterial populations during protein mineralization in soil.

However, if the collection of isolates studied consists of two distinct and unrelated populations, a dumbbell-shaped distribution of the isolates may occur in the factor space. Such a distribution is most likely to occur along the first component, which effectively finds the long axis of the largest variation. If there is a heavy loading in the first factor and if subpopulations can be recognized when the individual isolates are plotted against the first two factors, it might be worthwhile to make separate analyses for these subpopulations, as the heavy weight of the first factor in the combined material might obscure the meaning of subsequent components (Ivimey-Cook, 1969). The example in Fig. 4c is taken from the study of Ivimey-Cook (1969) on onion species, where the first twelve factors account for 65% of the total variance in the total population while the factors of the N and B populations account for 72% and 74%, respectively. However, the first factor in the combined population accounts for 16% of the variance, which is more than what is accounted

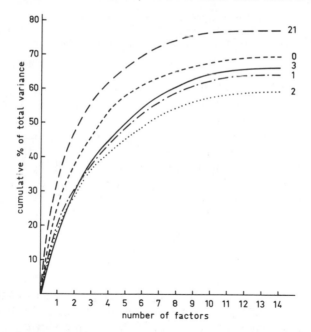

Figure 4b. Cumulated contributions of eigenvalues to total test variance for the first factors. Bacterial populations from soil at different times after protein addition. 0 days : $n = 19, N = 82$ 1 day: $n = 19, N = 92$; 2 days: $n = 19, N = 100$; 3 days: $n = 19, N = 100$; 21 days: $n = 19, N = 108$. From Rosswall (unpublished).

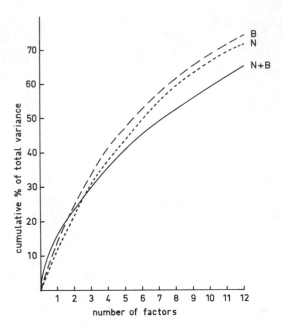

Figure 4c. Cumulated contributions of eigenvalues to total test variance for the first factors. Species of onions consisting of two distinct subgroups (N and B). N + B: $n = 64$, $N = 104$; N: $n = 61$, $N = 46$; B: $n = 58$, $N = 58$. Plotted from Ivimey-Cook (1969).

for by the first factors of the two subsets. Although the first part of the curve is steeper for the $(N + B)$ population, the diversity is large, as seen from the lower cumulative value for factor 12. It is also possible to determine whether there are distinct subgroups in the population by other methods (e.g., Gengerelli, 1963).

The next computational step is an estimation of the factor loadings (the eigenvectors or the latent vectors). Geometrically, the factor loadings are found by projection of the tests onto the new axes (see Fig. 2). Mathematically, they are obtained by use of a new matrix (V), which can be obtained from the R-matrix by

$$(R - \lambda I) \cdot V = 0$$

Thus, the sum of the squared factor loadings of all tests is equal to the eigenvalue of each single factor. Accordingly, the sum of the squared factor loadings of a test, summarized for the contribution of a test to all the factors, is called the communality (h^2) of the single test. A rough estimate of the communality, the limit of which is 0 to $+1$, is given by r_{max}, the maximum absolute value of the correlation of each test.

The contributions of the factors to the communality have also been used in an attempt at a cellular interpretation of the factors (Kocková-

Kratochvílová, 1972, 1976), with the largest contribution to communality in the first factor reflecting abilities directly controlled by the nucleus—such as cell form and protein synthesis—while factors regulated by the mitochondrion—such as assimilation of sugars and oxidation of other compounds—have positive weights in the second most prominent factor, etc. (Fig. 5). Actidione, which affects several processes in the cell, thus has heavy loadings in four of the first six factors since it acts on several levels of organization within the cell. Although this interpretation is open to discussion, it shows an interesting approach in trying to relate the results not only to individual tests or groups of tests but also to function. Perhaps such an approach can be used not only on the cellular level but also on the population level.

4.5. Rotated Factor Matrix

The structure of the factor matrix, extracted from the R-matrix, should be as simple as possible. This is achieved by rotation of the factors (the axes). The intention should be to get axes that are as meaningful as possible. The value of bipolar axes is sometimes quoted from the eco-

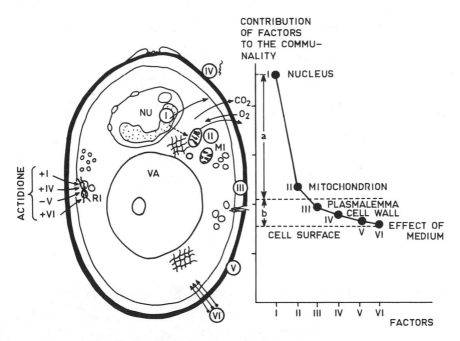

Figure 5. Contributions of the first six factors to the total communality in a factor analysis of 100 strains of *Saccharomyces* and 30 tests. An attempt has been made to interpret the factors on the basis of the structure and function of the yeast cell (Kocková-Kratochvílová, 1972, 1976).

logical point of view. Roughly half the loadings of a bipolar axis are positive, and roughly half are negative. Another desirable criterion is, for example, a fairly large number of zero factor loadings on all but one factor.

Generally, there are two types of rotation, viz., orthogonal rotation,
which is the most usual rotation procedure of principal-components
analysis, and oblique rotation, which is used in most factor-analysis
procedures. Sometimes the exact method of rotation is given, e.g.,
orthogonal "varimax" rotation or oblique "direction quartimin" rotation.

The rotated factor matrix results in estimates of the factor loadings
of the individual tests as well as the factor scores of the individual
isolates. It is thus possible to plot the positions of both the variables
(tests) and the cases (isolates) against the factors. An example of such a
plot is given in Fig. 6, in which the results from a study on soil bacteria are
shown (Soumare *et al.*, 1973). The positions of the individual isolates of
four genetic populations (*Pseudomonas, Bacillus, Arthrobacter*, and
Flavobacterium) are shown. The axes of the ellipses are the standard
deviations of the mean factor loadings on the first two factors. The mean
values of all the populations are significantly different from each other.

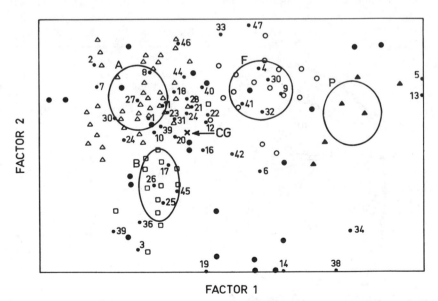

Figure 6. The positions of 76 soil bacterial isolates and 47 tests (the numbers refer to the list
of tests in Table VI) in relation to the first two principal factors after factor analysis. Four
populations were identified, and ellipses have been drawn around the mean values for the
individual isolates in each population. The axes of the ellipses were taken to be the standard
deviations for the isolates along the axes. P, *Pseudomonas* (▲); F, *Flavobacterium* (○); B,
Bacillus (□); A, *Arthrobacter* (△); ●, nonclassified strains (Plotted from data in Soumare *et
al.*, 1973). CG, center of gravity for all isolates.

Table VI. List of Tests Used in Describing Populations of Soil Bacteria[a,b]

---

1. Gram reaction
2. Presence of spheric cystites
3. Presence of spores
4. Production of insoluble pigments
5. Production of soluble pigments
6. Cytochrome oxidase
7. Catalase
8. Urease
9. Acid phosphatase
10. Starch hydrolysis
11. Gelatin hydrolysis
12. Resistance to KCN (0.5%)
13. Alkalinization of arginine anaerobically
14. Production of H_2S from cysteine
15. Production of keto-lactose from lactose
16. Reduction of NO_3 to NO_2
17. Mobility
18. Type of respiration
19. Fermentation of glucose
20. Oxidation of glucose
21.   arabinose
22.   xylose
23.   mannitol
24.   fructose
25.   maltose
26.   saccharose
27. Utilization of citrate
28.   acetate
29.   succinate
30. Lipolysis (Tween 80)
31. Dissolution of $CaHPO_4$
32. Resistance to penicillin
33.   streptomycin
34.   chloramphenicol
35.   NaCl (3%)
36.   NaCl (10%)
37. Growth at pH 9.0
38.   pH 4.0
39.   37°C
40.   4°C
41. Reduction of methylene blue
42. Ammonification
43. Indole production from peptone
44. Growth on mineral nitrogen
45. Amino acids needed for growth
46. Amino acids and vitamins needed for growth
47. Yeast extract or soil extract needed for growth

---

[a]From Soumare *et al.* (1973).
[b]Cf. Figs. 6 and 7.

FACTOR 1

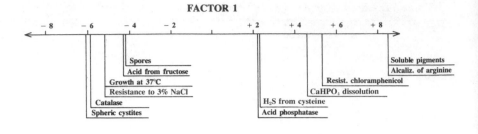

FACTOR 2

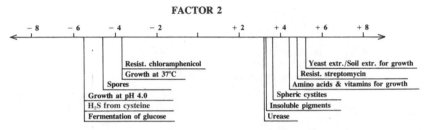

Figure 7. Graphical presentation of the heaviest factor loadings on the first two factors in Fig. 6. The units are arbitrary. (From data in Soumare *et al.*, 1973.)

The positions of the individual tests in relation to the first two factors are also shown; the list of tests appears in Table VI.

The factors can be functionally interpreted by examining the test with highest and lowest factor loadings according to the method proposed by Sundman and Gyllenberg (1967). The interpretation of the first two factors in Fig. 6 is presented in Fig. 7. The first factor is thus strongly loaded with tests for presence of soluble pigments and the ability to alkalinize arginine anaerobically, while there are heavy negative loadings for presence of spheric cystites, catalase, resistance to 3% NaCl, and growth at 37°C. Factor 2 is heavily loaded for tests with complex nutrient demands (No. 46 and 47), resistance to streptomycin, and presence of spheric cystites, while some of the negative loadings are the same as for factor 1, viz., presence of spores and growth at 37°C. It should be noted that the signs of the loadings of a given factor can be changed without loss of information—the factor can be "reflected" (Sundman and Gyllenberg, 1967). When comparing this with Fig. 6, it is evident that *Pseudomonas* and *Flavobacterium* have positive loadings in the two first factors, while *Arthrobacter* spp. are negative for factor 1 while positive for factor 2. It is thus possible to make a visual presentation of the position of different populations in a "physiological space." The radii of the ellipses are also measures of the heterogeneity of the different populations, and the distances between their centers of gravity are a quantitative estimate of their similarity.

The factors can be described in ecological terms—i.e., for carbohydrate metabolism, complex nutritional demands, and acid environment (Sundman, 1970)—or in more applied terms—i.e., the food deterioration factor described by Gyllenberg (1965) and Gyllenberg and Eklund (1967) with heavy loadings for proteolytic and lipolytic abilities. Some factors seem self-evident; e.g., factors with heavy loadings for presence of spores being a *Bacillus* factor, while penicillin resistance, soluble pigments, and possibility of denitrification indicate a *Pseudomonas* factor. On the other hand, oxidase activity and nitrate-reductase factor (Sundman 1968, 1970) are not as easily interpreted. Sundman (1970) suggested that the oxidase–nitrate-reductase factor can be an index for soil fertility since this factor was heavily loaded for a fertile grassland soil but had a low estimate for a forest humus.

Two types of factors can usually be defined: "specific" and "general" factors (Gyllenberg and Eklund, 1967). In a similar manner, tests can be "pure," i.e., occurring with a significant loading (e.g., ≥ 0.2) in one factor only, or "general," i.e., occurring in several factors.

The discriminating abilities between populations of factors obtained can be determined through variance analysis, and the factors that show significant differences between populations can be used in making factor profiles (Sundman, 1968). In a study of a forest humus, a grassland soil, and two field soils, Sundman (1970) noted that six of the first nine factors had a highly significant interpopulation variance ($p < 0.1\%$). Factor profiles of the four investigated populations are shown in Fig. 8. It should

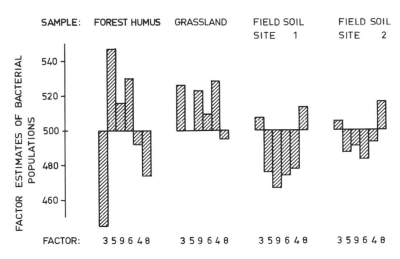

Figure 8. Factor profiles of bacterial populations from four soil samples. The factors showed significantly ($p < 0.1\%$) greater variations between than within populations. The factors are described as follows: 3, oxidase/nitrate reductase; 5, *Bacillus*; 9, humification; 6, acid environment; 4, complex nutritional demands; 8, pigment-correlated hydrolytic ability (from Sundman, 1970).

be noted that the two field-soil samples showed similar factor profiles, while the forest humus bacteria gave rise to an entirely different pattern.

Indications of the suitability of using factor profiles for population comparisons are also given by the fact that factors are similar for a population of fresh soil isolates and for isolates of the same origin which had been subject to several monthly laboratory culture passages in sand (Sundman, 1968). The factors obtained from isolates from a given soil thus seem to be stable. On the other hand, there are marked seasonal differences, and Rosswall and Torstensson (in prep.) found that populations of bacteria sampled in the field in the spring and in the autumn differed markedly.

These comparisons are purely qualitative, but it is also possible to determine the distance between the point centroids of the populations; Table VII presents data on the distances of the four populations described by Sundman (1970; Fig. 8). It can be seen that the field-soil populations have point centroids close to each other, and this distance can be regarded as a measure of stochastic variability (Niemelä and Sundman, 1977). It is thus possible to make a quantitative comparison of microbial populations based on the physiological ability of individual isolates.

The extent of the individual isolates in a given population can be determined by calculating the radius of the cloud of isolates according to the formula (Sundman and Carlberg, 1967):

$$r = 2\sqrt{\frac{\sum d^2}{n}}$$

where d is the euclidean distance from the individual isolate to the point of gravity and n is the number of isolates.

Sundman and Carlberg (1967) noted that the radii of bacterial populations showed a tendency to decrease with decreasing fertility of the soil (r for a compost being 2.18; forest humus spruce–birch stand, 1.88; forest humus lichen heath, 1.73; sawdust, 1.23). Soil fertility thus seems to influence the physiological versatility of the bacterial flora. Two of the factors showed consistent variation with fertility of the soil

Table VII. Euclidean Distances between the Point Centroids of Bacterial Populations from Four Different Soils [a,b]

	Forest humus	Grassland	Field site 1	Field site 2
Forest humus	0			
Grassland	115	0		
Field site 1	172	81	0	
Field site 2	127	40	10	0

[a]From Sundman (1970).
[b]The distances ($d^2 \times 10^2$).

samples. Factor III thus decreased in the order garden compost > forest humus > heath humus > decaying sawdust; factor III was a deep-growth–nonmotility factor (Sundman and Gyllenberg, 1967; Sundman, 1968). The opposite trend was found for factor IV, carbohydrate utilization. Sundman (1968) observed that protocatechuic acid-utilizing soil bacteria had low estimates for factor III and high for factor IV, thus resembling the native flora of decaying sawdust.

Another way of measuring homogeneity of mixed populations was suggested by Gyllenberg (1963). The characterization figure (C; see Section 3.3) summed for a group over all the tests will give a measure of the heterogeneity. If all tests are negative or positive, C equals the number of tests; if all tests show 50% ($+$) and 50% ($-$), C is equal to the number of tests divided by 2. Other indices for measuring inter- and intrapopulation variations have been discussed by Delabre et al. (1973).

5. Applications of Factor Analysis in Environmental Microbiology

5.1. Introduction

Increasing interest is being shown in techniques suitable for use in describing the impact of man-induced environmental change on the biotic structure and on biological processes. As the amounts of pollutants increase, ecosystems are frequently subjected to severe stress. This is a result of the introduction of new products into the environment (DDT or PCBs) as well as of the presence of increased amounts of naturally occurring substances at concentrations far above those usually found (SO_2 in the atmosphere or nitrogen and phosphorus compounds in water bodies).

Studies of the impact of environmental pollutants on microorganisms focus on any of the following main factors:

(a) Numbers of microorganisms (plate or direct counts)
(b) Groups of microorganisms (populations or functional groups)
(c) General indicators of activity (e.g., respiration or ATP concentration)
(d) Selected key processes (e.g., N_2 fixation or cellulose decomposition)

Few techniques seem to be available for studies of the possible impact on the population level. Factor analysis and principal-components analysis have, however, been used to evaluate the effect of a wide range of stress factors on populations of bacteria, viz., clear-cutting (Niemelä and Sundman, 1977), various types of crop husbandry (Gyllenberg and Rauramaa, 1966), cadmium pollution (de Leval et al., 1976), repeated pesticide application (Torstensson and Rosswall, 1977), acid precipitation (Rosswall, unpublished data), eutrophication (Persson and Rosswall, 1978), land cultivation (Sundman, 1970), and manure decomposition (Rosswall, 1976). Factor analysis has also been used in characterizing the site factors for use in microbiological studies (Hagedorn and Holt, 1975c).

5.2. Effect of Soil-Management Practices on Bacterial Populations

In their investigation of the influence of various soil-management techniques on the functional characteristics of the populations of soil bacteria, Gyllenberg and Rauramaa (1966) found a more homogeneous population in a field with a conventional rotation scheme and a more diverse bacterial flora in a field cropped to only cereals for several years with concomitant use of herbicides. Their investigation thus gave no indication that the present-day use of monocultures with heavy pesticide and fertilizer application results in a simple population structure. No investigations of this kind have, however, been made on soil fungal populations. The observations of Gyllenberg and Rauramaa (1966) were partially supported by Sundman (1970), who could not find evidence for a decrease in physiological versatility of the bacterial soil populations in two cultivated field soils when compared with that of a pasture soil.

These observations reflect the fact that many soil bacteria are ubiquitous, and Anderson (1977) has observed that there is generally "a very low level of generic and specific endemism of bacteria and fungi on a biogeographic scale." It is thus probable that the distribution of microorganisms between ecosystems and soil types is more even than that of most other organisms.

For example, although the physicochemical characteristics of a forest humus and the underlying mineral soil layers vary to a great extent, there do not seem to be any large differences in the physiological capacities of the bacterial populations from these soil layers (Niemelä and Sundman, 1977; Clarholm and Rosswall, unpublished data).

In Fig. 9 the factor scores of the individual isolates from different horizons of a podzolized forest soil are shown. There were no consistent differences between populations from mineral and organic layers, but there seemed to be a seasonal difference (factor 2) between the different populations. However, four and seven years after a clear-cutting, marked differences were observed between the populations of the humus and mineral layers (Niemelä and Sundman, 1977). In an examination of the first seven factors of Niemelä and Sundman (1977), one (No. 4) showed a wide range of values for the populations isolated from sites which had been clear-cut for different periods of time (Fig. 10). Whereas factor 4 clearly separated the population from the soil four years after clear-

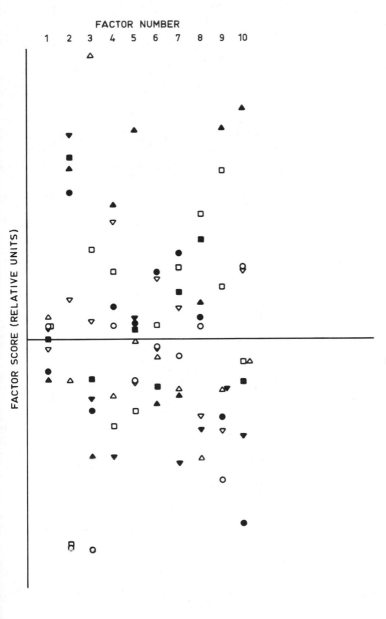

Figure 9. Graphical presentation of the average factor scores (arbitrary units) of soil bacteria populations from a podzolized forest soil at Jädraås (IhV), Central Sweden. (○) field layer, (☐) litter/humus layer, (△,▽) mineral soil layers. Closed symbols, July sampling; open symbols, October sampling. Clarholm and Rosswall, unpublished data.

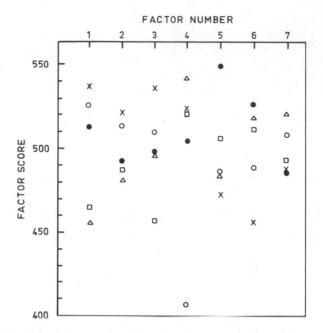

Figure 10. Graphical presentation of the average factor scores (arbitrary units) of soil bacterial populations from sites at various times after clear-cutting. (●), no cutting; (○), 4 years after clear-cutting; (×), 7 years after clear-cutting; (△, □), 13 years after clear-cutting (two samples) (Niemelä and Sundman, 1977).

cutting, factor 5 separated the population from the control plot. Factor 4 was characterized by positive loadings for $CaHPO_4$ dissolution and the ability to produce acid from sucrose, while it had a strong negative loading for nitrate reduction. Factor 5 showed strong positive loadings for lipolysis and proteolysis and negative for acid production from rhamnose. Clear-cutting had thus caused significant increases in proteolytic, lipolytic, and rhamnose-negative bacteria. There was a clear indication that the composition of the bacterial flora 13 years after clear-cutting tended to be similar to that of the control soil.

A factor-analytical approach has also been used to study changes in chemical properties of a soil after two levels of thinning (Gonzalez *et al.*, 1977). Thinning at 30% or 50% did not seem to affect the soil chemical properties after ten years.

5.3. Effects of Chemicals on Bacterial Populations

The effects of pesticides on soil bacteria have been investigated by the use of multivariate methods by Gyllenberg and Rauramaa (1966), Skyring and Quadling (1969b), and Torstensson and Rosswall (1977).

Gyllenberg and Rauramaa (1966) added the herbicide MCPA to soils in laboratory studies and followed the change in bacterial populations over time. They found a significant trend toward more homogeneous populations, but this was also true after additions of straw. In their investigation of bacterial isolates from rhizosphere and nonrhizosphere soils by principal-components analysis, Skyring and Quadling (1969b) did not find any significant differences between populations from soils treated with Vapam (sodium N-methyldithiocarbamate) and control soils.

Torstensson and Rosswall (1977) investigated the effects of repeated application over a 20-year period of two herbicides, MCPA and 2,4-D, but they were unable to find any effects on the functional characteristics of the bacterial populations as determined by factor analysis.

When treating results from 11 tests by means of factor analysis, de Leval *et al.* (1976) were able to show that cadmium-resistant aquatic bacteria (100 ppm Cd) formed a distinct subpopulation. The data presented in their report did not permit a more detailed analysis. They argue, however, that microorganisms can be used as biological indicators of pollution.

6. Conclusions and Perspectives

Principal-components analysis and factor analysis are not the only methods for formulating a functional description of mixed microbial populations. However, these methods offer the possibility of making qualitative as well as quantitative comparisons of populations of bacteria and fungi. The computational procedures have evolved rapidly but have aroused the interest of the microbiologist only to a limited extent. This is probably due in part to the tedious work of collecting the primary data. Recent developments allow rapid characterization of a large number of isolates with regard to a given set of physiological/biochemical tests. It is possible to develop these techniques even further and to automate the procedures. The use of quantitative estimates instead of merely qualitative ones will offer even greater possibilities for analyzing complex interrelationships between different physiological capabilities as well as between different strains and populations.

The primary isolation of the strains to be tested, either bacterial or fungal, is inevitably the time-consuming part of an investigation of this kind. There is always the criticism that only a small part of the native microflora can be analyzed by virtue of the selectivity of the isolation techniques. It is important that these limitations be kept in mind, but they should not prevent the use of the techniques. Only by the use of an array of different techniques—all deficient in one way or the other—is it possible to obtain a further understanding of the complex interrelationships between various functional groups of microorganisms. It should, however, be noted that some important microbiological transformations in nature, e.g., autotrophic nitrification, are not included in this type of functional analysis.

Although principal-components analysis and factor analysis do not seem to have been used to any extent on data on the physiological capabilities of microfungi, this should be a field for further investigation.

These multivariate techniques, and others, should prove to be powerful tools in attempts at further elucidating the complex microworld in natural environments.

ACKNOWLEDGMENTS

The senior author acknowledges support from the Swedish Coniferous Forest Project and grants from Carl Tryggers Stiftelse and the Swedish National Environment Protection Board.

We also wish to thank Drs. R.S. Clymo and V. Sundman for valuable comments on parts of the manuscript.

References

Adams, J. N., 1964, A critical evaluation of Adansonian taxonomy, *Dev. Ind. Microbiol.* 5:173.

Adanson, M., 1757, *Histoire Naturelle du Sénégal. Coquillages. Avec la relation abrégée d'un voyage fait en ce pays, pendant les années 1749, 50, 51, 52 et 53,* Banché, Paris.

Adanson, M., 1763, *Familles des Plantes,* Vol. 1, Vincent, Paris.

Alker, H. R., 1969, Statistics and politics: The need for causal data analysis, in: *Politics and Social Sciences* (S. M. Lipset, ed.), pp. 244–313, Oxford University Press, New York.

Anderson, J. M., 1977, The organization of soil animal communities, in: *Soil Organisms as Components of Ecosystems* (U. Lohm and T. Persson, eds.), *Ecol. Bull. (Stockholm),* 25:15.

Baron, D. N., and Fraser, P. M., 1968, Medical applications of taxonomic methods, *Br. Med. Bull.* 24:236.

Beech, F. W., Carr, J. G., and Codner, R. C., 1955, A multipoint inoculator for plating bacteria or yeasts, *J. Gen. Microbiol.* 13:408.

Beers, R. J., and Lockhart, W. R., 1962, Experimental methods in computer taxonomy, *J. Gen. Microbiol.* 28:633.

Blazevic, D. J., and Ederer, G. M., 1975, *Principles of Biochemical Tests in Diagnostic Microbiology,* Wiley, New York.

Bochner, B. R., and Savageau, M. A., 1977, Generalized indicator plate for genetic, metabolic, and taxonomic studies with microorganisms, *Appl. Environ. Microbiol.* 33:434.

Bonde, G. J., 1975, The genus *Bacillus.* An experiment with cluster analysis, *Danish Med. Bull.* 22(2):41.

Brisbane, P. G., and Rovira, A. D., 1961, A comparison of methods for classifying rhizosphere bacteria, *J. Gen. Microbiol.* 26:379.

Cattell, R. B., 1952, *Factor Analysis,* Harper, New York.

Cattell, R. B. (ed.), 1966, *Handbook of Multivariate Experimental Psychology,* Rand McNally, Chicago.

Clarholm, M., and Rosswall, T., 1973, A comparison of bacterial populations from tundra sites by means of a multipoint technique, *Swedish IBP Tundra Biome Project Tech. Rep. 10.*

Clarke, D. L., 1968, *Analytical Archaeology,* Methuen, London.

Cobb, R., Crawley, D. F. C., Croshaw, B., Hale, L. J., Healey, D. R., Pay, F. J., Spicer, A. B., and Spooner, D. F., 1970, The application of some automation and data handling techniques to the evaluation of antimicrobial agents, in: *Automation, Mechanization and Data Handling in Microbiology* (A. Baillie and R. J. Gilberg, eds.), *Soc. Appl. Bacteriol. Tech. Ser.* **4**:53, Academic Press, London.

Colwell, R. R., and Wiebe, W. J., 1970, "Core" characteristics for use in classifying aerobic, heterotrophic bacteria by numerical taxonomy, *Bull. Georg. Acad. Sci.* **28**:165.

Cooke, W. B., 1965, The growth of yeasts at 37°C, *Mycopath. Mycol. Appl.* **25**:195.

Corlett, D. A., Jr., Lee, J. S., and Sinnhuber, R. O., 1965, Application of replica plating and computer analysis for rapid identification of bacteria in some foods. I. Identification scheme, *Appl. Microbiol.* **13**:808.

Darbyshire, J. F., Wheatly, R. E., Greaves, M. P., and Inkson, R. E., 1974, A rapid micromethod for estimating bacterial and protozoan populations in soil, *Rev. Ecol. Biol. Sol.* **11**:465.

Debette, J., Losfeld, J., and Blondeau, R., 1975, Taxonomie numérique de bactéries telluriques non fermentantes à Gram négatif, *Can. J. Microbiol.* **21**:1322.

Defayolle, M., and Colobert, L., 1962, L'espéce *Streptococcus faecalis.* II. Etude de l'homogénéité par l'analyse factorielle, *Ann. Inst. Pasteur* **103**:505.

Delabre, M., Bianchi, A., and Véron, M., 1973, Etude critique de méthodes de taxonomie numérique. Application à une classification de bactéries aquicoles, *Ann. Microbiol. Inst. Pasteur* **124A**:489.

de Leval, J., Houba, C., and Remacle, J., 1976, Les microorganismes en tant que bioindicateurs de la qualité des eaux douces, *Mém. Soc. R. Bot. Belg.* **7**:129.

Driver, H. E., 1965, Survey of numerical classification in anthropology, in: *The Use of Computers in Anthropology* (D. Hymes, ed.), pp. 301–304, Monton, The Hague.

Fung, D. Y. C., 1976, Miniaturized microbiological techniques, in: *Rapid Methods and Automation in Microbiology* (H. H. Johnston and S.W.B. Newsom, eds.), pp. 169–172, Learned Information (Europe) Ltd., Oxford and New York.

Fung, D. Y. C., and Hartman, P. A., 1972, Rapid characterization of bacteria, with emphasis on *Staphylococcus aureus, Can. J. Microbiol.* **18**:1623.

Fung, D. Y. C., and Hartman, P. A., 1975, Miniaturized microbiological techniques for rapid identification of bacteria, in: *New Approaches to the Identification of Microorganisms* (C. G. Hedén and T. Illéni, eds.), pp. 347–370, Wiley, New York.

Fung, D. Y. C., and Miller, R. D., 1973, Effect of dyes on bacterial growth, *Appl. Microbiol.* **25**:793.

Fusaro, R. M., 1972, Inoculation technique for fungus cultures, *Appl. Microbiol.* **23**:174.

Gaarenstroom, P. D., Perone, S. P., and Moyers, J. L., 1977, Application of pattern recognition and factor analysis for characterization of atmospheric particulate composition in southwest desert atmosphere, *Environ. Sci. Technol.* **11**:795.

Garrett, S. D., 1946, A multiple-point inoculating needle for agar plates, *Trans. Br. Mycol. Soc.* **29**:171.

Gengerelli, J. A., 1963, A method for detecting subgroups in a population and specifying their membership, *J. Psychol.* **55**:457.

Gonzalez, A., Boudoux, M., and Huberg, G., 1977, Etude par l'analyse en composantes principales des propriétés chimique d'un sol forestier 10 ans après éclaircie, *Can. J. Soil Sci.* **57**:147.

Goodfellow, M., 1966, The classification of bacteria in a pine wood soil, Ph.D. Thesis, University of Liverpool.

Goodfellow, M., 1968, Properties and composition of the bacterial flora of a pine forest soil, *J. Soil Sci.* **19**:154.

Goodfellow, M., and Gray, T. R. G., 1966, A multi point inoculation method for performing biochemical tests on bacteria, in: *Identification Methods for Microbiologists, Part A* (B. M. Gibbs and F. A. Skinner, eds.), *Soc. Appl. Bacteriol. Tech. Ser.* **1**:117, Academic Press, London.

Gottlieb, D., 1961, An evaluation of criteria and procedures used in the description and characterization of the Streptomycetes. A cooperative study, *Appl. Microbiol.* **9**:55.

Green, J. H., and Iljas, S., 1969, Packing method for preserving agar plates during prolonged incubation, *Appl. Microbiol.* **17**:322.

Gyllenberg, H., 1963, A general method for deriving determination schemes for random collection of microbial isolates, *Ann. Acad. Sci. Fenn.* **A IV 69**:1.

Gyllenberg, H. G., 1965, Character correlations in certain taxonomic and ecologic groups of bacteria, *Ann. Med. Exp. Fenn.* **43**:82.

Gyllenberg, H. G., 1968, Significance of the gramstain in the classification of soil bacteria, in: *The Ecology of Soil Bacteria* (T. R. G. Gray and D. Parkinson, eds.), pp. 351–359, Liverpool University Press, Liverpool.

Gyllenberg, H. G., and Eklund, E., 1967, Application of factor analysis in microbiology. 2. Evaluation of character correlation patterns in phychrophilic pseudomonads, *Ann. Acad. Sci. Fenn.* **A IV 113**:1.

Gyllenberg, H. G., and Rauramaa, V., 1966, Density, activity, and composition of the bacterial flora with special reference to the employed techniques of crop husbandry, *Acta Agr. Scand.* **16**:39.

Gyllenberg, H. G., Woźnicka, W., and Kurylowicz, W., 1967, Application of factor analysis in microbiology. 3. A study of the "yellow series" of streptomycetes, *Ann. Acad. Sci. Fenn.* **A IV 114**:1.

Hagedorn, C., and Holt, J. G., 1975a, A nutritional and taxonomic survey of *Arthrobacter* isolates, *Can. J. Microbiol.* **21**:353.

Hagedorn, C., and Holt, J. G., 1975b, Differentation of *Arthrobacter* soil isolates and named strains from other bacteria by reactions on dye-containing media, *Can. J. Microbiol.* **21**:688.

Hagedorn, C., and Holt, J. G., 1975c, Ecology of soil arthrobacters in Clarion-Webster toposequences of Iowa, *Appl. Microbiol.* **29**:211.

Harman, H. H., 1972, How factor analysis can be used in classification. I. Mathematical part, in: *Yeasts—Models in Science and Technics. Proceedings of the First Specialized International Symposium on Yeasts* (A. Kocková-Kratochvílová and E. Minárik, eds.), pp. 273–295, Publishing House of the Slovac Academy of Sciences, Bratislava.

Harman, H. H., and Kocková-Kratochvílová, A., 1976, Use of factor analysis to classify strains of yeasts: Application to genus *Torulopsis* Berlese, *J. Math. Biol.* **3**:27.

Hartman, P. A., 1968, *Miniaturized Microbiological Methods, Advances in Applied Microbiology, Supplement 1*, Academic Press, New York.

Harris, P. J., 1963, A replica plate culture technique, *J. Appl. Bacteriol.* **26**:100.

Hedén, C. -G., 1975, The modular approach to the automation of the microbiological routines, in: *New Approaches to the Identification of Microorganisms* (C.-G. Hedén and T. Illéni, eds.), pp. 13–37, Wiley, New York.

Hedén, C. -G., and Illéni, T. (eds.), 1975a, *Automation in Microbiology and Immunology*, Wiley, New York.

Hedén, C. -G., and Illéni, T. (eds.), 1975b, *New Approaches to the Identification of Microorganisms*, Wiley, New York.

Hill, I. R., 1970, Multiple inoculation technique for rapid identification of bacteria, in: *Automation, Mechanization and Data Handling in Microbiology* (A. Baillie and R. J. Gilberg, eds.), *Soc. Appl. Bacteriol. Tech. Ser.* 4:175, Academic Press, London.

Hill, L. R., Turri, M., Gilardi, E., and Silvestri, L. G., 1961, Quantitative methods in the systematics of actinomycetales. II, *Giorn. Microbiol.* 8:56.

Hill, L. R., Silvestri, L. G., Ihm, P., Farchi, G., and Lanciani, P., 1965, Automatic classification of staphylococci by principal-component analysis and a gradient method, *J. Bacteriol.* 89:1393.

Hissett, R., and Gray, T. R. G., 1973, Bacterial populations of litter and soil in a deciduous woodland. I. Qualitative studies, *Rev. Ecol. Biol. Sol* 10:495.

Holding, A. -J., and Collee, J. G., 1971, Routine biochemical tests, in: *Methods in Microbiology* (J. R. Norris and D. W. Ribbons, eds.), vol. 6A, pp. 1–32, Academic Press, New York.

Ibrahim, F. M., and Threlfall, R. J., 1966, The application of numerical taxonomy to some graminicolous species of *Helminthosporium, Proc. R. Soc. London* B165:362.

Isenberg, H. D., and MacLowry, J. D., 1976, Automated methods and data handling in bacteriology, *Annu. Rev. Microbiol.* 30:483.

Ivimey-Cook, R. B., 1969, The phenetic relationship between species of onions, in: *Numerical Taxonomy* (A. J. Cole, ed.), pp. 69–90, Academic Press, New York.

Jayne-Williams, D. J., 1975, Miniaturized methods for the characterization of bacterial isolates, *J. Appl. Bacteriol.* 38:305.

Jensen, V., 1968, The plate count technique, in: *The Ecology of Soil Bacteria* (T. R. G. Gray and D. Parkinson, eds.), pp. 158–170. Liverpool University Press, Liverpool.

Johnston, H. H., and Newsom, S. W. B. (eds.), 1976, *2nd International Symposium on Rapid Methods and Automation in Microbiology,* Learned Information (Europe) Ltd., Oxford.

Joseph, S. W., Duncan, J. F., and Sindhuhardja, W., 1975, Multipoint inoculating apparatus for biochemical determinations in disposable trays, *Lab. Pract.* 24:413.

Kocková-Kratochvílová, A., 1972, How factor analysis can be used in classification. II. Biological part. in: *Yeasts—Models in Science and Technics. Proceedings of the First Specialized International Symposium on Yeasts* (A. Kocková-Kratochvílová and E. Minárik, eds.), pp. 297–303, Publishing House of the Slovak Academy of Sciences, Bratislava.

Kocková-Kratochvílová, A., 1976, Taxometric study of the genus *Saccharomyces* (Menyen) Reess. 3rd part: Small species, *Biol. Práce* 22(6):1.

Krassilnikov, N. A., 1959, Species significance of antibiotic components in actinomycetes, *Mikrobiologiya* 28:168.

Kühn, I., and Hedén, C. -G., 1976, From biotyping to metabolic fingerprinting, in: *Rapid Methods and Automation in Microbiology* (H. H. Johnston and S. W. B. Newsom, eds.), pp. 173–177, Learned Information (Europe) Ltd., Oxford.

Kvillner, E., 1978, Applicability and validity of some numerical methods in plant ecology. I. Principle component analysis, *Vegetatio* (in press).

Lederberg, J., and Lederberg, E. M., 1952, Replica plating and indirect selection of bacterial mutants, *J. Bacteriol.* 63:399.

Lighthart, B., 1968, Multipoint inoculator system, *Appl. Microbiol.* 16:1797.

Littlewood, R. R., and Munkres, K. D., 1972, Simple and reliable method for replica plating *Neurospora crassa, J. Bacteriol.* 110:1017.

Lockhart, W. R., 1967, Factors affecting reproducibility of numerical classifications, *J. Bacteriol.* 94:826.

Lockhart, W. R., 1970, Coding the data, in: *Methods for Numerical Taxonomy* (W. R.

Lockhart and J. Liston, eds.), pp. 22–33, American Society for Microbiology, Bethesda, Md.

Lockhart, W. R., and Liston, J. (eds.), 1970, *Methods for Numerical Taxonomy,* American Society for Microbiology, Bethesda, Md.

Lovelace, T. E., and Colwell, R. R., 1968, A multipoint inoculator for petri dishes, *Appl. Microbiol.* **16**:944.

Maltby, E., 1975, Numbers of soil microorganisms as ecological indicators of changes resulting from moorland reclamation on Exmoor, U. K., *J. Biogeogr.* **2**:117.

Middlebrook, G., Reggiardo, Z., and Taylor, G. R., 1970, Continuous dispenser for multiple-well serological plate, *Appl. Microbiol.* **20**:852.

Niemelä, S. I., and Gyllenberg, H. G., 1968, Application of numerical methods to the identification of micro-organisms, *Spisy Přírod. Fak. Univ. Purkyně Brno* **Ser. K43**:279.

Niemelä, S., and Sundman, V., 1977, Effects of clear-cutting on the composition of bacterial populations of northern spruce forest soils, *Can. J. Microbiol.* **23**:131.

Nikitin, D. I., 1973, Direct electron microscopic techniques for the observation of microorganisms in soil, in: *Modern Methods in the Study of Microbial Ecology* (T. Rosswall, ed.), *Bull. Ecol. Res. Comm. (Stockholm)* **17**:85.

Ochiai, A., 1957, Zoogeographical studies on the soleoid fishes found in Japan and its neighbouring regions. II, *Bull. Jpn. Soc. Sci. Fish* **22**:526 (in Japanese, English summary).

Palmer, W. J., and LeQuesne, S. E. (eds.), 1976, *Rapid Methods and Automation in Microbiology and Immunology: a Bibliography,* Information Retrieval, London and Washington, D. C.

Persson, I. -B., and Rosswall, T., 1978, Functional description of bacterial populations from lakes with varying degrees of eutrophication. Report to Swedish Environment Protection Board (in Swedish).

Piguet, J. D., and Roberge, P., 1970, Problèmes posés par la diagnostic automatique des batonnets gram-négatifs, *Can. J. Publ. Health* **61**:329.

Pike, E. B., 1965, A trial of statistical methods for selection of determinative characters from Micrococcacaeae isolates, *Spisy Prirod. Fak. Univ. Purkyne Brno* **Ser. K35**:316.

Pochon, J., and Tardieux, P., 1962, *Techniques d'Analyse en Microbiologie du Sol,* Editions de la Tourelle, Paris.

Pohja, M. S., and Gyllenberg, H. G., 1967, Applications of factor analysis in microbiology. 5. Evaluation of the population development in coldstored meat, *Ann. Acad. Sci. Fenn.* **A IV 116**:1.

Quadling, C., 1967, Evaluation of tests and grouping of cultures by a two-stage principal component method, *Can. J. Microbiol.* **13**:1379.

Quadling, C., and Colwell, R. R., 1964, Apparatus for simultaneous inoculation of a set of culture tubes, *Can. J. Microbiol.* **10**:87.

Remacle, J., and de Leval, J., 1975, L'application des indices de richesse et d'activité pour la charactérisation microbiologique des sols, *Rev. Ecol. Biol. Sol* **12**:193.

Ridgway Watt, P., Jeffries, L., and Price, S. A., 1966, An automatic multi-point inoculator for Petri dishes, in: *Identification Methods for Microbiologists Part A* (B. M. Gibbs and F. A. Skinner, eds.), *Soc. Appl. Bacteriol. Tech. Ser.* **1**:125, Academic Press, London.

Rosswall, T., 1976, The need for rapid methods and automation in environmental microbiology, in: *Rapid Methods and Automation in Microbiology* (H. H. Johnston and S. W. B. Newsom, eds.), pp. 131–135, Learned Information (Europe) Ltd., Oxford.

Rosswall, T., and Clarholm, M., 1974, Characteristics of tundra bacterial populations and a comparison with populations from forest and grassland soils, in: *Soil Organism and Decomposition in Tundra* (A. J. Holding, O. W. Heal, S. F. MacLean, and P. W. Flanagan, eds.), pp. 93–108, Tundra Biome Steering Committee, Stockholm.

Rosswall, T. and Torstensson, N. L. T., In prep., Effect of MCPA and 2, 4-D on physiological characteristics of soil bacterial populations.

Rowe, R., Todd, R., and Waide, J., 1977, Microtechnique for most-probable-number analysis, *Appl. Environ. Microbiol.* **33**:675.

Rypka, E. W., Clapper, W. E., Bowen, I. E., and Babb, R., 1967, A model for the identification of bacteria, *J. Gen. Microbiol.* **46**:407.

Schmidt, E. L., 1973, The traditional plate count technique among modern methods, in: *Modern Methods in the Study of Microbiol Ecology* (T. Rosswall, ed.), *Bull. Ecol. Res. Comm. (Stockholm)* **17**:453, Swedish Natural Science Research Council, Stockholm.

Seman, J. P., Jr., 1967, Improved multipoint inoculating device for replica plating, *Appl. Microbiol.* **15**:1514.

Silvestri, L., Turri, M., Hill, L. R., and Gilardi, E., 1962, A quantitative approach to the systematics of actinomycetes based on overall similarity in: *Microbial Classification* (G. C. Ainsworth and P. H. A. Sneath, eds.), *Symp. Soc. Gen. Microbiol.* **12**:333.

Skerman, V. B. D., 1969, *Abstract of Microbiological Methods,* Wiley–Interscience, New York.

Skyring, G. W., and Quadling, C., 1969a, Soil bacteria: principal component analysis of descriptions of named cultures, *Can. J. Microbiol.* **15**:141.

Skyring, G. W., and Quadling, C., 1969b, Soil bacteria: comparisons of rhizosphere and nonrhizosphere populations, *Can. J. Microbiol.* **15**:473.

Skyring, G. W., and Quadling, C., 1970, Soil bacteria: a principal component analysis and guanine-cytosine contents of some arthrobacter-coryneform soil isolates and of some named cultures, *Can. J. Microbiol.* **16**:95.

Skyring, G. W., Quadling, C., and Rouatt, J. W., 1971, Soil bacteria: principal component analysis of physiological descriptions of some named cultures of *Agrobacterium, Arthrobacter,* and *Rhizobium, Can. J. Microbiol.* **17**:1299.

Smith, D. A., 1961, A multiple inoculation device for use with fluids, *J. Appl. Bacteriol.* **24**:131.

Sneath, P. H. A., 1957, Some thoughts on bacterial classification, *J. Gen. Microbiol.* **17**:184.

Sneath, P. H. A., 1972, Computer taxonomy, in: *Methods in Microbiology* (J. R. Norris and D. W. Ribbons, eds.), vol. 7A, pp. 29–98, Academic Press, London.

Sneath, P. H. A., 1974, Test reproducibility in relation to identification, *Int. J. Syst. Bacteriol.* **24**:508.

Sneath, P. H. A., and Collins, V. G. (eds.), 1974, A study in test reproducibility between laboratories: Report of a *Pseudomonas* working party, *Antonie van Leeuwenhoek J. Microbiol. Serol.* **40**:481.

Sneath, P. H. A., and Johnson, R., 1972, The influence on numerical taxonomic similarities of errors in microbiological tests, *J. Gen. Microbiol.* **72**:377.

Sneath, P. H. A., and Sokal, R. R., 1973, *Numerical Taxonomy, The Principles and Practice of Numerical Classification,* W. H. Freeman, San Francisco.

Sneath, P. H. A., and Stevens, M., 1967, A divided petri dish for use with multipoint inoculators, *J. Appl. Bacteriol.* **30**:495.

Söderström, B. E., 1977, Vital staining on fungi in pure cultures and in soil with fluorescein diacetate, *Soil Biol. Biochem.* **9**:59.

Sokal, R. R., and Sneath, P. H. A., 1963, *Principles of Numerical Taxonomy,* W. H. Freeman, San Francisco.

Soumare, S., Losfeld, J., and Blondeau, R., 1973, Apports de la taxonomie numérique à l'étude du spectre bactérien de la microflore des sols du nord de la France, *Ann. Microbiol. Inst. Pasteur* **124B**:81.

Sundman, V., 1968, Characterization of bacterial populations by means of factor profiles, *Acta Agr. Scand.* **18**:22.

Sundman, V., 1970, Four bacterial soil populations characterized and compared by a factor analytical method, *Can. J. Microbiol.* **16**:455.

Sundman, V., and Carlberg, G., 1967, Application of factor analysis in microbiology. 4. The value of geometric parameters in the numerical description of bacterial soil population, *Ann. Acad. Sci. Fenn.* **A IV 115**:1.

Sundman, V., and Gyllenberg, H. G., 1967, Application of factor analysis in microbiology. I. General aspects on the use of factor analysis in microbiology, *Ann. Acad. Sci. Fenn.* **A IV 112**:1.

Taylor, J. B., 1974, Biochemical tests for identification of mycelial cultures of basidiomycetes, *Ann. Appl. Biol.* **78**:113.

Torstensson, N. T. L., and Rosswall, T., 1977, The effect of 20 years' applications of 2,4-D and MCPA on the soil-flora, in: *The Interaction of Soil Microflora and Environmental Pollutions,* vol. I, pp. 170–176, Instytur Uprawy Nawoźenia i Bleboznawstwa, Pulawy, Poland.

Véron, M., and Le Minor, L., 1975, Nutrition et taxonomie des *Enterobacteriaceae* et bactéries voisines. II. Résultats d'ensemble et classification, *Ann. Microbiol. Inst. Pasteur* **126B**:111.

Webb, L. J. Tracey, J. G., Williams, W. T., and Lance, G. N., 1970, Studies in the numerical analysis of complex rain-forest communities. V. A comparison of the properties of floristic and physiognomic-structural data, *J. Ecol.* **58**:203.

Wilkins, T. D., and Walker, C. B., 1975, Development of a micromethod for identification of anaerobic bacteria, *Appl. Microbiol.* **30**:825.

Wilkins, T. D., Walker, C. B., and Moore, W. E. C., 1975, Micromethod for identification of anaerobic bacteria, *Appl. Microbiol.* **30**:831.

Yule, G. U., and Kendall, M. G. 1950, *An Introduction to the Theory of Statistics,* 14th ed., Hafner, New York.

2

Limiting Factors for Microbial Growth and Activity in Soil

Y. R. DOMMERGUES, L. W. BELSER, AND E. L. SCHMIDT

1. Introduction

Over the last decades important progress has been made in our knowledge of microbial enzyme machinery and its manipulation, especially in the field of free-living and symbiotic nitrogen fixation, allowing the possible development of new microbial strains or new plant–microorganisms systems in order to improve the quantity and quality of crop yields. However, it must be remembered that the soil microorganism involved in plant associations functions in the complex environment made up of the soil and the lower atmosphere. Thus, the growth and activity of any given soil microorganism depend not only on its genetic characteristics but also on a complex of factors constituting its environment, which finally governs the expression of its intrinsic capabilities. Since the impact of environmental factors on soil microorganisms is often still poorly understood, it appeared necessary to review current concepts concerning this aspect of microbial ecology, focusing attention on the nature and role of limiting factors.

1.1. Concept of the Limiting Factor

A widely used physiological concept known as Liebig's law of the minimum states that "under steady state conditions the essential material

Y. R. DOMMERGUES • Centre Nationale de la Recherche Scientifique, ORSTOM, Dakar, Senegal. • L. W. BELSER and E. L. SCHMIDT • Department of Soil Science, University of Minnesota, St. Paul, Minnesota, U.S.A.

available in amounts most closely approaching the critical minimum needed by a given organism will tend to be the limiting one" (Odum, 1971). This concept is familiar to microbiologists. Thus, it is generally assumed that, in laboratory continuous culture of a given microorganism, the specific growth rate is controlled by only one limiting nutrient. Monod's theoretical model of microbial growth is based on this concept since, in the equation he proposed, the concentration of the limiting nutrient (S) appears explicitly in the function used to calculate the specific growth rate (μ) of the microorganism:

$$\mu = (\mu_m) \left(\frac{S}{K + S} \right)$$

where: μ is the microorganism's specific growth rate; μ_m is the microorganism's maximum specific growth rate; S is the ambient culture concentration of the limiting nutrient; and K is a constant. Some compounds may act as limiting factors only when their concentration is very low. Such is the case of oligoelements (such as molybdenum, copper, zinc, and cobalt) or vitamins.

Liebig's law of the minimum is restricted to chemical material (e.g., phosphorus) necessary for growth and activity. It was later extended to include other factors (e.g., temperature) and to factors that affect organisms because they are in excess (e.g., toxic compounds). It may also be extended to include biological factors. Since the growth or activity of any organism may be limited by either too low a level of a given factor or too high a level of the same factor, each has "an ecological minimum and maximum with a range in between which represents the *limits of tolerance*" (Odum, 1971). This concept, designated as the law of tolerance, is also well known by microbiologists. A classical example concerns the temperature ranges of the psychrophilic, mesophilic, and thermophilic organisms (Table I).

The law of tolerance may explain at least certain aspects of the

Table I. Temperature Ranges of Psychrophilic, Mesophilic, and Thermophilic Bacteria[a]

Bacteria	Minimum°C	Optimum°C	Maximum°C
Psychrophilic	0	15	30
Mesophilic	5–25	18–45	30–50
Thermophilic	25–45	55	60–90

[a]After Salle (1973).

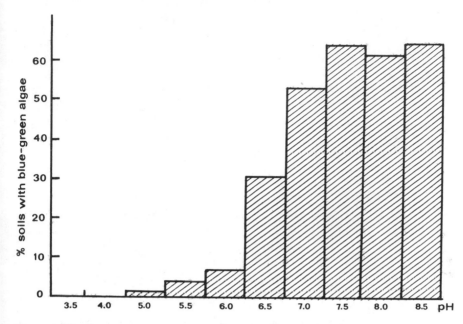

Figure 1. The percentage number of Scottish soils of different pH values showing the presence of blue-green algae after incubation for 30 days in the laboratory (after Fogg *et al.*, 1973).

distribution of microorganisms in the field. A good example concerns the blue-green algae and their preference for neutral or alkaline conditions. This preference, which is evident in the laboratory, appeared to be confirmed in the field since Stewart and Harbott (in Fogg *et al.*, 1973) found that blue-green algae were distributed mainly in soils of pH 7.0–8.5 (Fig. 1).

1.2. Application of the Limiting Factor Concept to Soil

Environmental conditions that are met in complex systems such as the soil–plant–atmosphere system, or the subsystem made up of the soil itself, differ widely from laboratory experimental conditions. Due to the multiplicity of the environmental factors and the variability of the expression of each factor in soil, the law of the limiting factor, as stated above, cannot be used without taking into account:

1. Interactions between environmental factors
2. Alterations of the range of tolerance
3. Variations of environmental factors both in time and in space

1.2.1. Interactions between Environmental Factors

The limiting effect of one factor is often strongly affected by others present. Returning to the example given above related to the distribution of blue-green algae, one should stress that

> It is difficult to distinguish whether the reported effects of pH are due directly or solely to changes in hydrogen-ion concentration or whether other physico-chemical factors are also involved. For example, at high pH levels the poor growth of blue-green algae may be due to soluble iron being precipitated out of solution as ferric hydroxide which thus becomes largely unavailable. Also, low pH levels could lead to an unavailability of molybdenum which is essential for nitrogenase and nitrate reductase. (Fogg *et al.*, 1973)

Moreover, a given limiting factor may affect a biological process indirectly through an intermediary agent, thus eliminating the synchronism between the expected effect of this factor and its actual effect. A good example is that of the effect of the limiting factor of light on nitrogen fixation by *Rhizobium* in symbiosis with legumes (Gibson, 1976).

1.2.2. Alteration of the Range of Tolerance

The limits of tolerance of a given organism may be reduced with respect to other ecological factors. For example, *in vitro* experiments show that most fungi have a wide range of tolerance for pH (3.0 to 8.0), but in soil the limits of tolerance are reduced by the factor competition by bacteria. The bacteria are mostly neutrophilic, so that the limits of pH tolerance for fungi in the soil range from 3.0 to 6.5. Moreover, organisms commonly do not exhibit the same optimal response in nature to a given factor as determined *in vitro*.

The range of tolerance of an organism may be extended for a given factor by virtue of interaction with other factors. The first example concerns an association between a plant and a soil microorganism. The presence or absence of mycorrhizal fungi on the root of conifers is currently thought to be "the most important operative factor controlling the extension or non-extension of tree species" (Langford and Buell, 1969). A second example refers to the range of tolerance of anaerobic diazotrophs in soil. It is well known that in laboratory cultures anaerobic diazotrophs such as *Clostridium butyricum* cannot fix N_2 in the presence of O_2, the latter gas acting as a limiting factor by inhibiting nitrogenase activity. However, when associated with oligonitrophilic bacteria, especially *Pseudomonas azotogenesis*, *C. butyricum* appeared to reduce acetylene actively in a seemingly aerobic environment (Line and Loutit, 1973). Thus, association of a given microorganism with another species can extend significantly its range of tolerance. This case is not a rarity, and the soil is probably a preferential system for associations of this type.

1.2.3. Variations of Environmental Factors in Time and Space

Variations in space are a consequence of the intrinsic heterogeneity of soil; variations in time result from regular and irregular changes of major environmental factors (day length, temperature) and successional changes during substrate utilization. It is well known that soil is made up of a mosaic of microhabitats (microsites), each of which is characterized by its specific set of physical, chemical, and biological factors (Hattori and Hattori, 1976; Marshall, 1975). Obviously such a heterogeneity impedes the resolution of limiting factors. Let us consider for example the problem of pO_2 as a limiting factor for denitrification in soils. Although *in vitro* studies using pure bacterial strains demonstrate that denitrification occurs only in the absence of O_2, experiments carried out with soil samples in contact with air indicate that nitrate and nitrite can be reduced to nitrous acid and N_2, thus suggesting that denitrification is not always a strictly anaerobic process. Such a conclusion is erroneous. The apparent contradiction between reported results vanishes if one admits that soil is made up of a juxtaposition of anaerobic (P) and aerobic (Q) microsites, the overall aeration depending finally upon the proportion of P and Q sites. The gas discontinuity in soil has recently led Flühler *et al.* (1976) to propose the concept that "aeration may be understood as a statistical expression of microscale heterogeneity."

The study of Flühler *et al.* (1976) was based on the assumption that variations of the environmental factor (O_2, in the case studied) are discontinuous. This as well may be a simplification. Most often gradients occur for a number of factors. They may concern very short distances (a few Angstroms, for example, in the vicinity of a clay particle), longer ones (a few millimeters, for example, in the rhizosphere), or a whole soil profile (Reddy *et al.*, 1976). Factors involved are as diverse as ion and organic substrate concentration, O_2 concentration, water tension, and temperature. At a precisely located site within such gradients, specific microbial strains may find their optimum environment; the examples of sulfate-reducing and diazotrophic bacteria thriving in the rhizosphere are good illustrations of such natural gradients (see Sections 5.1 and 5.4).

Since environmental factors such as light and temperature vary widely in time, it is not surprising to observe correlated variations in microbial growth and activity. Thus, diurnal variations have often been demonstrated for nonsymbiotic nitrogen fixation (e.g., Balandreau *et al.*, 1974) and for symbiotic fixation. In the latter case, nitrogen fixation,

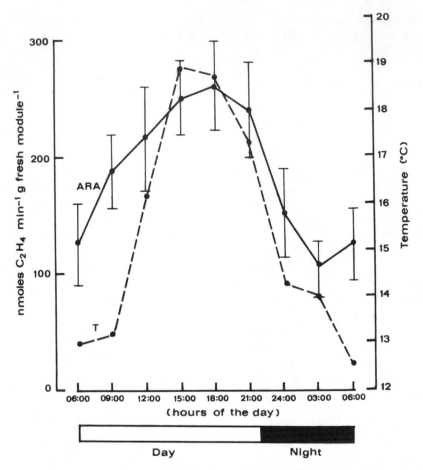

Figure 2. Diurnal changes in acetylene-reducing activity (●——●) of nodulated, 12-week-old seedlings of white clover; 95% confidence limits attached to data. Root temperature changes (●‑‑‑●) are shown (after Halliday and Pate, 1976).

measured by the acetylene reduction method, appears to run parallel with changes of temperature (Fig. 2) or other environmental parameters, especially light (Wheeler, 1971; Mague and Burris, 1972; Sloger *et al.*, 1975; Masterson and Murphy, 1975; Ayanaba and Lawson, 1977).

1.2.4. Modified Concept of Limiting Factors

We have emphasized the complexity of the network of factors implicated in the functioning of the soil or in the soil–plant–atmosphere system, thus calling into question the fate of the limiting-factor concept as stated in Section 2.1. Fortunately, whatever complex makes up a given system, only a few factors are operationally significant for a given organism placed in a given situation. Thus, ecology "can focus its attention, initially at least, on those environmental conditions most likely to be critical or limiting" (Odum, 1971). It is not desirable ". . . to make long uncritical lists of possible factors but rather to achieve these more significant objectives: 1. To discover by means of observation, analysis and experiment which factors are operationally significant and 2. to determine how these factors bring about their effect on the individual population or community, as the case may be" (Odum, 1971). The limiting-factor concept appears to be adaptable to the complex systems of the soil with the proviso that it be restated as follows: The growth and activity of a given microbial population in a complex system depends on the combination of several limiting factors acting collectively and interdependently. Thus, the activity of autotrophic nitrifying bacteria is essentially controlled by three limiting factors: pH, soil ammonium content, and O_2 concentration in the soil atmosphere and soil solution. The activity of denitrifying bacteria is generally governed by three limiting factors: amount of compounds acting as a source of electrons, soil nitrate content, and O_2 concentration in the soil atmosphere and soil solution. Similarly, nitrogen fixation (measured by C_2H_2 reduction) by soybeans was reported to be closely correlated with light intensity and air temperature (Sloger *et al.*, 1975).

The prerequisite of the success of the ecological approach is to recognize the limiting factors involved. This is not alway easy as far as complex systems are concerned. The earlier example of the limiting effect of pO_2 on denitrification is a good illustration of the type of difficulty that may be encountered in the exploration of limiting factors. After the main limiting factors have been determined, thanks to monovariate or, better, to multivariate analysis, it is necessary to carry out a quantitative analysis of the system. Here the objective is to evaluate the amount of the limiting factors present, to measure their normal and occasional variations, and to relate these data to variations in the microbial process to be studied. Such an approach is involved in the building of models (Section 6).

The situation is much less complex in the case of the extreme environment, where the limiting factor is unique and easily detected. Soil moisture tension and desiccation are obviously the limiting factors for microbial activity during the dry season in arid and semiarid soils; acidity can be easily evaluated in exceedingly leached soils or in newly drained sulfur soils; and the absence of energy-yielding substrate will be a dominating factor in tropical soils maintained in bare fallow.

2. The Soil Habitat

Microorganisms as they occur in soil are not uniformly distributed in a uniform environment. This is worth nothing since studies of factors limiting development of microorganisms commonly are derived under controlled conditions with a pure culture distributed uniformly on or in a uniform growth medium. Studies to examine microbial limitations directly in the soil or designed for extrapolation to the soil must consider the complexities of the diverse microbial population, of the soil environment, and of the interactions of microbes with that environment. Little is known with precision as to the details of these complexities because the classical methodology of soil microbiology has been clearly inadequate. Increased interest in the soil biological system during the past two decades, however, has not only pointed up the inadequacies of the old methodologies but has had a heuristic effect to encourage the introduction of new and innovative approaches. Although the complexities still far outweigh the research technology, prospects for realistic insights into the behavior of microorganisms in soil, if not bright, may at least be viewed with guarded optimism.

2.1. Heterogeneous Distribution of Microorganisms in Soil

The close association of soil microorganisms with the particulate fabric of the soil is clearly evident in a simple light microscope study of soil. Glass microscope slides recovered and stained after only a few days' burial in soil demonstrate in an interesting and convincing way the attachment of microorganisms to surfaces. Easily inferred from such preparations is the heterogeneous and discontinuous nature of the soil matrix itself. The photomicrographs taken many years ago by Starkey (1938) show the wealth of intriguing features that the buried slide can reveal. They point out some localization of bacteria on the glass in the vicinity of mineral particles, greater localization where bits of organic debris occur, and the consistently still greater density of microorganisms on those slides where the particles are recognizable as plant roots—the rhizosphere.

Whereas the buried-slide–ordinary-light-microscope technique did clearly document localization and heterogeneous occurrence of microorganisms in soil, it fell short of describing the geography of microorganism occurrence in the undisturbed soil. The capability to visualize the microorganism in relation to the microsite occupied is a quite recent development, involving the application of the scanning electron microscope (SEM). Unfortunately, the SEM has not been applied widely enough as yet as a probe of microbial microsites in soil, but the few studies made have been instructive. Of special interest are the spatial relationships in which bacteria were seen as microcolonies, with apparently great expanses of uncolonized space around the microcolony (Todd *et al.*, 1973). Even in the rhizosphere, where microorganisms were generally thought to cover virtually all of the root surface, SEM observations have shown that the area of the root occupied by microorganisms is only a fraction of that potentially available for colonization (Rovira and Campbell, 1974; Old and Nicholson, 1975).

2.2. Heterogeneous Physical Nature of Soil Habitat

The view that electron microscopy presents of microorganisms widely scattered about as microcolonies on soil particle surfaces is not a promising one with respect to the measurement of factors that limit their distribution and development. Because of the complex physical arrangement of the soil particles, it becomes difficult to impress a single experimental factor on a soil system with assurance that the factor is expressed uniformly throughout the soil and will reach the microsite of interest. The SEM shows promise for realistic description of the way things are at the microsite level, but it can provide only limited information about the dynamics of those microsites when limiting factors are applied or relieved.

Even before the SEM, it had been generally conceded that nearly all soil microorganisms are attached to particle surfaces. Microorganisms growing in the adsorbed state are commonly altered in response to environmental factors, respiration, and metabolic activities as compared to growth in nonparticulate circumstances. When adsorbed-versus-free comparisons have been made, the impact of adsorption has varied greatly between bacteria and fungi and among bacteria and has been affected markedly by the nature of the clays that aggregate to form adsorbing surfaces (Stotzky, 1972).

2.3. Microbial Diversity in Soil

Further complications stem from the diversity of the microorganisms of the soil. Growth habits of unicellular bacteria differ from those of filamentous bacteria and from those of fungi. This may not be a complication in some instances, as when major limiting factors are evaluated according to activity and the activity index is a general microbial parameter. Such was the case in the study by Stotzky and Norman (1961), in which the objective was to examine substrate, nitrogen, and phosphorus as limiting factors for the degradation of glucose in soil. Glucose was chosen as a substrate utilizable by the vast majority of soil bacteria, and CO_2 evolution was the indicator of its utilization. The authors concluded that the soil responses were in line with expectations based on pure culture studies. An entirely different problem is faced if growth response is to be followed independently of activity or is to be related to activity. Here the diverse microbes involved use the substrate at different rates, with different metabolic efficiencies, with different synthetic consequences, and with different growth habits, and no methodology exists to estimate associated growth.

Still another problem resides in more specific transformations—those carried out by a segment, and sometimes a small segment, of the soil population. Such transformations—nitrogen cycle events, pesticide breakdown, possibly the formation of a particular antibiotic, or the consequences of a specific microbe–plant interaction—are the ones that deserve priority in terms of their underlying microbiology. A case in point is found in nitrogen fixation, in which one of the current research needs has been highlighted (Evans, 1975) as the establishment of factors governing the type and population size of free-living, nitrogen-fixing bacteria associated with certain plants. In the free-living, nitrogen-fixing arena, *Spirillum*, for example, has attracted much attention because of reports that it fixes nitrogen in association with grass roots (Day and Dobereiner, 1976). If this relationship is to be evaluated with respect to its exciting potential for enhancing world food production, it will be necessary to relate the growth of *Spirillum lipoferum* to its nitrogen-fixing activity in the grass rhizosphere. Measurement of the nitrogen-fixing activity alone will not be adequate because other nitrogen fixers may be present unpredictably and unrecognized, while a real issue must be the way in which changes in *Spirillum* are related to nitrogen fixed. The methods to quantify *S. lipoferum* as it may occur in the rhizosphere, and as its response there to physiological factors may be reflected in nitrogen-fixing activity, are not yet at hand.

3. Methodology for Study of Limiting Factors in Soil

Mere identification of the factors likely to limit a given soil microbial process is easily inferred once the general microbiology of the process is known. The chasm between this level of understanding and one that provides a basis for predicting the impact of major factors on the way a microbial process will proceed in the soil is vast and presently unbridged. Predictive capability cannot be achieved without the availability of methods that will allow for adequate quantification of the process and of its microbial causation. The restrictions thus far stem from the inadequacies of conventional methodology.

3.1. Process Chemistry and General Microbial Indices

One approach has been to measure the process chemistry alone and either disregard the microbiology base or merely infer microbial activity from the chemical data. Since the chemistry involved in following a process is often relatively simple, it is not surprising that the inevitably complex microbiology is avoided. Estimates of microbial activity based on the process alone may be extremely valuable, as has been the case for symbiotic nitrogen fixation in legumes. A great deal, for example, has been learned about major environmental factors limiting fixation under field conditions with only yield, dry weight, or total plant nitrogen as the indirect indicator of the microbial activity. Here the relationship is so tightly specific that quantifying the plant response often provides a useful measurement of the microbial activity involved. If the problem is switched to determine how to predict or control free-living nitrogen fixation, it becomes necessary to establish which of the nonsymbiotic fixers are involved in a given circumstance. It is necessary to relate microbial response to fixing activity because the different organisms will have different limiting factors. Thus, free-living nitrogen fixation may now be assessed by some sensitive instrumentation in various environments and specific circumstances (Knowles, 1977), but the data will be of little predictive value until the circumstances are defined with respect to the microbiology of the process.

Another common approach has been to use some index of microbial development such as dilution plate counts, most probable number (MPN), or respiratory activity to correlate with a microbial process. Usually the process is a fairly general one, such as cellulose decomposition or ammonification, involving diverse microorganisms. The plate count has undoubtedly been the microbial indicator used most extensively and probably with the most frustrations because of its severe limitations (Schmidt, 1973b). Speaking of the various soil respiration techniques, Stotzky (1965) probably reflected a still-held consensus in concluding that CO_2 evolution is the most appropriate approach despite its numerous limitations. But any procedure attempting to couple the microorganism and the process faces severe challenges in view of the complexity of the soil and the diversity of the soil population. Some interesting new possibilities have developed in recent years in connection with the examination of various environments including soils. It is worthwhile to take note of these with respect to the potential they may have in relating the growth and activity of the microbial cell to some biochemical process in the soil.

3.2. Indicators of Microbial Activity

3.2.1. Isotope Tracers

General heterotrophic activity measurements have been greatly refined with the advent of isotope labels. Use of ^{14}C-labeled substrates provides a means of following decomposition with great precision and for observing the effects that result when major factors are manipulated. Special radiorespirometers have been devised to measure continuously the [14]CO$_2$ released over a period of time in a ^{14}C-substrate mineralization experiment, and these respirometers have been used to a limited extent in soil studies. The potential of labeled carbon substrate for kinetic analysis of soil has been discussed by Mayaudon (1971) and is illustrated by his radiorespirometric data showing the limiting effects of pH on the mineralization rate of [^{14}C]glucose in a meadow soil (Table II). A number of practical and theoretical considerations in the use of ^{14}C-substrates for heterotrophic activity are presented by Wright (1973). Most tracer studies applied to soils have used labeled glucose as the substrate—a convenient and useful choice to meet certain objectives, but one which is uninformative with respect to the major substrates that reach the soil. Some very important possibilities have been opened up just recently to expose the transformations of lignin and lignocellulose to study in natural environments where the microbial ecology of these major substrates is virtually unknown. Experimental programs are underway using synthetic ^{14}C-labeled lignins (Hackett *et al.*, 1977) and naturally occurring but specifi-

Table II. Effect of pH on the Variation of Maximum Mineralization
($A_{eq.}$) and Initial Mineralization Rate (V_{in}) in Fresh,
Sieved Meadow Soil after the Addition of [3,4-^{14}C]Glucose [a,b]

pH	$A_{eq.}$	V_{in}
5.0	24.0	3.15
5.5	26.0	3.91
6.0	27.0	4.71
6.5	29.0	5.78
7.0	29.5	6.68
7.5	30.0	4.51

[a]Experimental conditions: ^{14}C-substrate: 50 μl/0.2 μCi/3.6 μg C added to 1 g soil (dry weight). V_{in} expressed as μg C ([14]CO$_2$)/min/100 g soil; $A_{eq.}$ expressed as % [14]C.
[b]After Mayaudon (1971).

cally labeled lignocelluloses containing ^{14}C in either their lignin or cellulose components (Crawford *et al.*, 1977). The techniques are now available for the first time to examine the factors that affect and limit the rate of important natural substrate turnover in nature. These data are needed badly for the construction of simulation models.

3.2.2. Acetylene Reduction

Because of its obvious importance to food production and soil fertility, the process of nitrogen fixation has been studied extensively with regard to factors that limit its exploitation. Of the many methods used as indices of nitrogen-fixing activity, ranging in sophistication from yield data to isotope labeling and analysis (Hardy and Holsten, 1977), the most recent, acetylene-reducing activity, has attracted great interest. The sensitivity (1000–10,000 times that of ^{15}N methods), rapidity, and convenience of the assay have led to its widespread adoption in the decade since its introduction; it has been said to represent one of the most important developments in nitrogen-fixation research (Hardy *et al.*, 1968). In addition to the unquestioned value of the assay for research at the biochemical and molecular level, the technique has been used with great success for the detection of nitrogen-fixing competence in bacteria and the detection of nitrogen-fixing associations. However, problems other than detection confront the researcher whose aim is to determine the factors that exert control of nitrogen fixation in field soils. An important objective in symbiotic legume-fixation research is to determine the factors that limit the process and to determine how best these limits may be relieved. Yield data, Kjeldahl *N*, etc., cannot address all such problems. The acetylene-reduction assay may be extremely useful in this context (Hardy *et al.*, 1968; Hardy and Holsten, 1977) as an additional indicator of the microbial activity involved. But clearly the method must be used cautiously in the field since problems of sampling are severe, diurnal variations occur, unpredictable energy loss through H_2 evolution may occur (Schubert and Evans, 1976), and various other shortcomings pertain (Ham, 1977).

Factors affecting the important soil transformations of nitrification and denitrification are also studied frequently simply by following the process chemistry. This approach is compromised in both instances by possible interactions between the two processes or with other nitrogen-cycle events during the experiment. The more natural the circumstances studied, the more difficult it becomes to isolate a given process for study. For denitrification at least, a new approach seems capable of providing the isolation needed for short-term experiments even under field conditions. This possibility was noted almost simultaneously by Balderston *et al.* (1976) and Yoshinari and Knowles (1976) in connection with observations that N_2O formed by pure cultures of denitrifying bacteria was not reduced in the presence of C_2H_2. Since the C_2H_2 does not affect the nitrate- and nitrite-reduction pathways, the accumulation of N_2O is a conveniently quantifiable index of denitrification. A big advantage results since the further reduction of N_2O in denitrification sequences results in products that are difficult to quantify.

Balderston *et al.* (1976) reported preliminary experiments to demonstrate the potential of the acetylene-reduction technique as an indicator of denitrification in a marine sediment, and Yoshinari *et al.* (1977) reported experiments carried out in soil. The latter authors found that incubation of soil in the presence and absence of C_2H_2 permits assay of both denitrification and nitrogen fixation and provides information on the mole fraction of N_2O in the products of denitrification. The limiting effects of available energy on both denitrification and nitrogen fixation occurring in anaerobic moist soil are shown in Fig. 3.

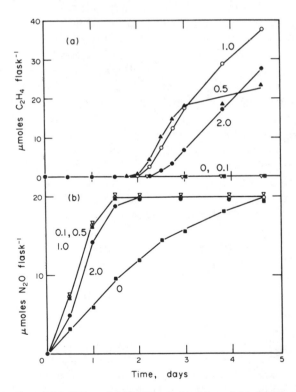

Figure 3. Production of (a) C_2H_4 and (b) N_2O by anaerobic moist soil in the presence of 0.1 atm C_2H_2, 40 μmol nitrate per flask and 0, 0.1, 0.5, 1.0, and 2.0% glucose as indicated (after Yoshinari *et al.*, 1977).

3.3. Indicators of General Microbial Numbers or Biomass

Even though end-product analysis or substrate disappearance alone may provide valuable insights into soil microbial processes and their major regulatory factors, the ideal involves the coupling of process chemistry to microorganism response. Different microorganisms or different suites of microorganisms may be responsible for the same process under different conditions, and the transformations of interest may overlap and interact to obscure the process chemistry. By knowing what microorganisms are involved and their population or enzymatic dynamics relative to the process, extrapolation from the specific to the general will be more secure.

The more generalized processes—those participated in by many different kinds of microorganisms or by one or more diverse "physiological groups"—present difficult problems. Plating or solution-culture methods suffer from selectivity and artificiality; direct microscopic methods are subject to grave limitations but appear far more promising for quantifying relationships between microbe and process. Electron microscopy eliminates the resolution problem that complicates distinguishing between microorganisms and inert artifacts in light microscopy but introduces other problems. The great resolution has its premium in the extremely small area examined, so that where quantification may be attempted, as with the SEM, sample variability is a dominant consideration. SEM has considerable promise, however, in estimating microbial biomass in soil, but it has not been used in this context, probably because of instrument cost, sample preparation problems, and sampling difficulties. Transmission electron microscopy (TEM) is useless for quantifying microorganisms in soil but possibly may be of some value in the rhizosphere (Foster and Rovira, 1976). However, the sophisticated and lengthy procedures involved in sample preparation and difficulties with respect to microscopic field orientation and size do not add to the attractiveness of the TEM methodology.

3.3.1. Direct Microscopy

Prospects for estimation of microbial response in soil by light microscopy have been greatly enhanced by methodologies combining membrane filtration and fluorescent dyes. Procedures outlined by Babiuk and Paul (1970) using the fluorochrome fluorescein isothiocyanate (FITC) have proven useful for enumerating microorganisms during soil transformations (Shields et al., 1973). Increasing availability of microscopes allowing visualization by reflected fluorescence (epifluorescence) makes the use of nonspecific fluorochrome stains still more interesting. With such instrumentation, samples may be dispersed, passed quantitatively through a membrane filter of appropriate porosity to retain bacteria, and stained with the fluorochrome, and the fluorescing cells can be counted relative to the amount of soil represented.

The epifluorescence total count protocols are now best advanced for aquatic systems. Hobbie *et al.* (1977) outlined procedures, precautions, and shortcomings for membrane epifluorescence examination of samples stained with acridine orange. The essentials of the approach as well as the precautionary considerations are applicable to soils (E. L. Schmidt, unpublished observations) whether using the acridine orange procedure of Hobbie *et al.* (1977) or FITC following procedures outlined by Fliermans and Schmidt (1975b). The still simpler acridine orange staining procedure for enumeration of total bacteria in aquatic samples proposed by Ramsey (1977) also appears to be excellent for soil samples as based on limited data (E. L. Schmidt, unpublished observations).

3.3.2. Biomass Estimates by Indicator Chemical Constituents

Numerous types of biomass estimates have been proposed based on the occurrence of chemical constituents that are consistent features of microorganisms. One such indicator of microbial biomass is adenosine triphosphate (ATP). Initially proposed for analysis of water and sediments, it has found widespread usage in such materials only despite many drawbacks and controversial aspects. Some early enthusiasm was generated (Ausmus, 1973), but the problems of differing ATP contents for different microorganisms, of varying ATP with physiological state in the same organism, and of ATP extraction from soils indicate that the approach has no practical utility for measuring dynamic aspects of the soil population.

Muramic acid was suggested as an indicator of microbial biomass in soil by Millar and Casida (1970). Further development of the assay and consideration of its application in estuarine and marine samples was undertaken by King and White (1977). These workers found the method promising; it is, however, time-consuming, requiring about a week to do four or five samples, requires large sample size, and is technically complex in view of the purification requirements. Should some of these difficulties be resolved, the muramic acid assay would be worthy of reexamination.

A new indirect method (Watson *et al.*, 1977) for estimating bacterial biomass in marine environments has not yet been attempted for soils. The test is performed by analyzing for lipopolysaccharide (LPS), a compound which occurs only in gram-negative bacteria. It will be of considerable interest to see whether this specific and sensitive indicator can be applied to soil where, as in the ocean, the gram-negative bacteria greatly predominate. As so often happens, the highly particulate nature of the soil environment may confound the attempts.

3.4. Enumeration of Specific Microorganisms

Certain of the microorganisms in soil are linked closely, sometimes obligatorily, to specific processes that they catalyze. When ammonia is present in a soil and certain conditions prevail, that ammonia will be oxidized; if conditions are not suitable the ammonia will not be oxidized, but when it is, the activities of a small, well-circumscribed group of autotrophic bacteria—the nitrifiers—are responsible. Somewhat parenthetically, it should be noted that, despite all claims to the contrary, no instance of nitrification in nature by a heterotrophic microorganism has yet been demonstrated unequivocally. Again, when special circumstances occur, a certain relatively few kinds of bacteria fix atmospheric nitrogen; here altered circumstances may involve other nitrogen-fixing forms but still of limited diversity. Such classical examples of microbe–process specificity could be augmented by less well known instances of attack on refractory pesticide molecules or perhaps the production of an antibiotic substance. In all such instances, it is desirable to study the process and the microorganism together in limiting factor experiments.

3.4.1. Selective Media

A great many media have been devised with highly developed selective properties for the isolation of specific microorganisms; however, those that can be used to quantify certain soil microorganisms with adequate efficiency are unusual. The culture techniques fail to mimic the natural environment well enough, or are imprecise, or both. The free-living nitrogen-fixing genus *Azotobacter* is among the few soil bacteria that should be well suited to selective plating based on simply constituted, nitrogen-free media. Such media do in fact work reasonably well and may be used to follow the fate of particular *Azotobacter* inoculated at high densities relative to the soil population, as was done with *Azotobacter paspali* (Brown, 1976). Selective plating is less effective at the low population levels of *Azotobacter* usually found in soil, since plates at low dilutions develop high numbers of oligonitrophiles which obscure the *Azotobacter* (Dunican and Rosswall, 1974). Hegazi and Niemelä (1976) suggest a membrane-filter counting technique for soil with low *Azotobacter* density; they note also the poor precision and erratic counts found with the MPN technique. Ideally suited as *Azotobacter* may appear to be for selective plate enumeration, there remains an urgent need for a reliable method of assessing populations of it and other aerobic nitrogen-fixing bacteria (Postgate, 1972).

An increasing array of antibiotics and inhibitors that may be added to media to increase selectivity gives some promise of extending the scope of specific microorganisms that might be enumerated in soil. This approach has been successful for selective plating of fluorescent pseudomonads (Simon and Ridge, 1974) and agrobacteria (New and Kerr, 1971) but thus far has been useless for rhizobia (Pattison and Skinner, 1973). The possible introduction of antibiotic-mutant strains of rhizobia into soil and subsequent enumeration by selective plating on appropriate antibiotic-containing media was first proposed by Obaton (1971), but thus far the method can be applied only to a very few soils.

3.4.2. Immunofluorescence

Easily the most exciting possibility for studying the population dynamics of a specific microorganism in relation to its biochemical activity in the soil is inherent in fluorescent-antibody (FA) techniques. Antibodies are prepared against a particular microorganism of interest and labeled with a fluorescent marker; this "fluorescent antibody" is then used as a stain of a microscopic specimen of the natural environment to detect, with great specificity and sensitivity, the organism of interest. Detection is afforded by an appropriate microscope. The approach, complications, and precautions have been described by Schmidt (1973a). FA techniques were extended by Bohlool and Schmidt (1973b) and Schmidt (1974) to the quantification of a specific bacterium in complex environments. The quantitative approach involves dispersion of a soil sample to release bacteria, flocculation of dispersed soil to leave the bacteria in suspension, and distribution of the suspended bacteria onto a membrane filter, which is then stained with FA and examined with epifluorescence microscopy. Figure 4 shows the growth curves of two strains of soybean rhizobia to illustrate the capability of estimating numbers of one bacterium in a context of the total mixed population in soil.

A later refinement was introduced (Fliermans and Schmidt, 1975a) which combined immunofluorescence and autoradiography to allow simultaneous observation and recognition of a particular bacterium while assessing its activity with respect to a given isotope. The bacterium is recognized by FA, and its metabolic activity is interpreted from the silver grain development about the same cell. These combined techniques appear to have the potential of relating biomass activity and process chemistry to a degree unmatched by any technique used at the present time.

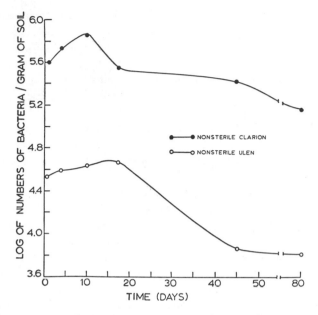

Figure 4. Growth curves of *Rhizobium japonicum* 110 in field soil samples of Clarion and Ulen soils, based on direct counts of *R. japonicum* 110 by the quantitative FA-membrane filter technique (after Schmidt, 1974).

4. Major Limiting Factors

The list of possible limiting factors in the soil environment considered here is far from exhaustive. We have, for example, omitted dealing with the clay microsite in relation to survival and activity of soil microorganisms (cf. Marshall, 1975; Hattori and Hattori, 1976) because so little is known as yet of its significance. Since it is impractical to review all factors possibly implicated in limitation, attention was restricted to some key factors studied not only in the laboratory but also in the field, which are well recognized as capable of limiting microbial growth.

4.1. Energy

Like other living organisms, microorganisms require energy for their growth and metabolic activity. The energy used is derived either from solar radiation or from the oxidation of inorganic or organic compounds. According to Payne's review (Wiebe, 1971), "an average of 0.118 g of cells is very likely to be generated aerobically or anaerobically from any sort of microbial culture for every kilocalorie of energy removed from the culture medium by the growing cells." This relation underlines the importance of the energy factor as an agent of regulation of microbial growth and activity.

The primary source of energy for ecosystems is solar radiation. Only photosynthetic microorganisms (algae and photosynthetic bacteria) are able to convert it directly for their own use. With the exception of some specific soil types where photosynthetic organisms—especially algae—are abundant (e.g., paddy soils), the bulk of the microflora is made up of chemotrophs, that is, of microorganisms which obtain their energy from the oxidation of either inorganic compounds (chemolithotrophs) or organic compounds (chemoorganotrophs). Since chemoorganotrophs are dominant, it is easy to predict that the soil content of energy-yielding organic compounds should be an important limiting factor. In fact, this factor plays a prominent role governing the growth, the activity, and the distribution of the major part of the soil microflora.

In the soil, the major sources of organic energy-yielding compounds to free-living microorganisms are of plant origin. They may be incorporated spontaneously (e.g., surface litter, root litter, and root exudates) or artificially (e.g., composts or diverse organic manures). A striking example of the limitation of microbial populations due to the lack of organic energy-yielding compounds is that of semi-arid psamment soils. In such soils, the input of organic compounds by plants is limited as a consequence of the relatively low productivity of the plants; on the other hand, biodegradation of organic compounds of plant origin during the rainy season is rapid. Thus, microorganisms very rapidly use up their energetic substrate, and the population numbers tend to be very low. Maintaining humid tropical soils in bare fallow may lead to the same result.

Microorganisms associated with plants, especially *Rhizobium*, actinomycete-like diazotrophs, and mycorrhizal fungi, obtain their energy from the host-plant and thus rely on the photosynthetic ability and photosynthate distribution of the plant. For such systems, light is obviously a limiting factor.

4.2. Light

4.2.1. Light Deficiency

A clear example of the effect of light on microorganisms is related to the legume nitrogen-fixing system. The large requirement for ATP, which has been demonstrated *in vitro* for nitrogenase activity, as well as the considerable energy requirements for assimilating combined nitrogen suggests that photosynthesis and photosynthate distribution frequently control the level of nitrogen fixation by legumes. A direct relationship has been reported by many authors between light intensity, nodulation, and nitrogen fixation (cf. reviews of Lie, 1974, and Gibson, 1977). Insufficient light was shown to act by limiting the supply of carbohydrates to the *Rhizobium*, but light effects are not restricted to photosynthesis since evidence exists that some may be attributable to other mechanisms. By exposing plants to light of different wavelengths, Lie demonstrated a requirement for red light: nodulation was excellent in red and poor in blue light (cf. review of Lie, 1974). Moreover, photomorphogenetic effects on nodule initiation and development induced by light should not be ignored (Gibson, 1977).

4.2.2. Light Excess

If light intensity is limiting when too low, conversely it may be detrimental when too high. High light intensities are thought to be responsible for the "nitrogen hunger" period observed with different legumes. When legumes are exposed to high light intensities, nodule development is retarded, probably because the amount of nitrogen available for the nodule is reduced due to a large increase in the level of carbohydrates in the plant tissues. Shading such plants or providing them with combined nitrogen initiates rapid development of nodules (Fred *et al.*, 1938).

Reynaud and Roger (1976) studied the diurnal variations of acetylene-reducing activity (ARA) of an *Anabaena* sp. bloom in Senegal by shading the bloom with 1, 3, or 5 screens. Light intensity was reduced, respectively, to 60%, 22%, and 7% of that of the control (no screen), and ARA was significantly increased when the bloom received only 22% of the incident sunlight, suggesting an inhibitory effect of high light intensities on ARA (Fig. 5). The limiting effect of excessive light intensity upon the ARA of blue-green algae explains why their biomass is very low in Senegalese paddy fields at the beginning of the growth cycle. The light intensities occurring in Senegal at this time are 70,000 lux at 1300 hr; this increases progressively as the plant cover develops.

4.3. Temperature

Like other organisms, microorganisms are very sensitive to temperature changes; in other words, temperature is a key limiting factor. It must be kept in mind, however, that (1) each species and often each strain has its own minimum, optimum, and maximum temperature, and (2) interactions with other factors occur in the field, which may induce alteration of

Y. R. Dommergues, L. W. Belser, and E. L. Schmidt

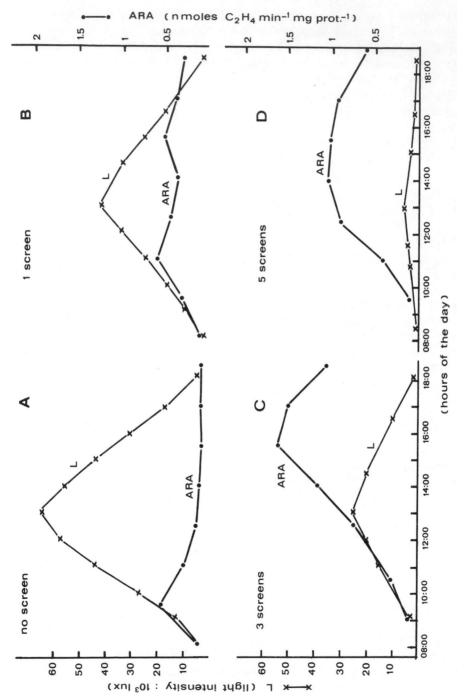

Figure 5. Diurnal variations of acetylene reduction by an *Anabaena* bloom placed under artificial cover (after Reynaud and Roger, 1976).

the response curve to temperature. Moreover, when symbiotic systems are dealt with, one should take into account the effect of temperature on each of the partners and on the association itself. In his recent review, Gibson (1977) draws attention to the possibility of compensation by the symbiotic legume system for adverse effects of moderately low or high temperatures. For example, under moderately high temperatures, nodule numbers and/or nodule weight may be higher than under "optimum" temperature conditions. The greater volume of bacteroid tissue formed under these conditions is interpreted as resulting from the ability of the system to compensate for unfavorable temperature conditions.

4.4. Water Tension

Soil water status is often described in terms of moisture content, but it is much more important to know the actual availability of water. Water availability is expressed as water suction or water tension, the corresponding usual unit being the "bar." The higher the water tension, the less available the water for organisms living in the soil.

The tolerance of soil microorganisms to high water tension varies widely with the species. Such differences in tolerance to moisture stress lead to the predominance of microorganisms that possess a wider range of tolerance to this limiting factor. Generally speaking, bacteria are more exacting than fungi and actinomycetes, a characteristic which explains why fungi and actinomycetes become prevalent as soils dry (Conn, 1932; Dommergues, 1962; Kouyeas, 1964). Another consequence of differential response among soil microorganisms to high water tensions is the accumulation of ammonium nitrogen as soils dry. Under such conditions of high water tension, the activity of the diverse ammonifying organisms is not limited, while that of nitrifying bacteria—the water requirements of which are higher—is inhibited. The accumulation of ammonia that can accompany the mineralization of organic matter at low levels of available moisture is illustrated in Fig. 6.

The limiting effects of low moisture availability may be different depending on the parameter measured. An illustration of this is found in the report of Wong and Griffin (1974), who studied the growth and antibiotic production of 18 streptomycetes in agar culture as a function of osmotic potential. Not only did the isolates vary markedly in individual responses, but the conditions that favored antibiotic formation were distinctly different from those that favored growth. Maximum antibiotic production occurred at high water tension (20–35 bars) when the growth rate was less than half the maximal. Limiting factors other than water tension are, of course, frequently expressed very differently in terms of activity response as compared to growth response.

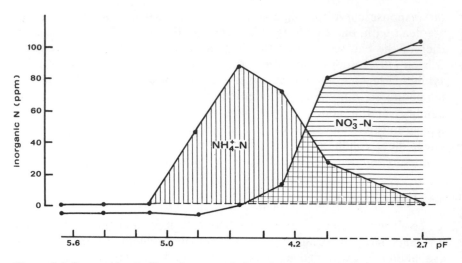

Figure 6. Influence of soil pF on the accumulation of ammonia and nitrate nitrogen. The process of ammonification predominates over nitrification in the pF range of 4.2 to 5.0 (100 bars tension) in this black clay tropical soil of Senegal (after Dommergues, 1977).

4.5. Oxygen

The O_2 content of the soil atmosphere and soil solution is governed by the soil structure and water content; hence, these soil characteristics play a key role as limiting factors for the activity and growth of aerobic and anaerobic free-living microorganisms. Soil O_2 content may also affect symbiotic systems, in particular nitrogen-fixing systems and mycorrhizal associations. As far as the latter system is concerned, it is well known that mycorrhizal initiation can be impeded by low soil pO_2, which derives from the fact that mycorrhizae are strongly aerobic (Harley, 1969) and respire actively (Marks and Kozlowski, 1973). An important practical consequence of the inhibitory effect of low pO_2 on mycorrhizal development is that good soil aeration is essential for successful inoculation (Mikola, 1973). Artificial drainage may be necessary in waterlogged soils in order to improve aeration, thus allowing the extension of the mycorrhizal infection.

4.6. pH

It has already been stressed that the soil differs from laboratory media by the fact that it is a "heterogeneous discontinuous and structured environment composed of particles of various sizes" (Hattori and Hattori, 1976) and thus made up of microhabitats of variable sizes that differ in their own specific characteristics, such as pO_2 (cf. Section 4.5), quality and quantity of substrates, and, of course, pH.

The pH of a given soil sample as it is usually measured is an overall value. Actually soil is formed of microhabitats, each with its own pH. Thus, a soil sample of pH 5.5 may contain microhabitats of pH 7.0 or more and microhabitats of pH lower than 5.5. In such a soil, nitrification may occur even though it is known to be inhibited in a culture medium in which the pH is lower than 6.0. This is actually the case in alluvial tropical soils of pH 5.5 where nitrification is very active. The investigator must proceed cautiously when studying the limiting effect of pH in soil, so as to consider the possible occurrence of a statistically significant proportion of microhabitats that depart from an average pH.

4.7. Mineral Nutrients

Microorganisms behave as active competitors with plants for certain mineral nutrients but are more tolerant than plants to low levels of mineral nutrients. Examples of limitation of microbial growth or activity due to mineral deficiencies have been seldom reported except in the case of extreme environments such as some acid peat soils. In such unusual conditions, phosphorus and calcium limitations appear to be responsible for the low rate of organic matter decomposition (Kin and Dommergues, 1970, 1972); however, in this system it is difficult to separate the limiting effect of phosphorus and calcium deficiencies from that of acidity. Conversely, there are many examples of the effect of excessive levels of inorganic compounds. The problem of the inhibition of nitrogen fixation by soluble combined nitrogen will be dealt with in Section 5.1. Another, though much less studied, example is the effect of high salt concentration on some processes such as *Rhizobium* infection of legumes (Dart, 1974).

4.8. Biological Factors

The consequences of microbial interactions undoubtedly limit the development and activity of certain microorganisms in soil. This is surely the case for those generalized soil processes that involve the growth and activities of many different individual species. The communities that develop in the context of common available substrate and physical proximity change as the substrate is changed; limiting factors are imposed on some as they are replaced by newcomers to the succession. There is ample opportunity for biological factors in the sense of specific growth products, toxins, and lytic agents to function as aggressive devices for the newcomer or as offensive attributes for the functioning microorganisms. Although intuitively logical, the mechanisms and significance of limiting factors of these as well as other biological interactions are virtually unknown.

The question of the significance of antibiotics in natural soil remains as when reviewed 20 years ago by Brian (1957)—highly equivocal. A few fungal antibiotics were detected in extracts made from decaying fresh organic matter, but the limitation of methodology remains. If antibiotics are produced to limit a competitor, the production is at the microsite level, the concentration extremely low, and the persistence of the molecule probably very brief. Equally uncharacterized and unevaluated is the question of specific stimulants and inhibitors other than antibiotics. Rhizosphere dynamics have been examined with this in mind for many years in plant pathological studies. The rhizosphere clearly stimulates the growth and activities of microorganisms, with substrate the main stimulating, regulating, and limiting factor so far as is known. As noted by Deverall (1977), there is little reason to believe that susceptible roots are exceptionally stimulating to parasitic fungi in the soil or that resistant roots are repressive.

Toxins formed by plants have been implicated in limiting certain specific soil processes. Most notable of these is nitrification, regarding which Rice and Pancholy (1973) have hypothesized that climax vegetation inhibits the nitrifiers by means of tannins and root products. Evidence in support of the hypothesis is fragmentary, and there is need to study nitrification and nitrifiers in rhizospheres with improved methodology. Specific inhibition of nitrifiers has not been confirmed. Recent evidence (Nakos, 1975) indicates that lack of nitrification in a forest soil is due to the absence of nitrifiers and not to plant toxins since concentrated extracts of the nonnitrifying soil when added to a nitrifying soil had no effect on nitrification. Bohlool and Schmidt (1977) found that high concentrations of tannin failed to inhibit three genera of ammonia oxidizers and two strains of nitrite oxidizers in pure culture experiments.

Nitrifiers may well be limited, however, by volatile substances of microbial origin under appropriate circumstances in the soil. Bremner and Bundy (1974) found that nitrification was highly sensitive to several volatile sulfur compounds, especially carbon disulfide. Such inhibition could occur in soils at microsites characterized by restricted gas exchange and the decomposition of sulfur-containing amino acids. Volatiles deserve further attention as possible limiting factors for soil microorganisms. Although carried out in culture media rather than in soil, Moore-Landecker and Stotzky (1973) observed that nine species of bacteria and an actinomycete emitted volatiles that induced morphological abnormalities in common soil fungi.

Bacteriophages have often been considered as potential limiting factors in soil ecosystems. Some brief attention is given to bacteriophage in respect to the rhizobia (see Section 5.1.2), but very little is known regarding these and other possible limiting factors of biological origin such as predation and bacteriocins.

5. Factors Limiting Growth and Activity for Selected Processes

5.1. Nitrogen Fixation

The process of biological nitrogen fixation has long been recognized as fundamental to the operation of the entire biosphere. Important advances of the past few decades have contributed greatly to a better understanding of some aspects of the process, while more recent awareness of the need for still much better understanding has developed explosively against a backdrop of world food shortage, petroleum-based energy shortage, and concern for the environment. Suddenly the need to identify the factors that limit biological nitrogen fixation is starkly apparent as a first step in determining how to enhance the process at a practical level.

Biological nitrogen fixation is now under virtually continuous review. Recent summations have discussed fixation by free-living prokaryotes in various environments (Knowles, 1976, 1977) and by symbiotic nitrogen fixers (Gibson, 1976, 1977) with emphasis on the environmental factors influencing their nitrogen-fixing activity. Consideration here will focus only on the legume–rhizobia symbiosis, with the objective of highlighting some of the limitations that pertain before the symbiosis is established. This emphasis is justified in view of the fact that the legume symbiosis is the best understood and the most practically important of the fixation systems and offers the best prospects for the manipulation of limiting factors; moreover, the limitations that relate to rhizobial growth and survival in the soil and to colonization in the rhizosphere are poorly identified, little studied, and clearly crucial.

5.1.1. Soil Physical–Chemical Factors Limiting Nodule Initiation

The physical and chemical parameters have been considered extensively but usually in relation either to the performance of the symbiosis or to the rhizobia as inferred from growth response of inoculant cultures to laboratory manipulations. Study of rhizobia directly in the soil is needed, but the subject has been badly neglected because of methodological difficulties. Moisture stress and both low and high temperatures may have adverse effects on nodule initiation, nodule leghemoglobin content, and rhizobia survival (Gibson, 1976), even though rhizobia are mesophilic bacteria that function in subarctic, temperate, and tropical regions. The question of tolerance to interacting factors of high temperature and desiccation is particularly important to the enhancement of nitrogen fixation in semi-arid and tropical agriculture. Data are presently too incomplete to provide useful generalizations. Soil pH is generally conceded to have marked effects on the survival of rhizobia, and especially so in the acid range. In an important study by Ham *et al.* (1971), pH emerged as a likely major factor interacting with other soil factors associated with the dominance of certain serotypes of *Rhizobium japonicum* in Iowa soils. It is worthwhile to gain more information on the part played by pH on the population dynamics of rhizobia because such information might lead to some control of strain dominance. If different strains respond to changes in soil pH and related factors and such changes are reflected in nodulation, soil management by liming or fertilization could prove a practical means to foster the more desirable inoculant strains.

One of the most significant of the factors that may limit legume nitrogen fixation is the presence of combined nitrogen in soil. Combined nitrogen at low concentrations generally inhibits nitrogenase function, but not growth, of nitrogen-fixing bacteria. Effects of combined nitrogen in the soil usually are expressed in both the inhibition of nodulation and the functioning of the nodules. Various hypotheses have been suggested to account for a specific inhibitory mechanism for combined nitrogen, especially nitrate, and these have been summarized by Dixon (1969). Munns (1977) noted that there are exceptions to the nitrate inhibition response that have confused understanding of its physiological basis. It is clear that better understanding of this well-identified limiting soil factor is needed, for nitrate effects may be expressed at concentrations on the order of 1 mM. The available nitrogen status of a soil may thus be so low as to limit crop production but still high enough to interfere with the use of biological nitrogen fixation as a device for restoring combined nitrogen to the soil. Studies to date have concentrated on the symbiotic response and have largely overlooked possible effects of nitrate on the free-living rhizobia as they interact in the rhizosphere of the developing host plant.

Certain soil management practices can be limiting to the rhizobia. Those of most obvious concern are the fungicides or other pesticides so applied as to contact rhizobia inoculants. Some fungicides and insecticides can be detrimental to inoculants, but undesirable effects of specific pesticidal chemicals on soil rhizobia are still poorly documented. Much the same special situation applies to the general status of the soil and its management with respect to major plant nutrients and trace nutrients. Any of a large number of these may be limiting in a given *Rhizobium*–legume association, but none is likely to be of widespread importance as a limiting factor.

5.1.2. Soil Biological Factors Limiting Nodule Initiation

A very wide range of biological interactions of possible importance to rhizobia in the soil or rhizosphere is suggested in the literature. Little of this work relates directly to the natural environment, and no specific interactions have been resolved; the emphasis has been on inhibitory and antagonistic effects demonstrable in artificial culture media. Until experiments are carried out with the rhizobia and the alleged antagonist in the natural environment, all claims for "antibiotic" and "antagonistic" substances as limiting the rhizobia must be viewed with caution.

The possibility that attack by bacteriophage might be a factor limiting the survival of rhizobia was first suggested more than 40 years ago. Phages specific for rhizobia apparently are of common occurrence in soil and nodules. Kowalski *et al.* (1974), for example, found bacteriophages lytic to several host strains of *R. japonicum* in nearly all samples of soil and nodules examined. Rhizobia are known to form bacteriocins, but the role these interesting, largely protein, cellular inclusions may play in nature is unknown. Because bacteriocins are selectively bactericidal and strain specific, it is possible that certain strains of rhizobia maintain dominance in the soil and rhizosphere by virtue of bacteriocin activity directed against introduced strains. Bacteriocins often appear to be defective temperate bacteriophages. An interest in the ecological significance of the bacteriocin system may serve to renew interest in the general question of bacteriophage as a factor limiting the establishment of rhizobia in soil.

Predation of the rhizobial endophyte is another possible limiting factor to the initiation of nodulation. The concept recently received attention from Danso and Alexander (1975). Using high population densities of prey *R. meliloti* in a solution system with various protozoa, it was found that predation ceased when prey density dropped to about 10^6-10^7 cells per millilter. With the lower populations of rhizobia likely to be found equilibrated in soil and the far less ideal conditions there for predator mobility, it seems quite unlikely that protozoan predation would be specific enough or extensive enough to constitute a limiting factor for the rhizobia. The suggestion that the predatory bacterium *Bdellovibrio bacteriovorus* may be a factor in the decline of rhizobia was first made by Parker and Grove (1970). Keya and Alexander (1975) considered predation by *B. bacteriovorus* and found this not to be a factor in the decline of rhizobia in nonsterile soils. Further examination of the possible predation effects of this organism on rhizobia might be worthwhile since *Bdellovibrio* is thought to be widely present in soil and since enumeration techniques capable of adaptation to soil are at hand (Fry and Staples, 1976).

5.1.3. Intrinsic Factors Limiting the Endophyte

Virtually every study of the rhizobia attests to intrinsic differences among the species and strains examined. Of major concern to the effort of removing limiting factors that stand in the way of enhanced legume production are those properties of the *Rhizobium* that contribute to an efficient and effective nitrogen-fixing relationship with an important host legume cultivar. Certain other attributes relate to the aggressiveness and effectiveness with which the *Rhizobium* adjusts to free-living growth in the soil and responds to the presence of the root of a compatible host. Nodule initiation depends on these properties of the rhizobia and the manner in which some of them at least interact with equally significant, and equally uncharacterized, properties of the plant cultivar. According to Burton (1976), the qualities desired in rhizobia are (1) a high degree of infectiveness for the particular host, (2) the ability to fix large amounts of nitrogen, and (3) compatible effectiveness for a broad spectrum of host species and cultivars.

Effectiveness is dependent on the amount of nitrogenase synthesized by the endophyte and the functional response of the enzyme to the conditions in the nodule. The latter conditions depend greatly on the genetics and biochemistry of the host plant (Evans and Barber, 1977) and are of immense importance in maximizing the contributions of biological nitrogen fixation. Certain strains achieve greater effectiveness by somehow combining with the host to develop large nodules or large numbers of functional nodules. These properties are selected for under greenhouse conditions and evaluated in field trials and do not necessarily react similarly or consistently to both conditions. A newly recognized limiting factor in nodule function is the capacity to recycle hydrogen. The nitrogen-fixing reaction generates H_2 and constitutes a loss of energy supplied to the nitrogenase. Schubert and Evans (1976) reported that the majority of nodulated legumes examined lost 30–60% of the energy supplies to nitrogenase as evolved H_2. The energy is not necessarily lost, as the H_2 can frequently be activated by the enzyme hydrogenase. Genetic information for hydrogenase synthesis apparently resides in the *Rhizobium* strain but not to the same extent in all. Evans and Barber (1977) point out the desirability of relieving this limiting factor as soon as possible by use of appropriate seed inocula. Unfortunately, the mere selection of the most appropriate strains possible based on hydrogenase synthesis, nitrogenase synthesis, or any of the undefined attributes of efficiency for their use as a legume inoculant does not solve the problem. There remains the problem of competitive ability of the selected strain or strains.

Competiveness has been described by Burton (1976) as connoting the ability to induce nodulation in the presence of an abundance of invasive strains in the rhizosphere; he has equated competitiveness with the more commonly used term *infectiveness*. Competitive success, however, reflects more than the ability to compete with other rhizobia, for it involves as well the ability to establish in the soil and rhizosphere in competition with a total microbiota overwhelmingly more numerous than the rhizobia present. The widespread reality that makes competitiveness a limiting factor is that introduction of new inoculant strains in the face of a competent naturalized population of soil rhizobia results in a surprisingly low percentage of nodules formed from the applied inocula. Thus, mere inoculation with the new ammonia-insensitive, H_2-recycling, nitrogenase-efficient strain will not necessarily insure that these attributes are brought into play; the odds are that the more competitive and less desirable indigenous rhizobia will appear in the nodules.

The elements of competitive success are not known. The same unknown factors that enter into the capacity of certain rhizobia to establish in a soil in competition with the complex biota probably come into play more intensively in the rhizosphere and are augmented there by additional factors. Away from the root, competition is likely to resolve around the ability of rhizobia to utilize a given substrate in rivalry with other bacteria. In the legume host rhizosphere, there may be specific substrate conditions that favor proliferation of rhizobia, but these conditions are not yet classified and probably favor rhizobia in general along with many rhizosphere bacteria other than rhizobia.

Competition for rhizosphere substrates is a prelude to competition for "nodule sites," whatever those are, on the developing legume root. A recent hypothesis (Bohlool and Schmidt, 1974) to account for the great specificity of the interaction between the rhizobia and legume partners involves mutual recognition mediated by a lectin at the nodule site. In view of emerging experimental support for this hypothesis, one of the limiting factors in the competitive success of a particular strain may be its capacity to present the appropriate sugar on its cell surface in sufficient abundance to insure recognition at the nodule site. Still another attribute of competitiveness was suggested by the discovery of an unusual morphological feature of rhizobial cells (Bohlool and Schmidt, 1976). The feature, an extracellular polar structure found at the very tip of *R. japonicum*, has a functional role in attachment. Strains vary with respect to the incidence of this structure, but this has not yet been related to competitiveness.

5.1.4. Prospects for Relieving Limitations to Nodule Initiation

Genetic manipulations have deservedly been well publicized for their potential to release nitrogen fixation from a variety of current constraints. Plasmid-mediated transfer of genetic material between strains of *Rhizobium leguminosarum* has demonstrated the feasibility of this approach (Beringer and Hopwood, 1976). It is likely that the initial emphasis will continue to concentrate on the need to better understand the genetic material of the rhizobia, to match symbiont and host genetic features, and to mobilize characteristics that bear on nitrogen-fixation efficiency. No account has as yet been taken of the need to insure the survival and competitive success of any "superstrain" that may be developed. Insensitivity to soil nitrogen is about the only competitive attribute identified (Evans and Barber, 1977); genetic features of possible importance to the success of rhizobia in rhizosphere competition should be explored as soon as possible.

Selection of rhizobial strains of desirable properties has always concentrated on nitrogen-fixing efficiency as measured in greenhouse assays. The considerable potential of strain selection and mutant selection with respect to properties pertinent to nodule initiation have been largely ignored. Some initial success along these lines is suggested by the acquisition of some fungicide-resistant rhizobia (Odeyemi and Alexander, 1977). The objective of this work was to obtain strains of rhizobia that can be applied directly to fungicide-treated seeds.

The study of competitiveness and the identification of its attributes are likely to benefit from new methodological capabilities. Advances in fluorescent-antibody techniques (Schmidt, 1974) have provided an approach for the examination of survival and competitive interactions directly in the soil to provide a tremendous advantage over older methodology involving indirect analysis based on host-plant infection. Some use of the FA approach was illustrated in the persistence and competition study of Bohlool and Schmidt (1973a), and possibilities for extension by virtue of a quantification capability were evident in the report of Schmidt (1974).

Despite their importance, the rhizobia have been neglected with respect to cellular morphology, especially as free-living bacteria. Examination of cellular details of *R. japonicum* has recently taken note of several features that are distinctive and intriguing (Tsien and Schmidt, 1977). Figure 7 illustrates the surprising degree of differentiation found in *R. japonicum.* The morphological features observed may comprise attributes of competitiveness evolved to insure arrival of the rhizobia at the nodulation site, recognition at the root surface, attachment, and initial invasion as preludes to the initiation of its ecological niche, the nodule.

Whereas control of soil populations of rhizobia and the acquisition of sufficient knowledge to manipulate the soil adaptation of desirable strains remain a long-range objective, the application of legume seed inoculants is of great immediate and practical importance. Inoculant production technology is highly developed but geared to the developed nations. The methods of inoculant application as well are best adapted to agricultural technology that can recognize and accommodate the biological nature of the inoculant preparation. Pelleting of the rhizobia in a matrix around the

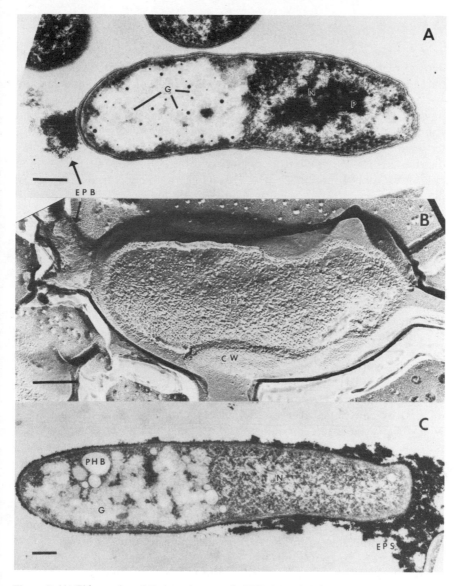

Figure 7. (A) Thin sectioned *R. japonicum* strain USDA 31, showing inclusion polymers and granules of glycogen (G) accumulated at one end of the cell and cytoplasmic material, nucleoplasm (N), and polyphosphate (P) at the other end. Extracellular polar body (EPB) present at one pole of the cell at the polymer end. (B) Freeze fractured USDA 31 cell showing outer fracture face (OFF) of cytoplasmic membrane and cell wall (CW). The extracellular polar body (EPB), surrounded with some microfibrils, is attached firmly at one pole of the cell. (C) Thin sectioned USDA 138 *R. japonicum* stained with ruthenium red during fixation. Inclusion polymers, granules of poly-β-hydroxybutyrate (PHB), and glycogen (G) at one end of cell. Extracellular polysaccharide (EPS) shown as darkly stained ruthenium red-positive accumulations around cytoplasmic end. Bar equals 0.2 μm (after Tsien and Schmidt, 1977).

seed has been under technological development for a considerable period and now affords some prospects for truly reliable preinoculation procedures (Brockwell, 1977). Many new combinations of rhizobia and carriers are under investigation in the hope of achieving an effective and reliable inoculant that could be applied by the small farmer in a poorly developed country. Y.R. Dommergues and colleagues (unpublished observations), for example, devised an inoculant pellet for testing in Senegal. They distributed *R. japonicum* in a matrix of acrylamide that was then polymerized and prepared in cubes. In preliminary trials, the cube inoculant increased nodulation over that obtained with liquid control preparations.

5.2. Nitrification

The process of nitrification occurs in a spectrum of richly diverse terrestrial and aquatic environments; it is important to agriculture, waste disposal, nutrient cycling, and the quality of the environment. So far as is known, the activities of two small groups of autotrophic bacteria are responsible for the process throughout nature. One group, the ammonia oxidizers, initiates the process with the formation of nitrite; a second group, the nitrite oxidizers, completes the process by converting nitrite as promptly as it is formed (nitrite rarely accumulates in nature) to the end product of the process, nitrate. The nitrifying bacteria cannot be studied directly by conventional microbiological techniques. Direct plating, even with strictly inorganic media, is useless because the organic materials introduced with the inoculum permit growth of nonnitrifying heterotrophs, and nitrifying chemoautotrophs, if present, are unrecognizable. Isolation is difficult, requiring tedious enrichment culture procedures. The usual approach to the microbiology of the process is indirect, using some modification of the MPN procedure. The MPN may have great statistical uncertainty and be imprecise even under the ideal conditions of pure culture (Rennie and Schmidt, 1977). Work by Fliermans *et al.* (1974) first demonstrated the potential use of FA techniques as a wholly new approach that has already proved useful for the examination of nitrification in soils, as noted below.

More thorough understanding of nitrification is patently needed in order to control the process in its many applications, such as nitrogen management in field crops, nitrogenous waste disposal in sewage systems, and prevention of groundwater and lake contamination with nitrate. Information on limiting factors serves a critical need for promoting nitrification in some circumstances and preventing it in others. When sufficient information of this type is at hand, it will be possible to construct models with useful predictive capabilities. Prospects for this are still somewhat remote, but the need is increasingly evident.

5.2.1. Major Factors Limiting Nitrification

The relationship between substrate and microbial activity in nitrification is obvious and clear-cut in view of the autotrophic nature of the nitrifiers and their dependence on either ammonia or nitrite as a specific energy source. This dependence is probably completely obligatory in nature, for although certain organic substrates may replace nitrite for *Nitrobacter* (the only chemoautotrophic nitrifier known to use an organic energy source), the organism grows more slowly in the absence of nitrite and has very little nitrite-oxidizing ability (Bock, 1976). A variety of organic compounds may stimulate nitrifiers, but there is no indication that the organic compound is serving as an energy source, even in part, under these conditions (Clark and Schmidt, 1967; Bock, 1976).

Threshold substrate concentrations for soil nitrification are not known, but judging from seawater isolates (Carlucci and Strickland, 1968), growth rates at very low concentrations of ammonia or nitrite are not greatly different from those at very high levels. There is a real need for further critical examination of the extent to which substrate concentration influences nitrification in soil at the lower, more realistic end of the scale. Ammonia and nitrite concentrations may become too high according to pure culture studies, but the relevance of this to soils is probably nil in the absence of other limiting factors, except where fertilizer application may lead to localized inhibitory ammonia concentrations. Substrate does, of course, limit the second stage of nitrification since nitrite does not normally accumulate, and its presence depends on the rate of ammonia oxidation. The fact that nonnitrifying situations occur in soils as indicated by the accumulation of ammonia and the absence of nitrate indicates that limiting factors other than substrate control nitrification (e.g., Fig. 6).

The factors of temperature, pH, aeration, and their interplay can limit nitrification. Tolerance of the nitrifiers to both temperature and aeration is surprisingly good. Nitrification occurred in a manure oxidation ditch at 4°C, and isolation of *Nitrobacter* was accomplished from a hot spring at 65°C (C. B. Fliermans and E. L. Schmidt, unpublished observations). In another study, nitrification down to 0.3 ppm O_2 was noted (Gunderson *et al.*, 1966). In highly organic soils featuring active total microbial activity, it may well be that nitrification is favored by low temperatures as a result of decreased demand for ammonia by heterotrophs and increased solubility of O_2.

A well known and major limiting factor in nitrification is pH. Acid forest soils commonly do not nitrify; however, these and other acid soils are usually found to nitrify well following the addition of lime (Chase *et al.*, 1968). The effect of acidity may be more an expression of aluminum toxicity (Brar and Giddens, 1968). Some acid soils between pH 4 and 5 nitrify slowly (Chase *et al.*, 1968), and this raises questions that have never been adequately resolved regarding the possibility that nitrifying populations of acid environments may be qualitatively different from the usual nitrifiers.

No micronutrient deficiency has ever been reported to limit soil nitrification, but a possibility exists that certain trace metals may inhibit the process. Skinner and Walker (1961) demonstrated the sensitivity of *Nitrosomonas europaea* to Ni, Cr, and Cu at 0.1–0.5 ppm. The pertinence of such metals as limiting factors was pointed out by Wilson (1977), who noted that heavy metals are often present in domestic sewage and industrial sludges in such high concentrations as to merit consideration with respect to their effects on nitrogen mineralization in soils receiving the sludges. He studied the effects of Zn and found that 100 ppm did limit nitrification in two of three different soils. Toxic micronutrients represent another aspect of soil nitrification that bears some emphasis.

5.2.2. Research Approaches Needed

A great deal of the research on nitrification has dealt with either the process chemistry alone or the process chemistry in conjunction with microbiology in enrichment situations. As noted before, the chemistry of nitrification is easily followed, but the microbiology has been sharply limiting, even under enrichment conditions. Great gaps exist in information relative to the normally low substrate conditions that exist in most soils, where available organic carbon is low, ammonia trickles in, and nitrate trickles out. In these situations as well as those more enriched, the gaps need to be filled by approaches that couple the process to the microbiology.

Limiting-factor studies that deal in process chemistry alone relate to the particular complex of environmental conditions under which the measurements were made and provide a poor basis for generalization. The case for coupling the microbiology and the chemistry rests on several points. The interaction of factors is integrated in nature in terms of the population response: the nitrifiers will reflect the interaction but the chemistry need not. Nitrate, for example, may be lost in denitrifying microsites contiguous with those forming nitrate. Only recently has the possibility of concurrent nitrification and denitrification been documented (Belser, 1977). Moreover, the nitrifying population involved in a given set of soil circumstances may not be the same population operative in other soils or in response to altered environmental factors in the same soil. Very little is known about the diversity of nitrifiers. The unfortunately still too common oversimplification, that of equating *Nitrosomonas* and *Nitrobacter* to nitrification, is being questioned, but the question of just how much diversity is to be encountered among the autotrophic nitrifiers is only now being addressed. The correlative question to diversity is that of activity. Microbial growth and activity need to be coupled to process chemistry under controlled conditions to determine what different kinds are involved under different circumstances, how many of each are involved, and how they differ in activity as affected by energy substrate concentration, environmental factors, and inhibitors.

It appears that a number of techniques are adequately developed to merit integrated approaches to study of the limitations that regulate nitrification in nature. The model by Laudclout *et al.* (1976) points the way for evaluating major factors and particularly for the quantification of interactions among the factors. The experimental system used by Laudelout was artificial and the microbial dynamics were not studied, but the possible contribution of mathematical analysis to the total problem is well illustrated.

Prospects for including the microbiological base and relating it to the process even in the soil have been heightened by advances in the FA technique. The data in Fig. 8 taken from Rennie and Schmidt (1974) show the relationship between nitrate production and *Nitrobacter* populations in an acid soil. The populations as measured by FA differ greatly from those estimated by MPN, a procedure which was particularly inefficient for the acid soil. The number of *Nitrobacter* produced, counted by FA, followed the production of nitrate. An increase in population of about 10^6 cells per gram was observed as the result of the oxidation of 100 ppm nitrite. Kinetic analysis of such data is clearly capable of providing additional information on the growth and activity of *Nitrobacter* in soils. In order to capitalize on the evident potential of this approach, more must be learned about the twin problems of nitrifier diversity in soil and nitrifier growth and activity in soil. Then with various factors impressed

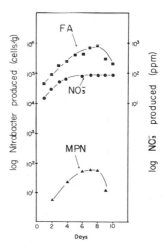

Figure 8. Net production of nitrate and *Nitrobacter* populations (FA and MPN counts) as a function of time following addition of 100 ppm nitrite-nitrogen to an acid (pH 4.7) Hubbard sandy loam soil (after Rennie and Schmidt, 1974).

on the soil, perhaps the impact of these can be assessed in terms of both the nitrifiers and the process that depends on their activity.

5.2.3. Nitrifier Diversity and Activity

Mention was made earlier of the facts that certain heterotrophic microorganisms are known to nitrify and that, although these are sometimes invoked to account for nitrifying activity in nature, there is no evidence that any of the heterotrophs that can nitrify in pure culture in fact ever do so in nature. Those microorganisms that have been implicated in nitrification under natural circumstances are all chemoautotrophic bacteria. The list is surprisingly short: five genera and seven species of ammonia oxidizers, and three genera and three species of nitrite oxidizers. Of these, virtually all of the cultural and biochemical studies have been performed with *Nitrosomonas* and *Nitrobacter*; just how legitimately the data may be transposed to the ecological situation is highly questionable. Are the forms the same in soils, waters, and sewage? Are those responsible for nitrification the classical autotrophs, are some as yet undescribed autotrophs, or are some not autotrophic at all?

In a first look at diversity among the nitrifiers by using FA, Fliermans *et al.* (1974) found surprising uniformity. All autotrophic nitrite oxidizers isolated from a wide range of habitats could be recognized serologically as either the winogradsky or agilis serotypes of *Nitrobacter winogradskyi* (current nomenclature for those traditionally referred to as *N. winogradskyi* and *N. agilis*). This limited diversity among the nitrite oxidizers now seems more apparent than real as different isolation techniques have yielded strains that belong to neither serotype (P.M. Stanley and E.L. Schmidt, unpublished observations).

The ammonia-oxidizing bacteria probably have a greater diversity than the nitrite oxidizers. Early data on this question are in press (Belser and Schmidt, 1978) and are included here as Table III. Examination of isolates obtained from a variety of soils suggested that three genera appear to be common in soil and widely distributed. The groupings into genera as listed in Table III were according to gross morphology, but within each group each isolate fell into a different serotype. How many different species are involved and the extent to which these serotypes differ from one another in nitrifying activity, growth rates, and response to environmental factors are still to be determined.

The autotrophic nitrifying population has already proven to be more diverse than may have been expected based on current taxonomy. Coexistence of different nitrifiers in the same soil is probably common; recently from a single Minnesota soil sample a *Nitrosospira*, a *Nitrosolobus*, and a *Nitrosomonas* were isolated in pure culture (Belser

**Table III. Serological Diversity among Three Genera of Autotrophic
Ammonia-Oxidizing Bacteria as Determined by Immunofluorescence** [a]

Culture [c]	Immunofluorescence Reaction [b]							
	ATCC FA	EK FA	Tara 7-15 FA	SI FA	Fargo FA	BNF FA	SP-1 FA	SP-30 FA
Nitrosomonas								
ATCC	4+	—	—	±	±	—	—	—
EK	—	4+	1–2	±	—	—	±	—
Tara 7-15	—	1+	4+	—	—	—	±	—
SI	—	—	—	4+	—	—	—	—
Nitrosolobus								
Fargo	—	—	—	—	4+	±	±	—
BNF	—	—	±	—	—	4+	—	1+
Nitrosospira								
SP-1	±	—	±	±	±	—	4+	—
SP-30	—	—	—	—	—	±	—	4+

[a] After Belser and Schmidt (1978)
[b] Symbols: —, no fluorescence; ±, trace; 1+ to 4+, increasingly good fluorescence.
[c] ATCC, *N. europaea*, American Type Culture Collection; EK, Morocco soil isolate; Tara 7-15, Fargo, Minnesota soil isolates; BNF, SP-1, SP-30, isolates provided by Dr. N. Walker, Rothamsted Experimental Station, England.

and Schmidt, 1978). The likelihood is that with improved methodology and accelerated interest in the nitrifiers, a picture of morphologically and physiologically diverse bacteria will emerge to replace the current *Nitrosomonas–Nitrobacter* stereotype. This will complicate the microbiological analysis problem but will contribute to a much better understanding of nitrification response to differing environmental conditions.

During growth of microbial populations where substrate metabolism and population increase can be followed, activity per cell and yields can be readily calculated. For growth of *Nitrobacter* in Hubbard soil (Fig. 8; data from Rennie and Schmidt, 1977), an activity of 2.2 ± 0.3 pg N/hr per cell and a yield of $4.4 \pm 0.4 \times 10^3$ cell/μg N can be calculated. The population was growing with an apparent generation time of 76 hr. In pure culture, *N. winogradskyi* growing with a 15 hr generation time has an activity of 0.16 pg N/hr/cell and a yield of 3.7×10^5 cells/μg N (calculated from data of Rennie and Schmidt, 1977).

The question arises, "How can one explain the difference between growth in soil and that in pure culture?" Since substrate is in excess in pure culture, the activity should be maximal. However, when compared to the activity in soil, it is only 7% of the value calculated for soil. Although it is possible that cells in soil have a larger activity per cell due to large biomass (Belser, 1977), it is more likely that the difference in activities observed between soil and pure culture is due to the inability to desorb cells from soil particles, thus counting only a fraction of the cells present. For example, if the counting procedure counted only 7% of the cells present in soil, then the activity per cell in soil and pure culture would be the same. It seems reasonable to assume that, if the cells in pure culture and soil are morphologically the same (i.e., contain the same amount of enzyme), the maximal activity per cell should be the same.

Unless some factor other than substrate limits the rate of oxidation in soil, the activity in soil should be maximal, since substrate was in excess during the growth study just considered. If the activity is maximal, then why does the population grow so slowly, with cells produced so inefficiently? The inefficiency of cell production can be seen by comparing the yields. If it is assumed that only 7% of the cells are counted in soil (i.e., the actual population is 13 times higher than measured), as suggested, six times as many cells are produced in pure culture as are produced in soil. Although some decrease in yield would be expected for increased maintenance per longer generation time (Pirt, 1965), this would be not nearly enough to account for so large a difference in yields. Thus, if the activity per cell in soil were maximal, there would appear to be some uncoupling between oxidation and growth.

On the other hand, if the activity per cell in soil were not maximal, one would have to assume that the counting efficiency was even less than the 7% used previously. If only 1% of the indigenous soil population was counted, the yield in soil would be similar to that measured in pure culture, but the activity per cell would only be 17% of that in pure culture. If this interpretation is correct, then something other than nitrite is limiting the activity per cell. The reduced activity could account for the slowed growth.

The enumeration problem discussed here may be due in part to the fluorescent-antibody stains not covering all the immunological strains of *Nitrobacter* present in soil. However, even if this is true, the basic questions are still the same. When substrate is in excess, is the activity per cell in soil maximal and the same as the maximum activity per cell measured in pure culture? If the activity in soil is maximal, than why do populations grow so slowly and produce cells so inefficiently? On the other hand, if the activity per cell is not maximal in the presence of excess substrate, then what factors limit the enzymatic activity?

5.3. Denitrification

It was noted earlier (Section 3.2.2) that a new approach to the study of denitrification offers considerable promise for the direct quantitative evaluation of the process, perhaps even under field conditions. While still only very preliminary, the possibility is of much interest because satisfactory direct methods have not been available. The usual approach has involved measuring the deficit of either total nitrogen or applied tracer nitrogen in greenhouse, lysimeter, or field experiments to obtain indirect evaluations of denitrification in different ecosystems. A summary of reported data reflecting this approach is given in Table IV.

Such evaluations indicate that denitrification under cropland conditions is widely variable since denitrification losses range from 0 to 40% of nitrogen applied as fertilizer. In paddy and pasture soils, losses are seldom lower than 20–25%. The increased cost and shortage of fertilizer nitrogen especially in developing countries must prompt soil microbiologists to gather more information concerning the factors that could limit denitrification in soils.

Basically denitrification is the process whereby nitrate, through the agency of nitrate reductase and cytochromes, is reduced to nitrite, nitrogen, or nitrous oxide during anaerobic respiration, with the gaseous products of denitrification escaping to the atmosphere. Denitrification can be carried out when denitrifying bacteria find simultaneously, in a given site, anaerobic conditions, a source of energy and electrons, and nitrate, which is the electron acceptor. In other words, denitrification can be limited by only *one* of the following factors: presence of O_2, lack of energy and electron-yielding substrate, or lack of nitrate. Other factors may be involved, especially pH and temperature, but these will not be considered here since their impact on denitrification is less marked in most croplands.

5.3.1. Anaerobiosis

Because of the heterogeneous distribution of microsites within an apparently well-aerated soil, denitrification may occur in anaerobic sites, especially in the core of aggregates, within plant residues, or in the

Table IV. Nitrogen Loss, Recovery by Crops, and Balance[a]

Ecosystem	N loss in denitrification (% of applied N)	Crop recovery of N (% of applied N)	N balance (% accounted for)
Cropland	0–40	50–75	60–111
Pasture	25–35	50–70	70–90
Forest	—	—	(79–90)
Paddy	20–50	26–45	—
Waters	—	—	—

[a]After Hauck (1971).

rhizosphere of many plant species. Air diffusion into these different sites depends on physical factors such as *moisture, structure* (inter- and intraaggregate porosity), and *temperature*. It also depends on *biotic factors*, especially the rate of O_2 uptake by aerobic microorganisms and by plant roots, and the rate of diffusion of air out of some root systems (e.g., rice). As some of these factors may be manipulated, one may limit the expression of denitrification in soil by managerial practices favoring air diffusion, thus reducing the total volume of anaerobic sites.

5.3.2. Energy and Electron-Yielding Substrates

In anaerobic sites, denitrification is controlled largely by the supply of energy and electron-yielding substrates. In soils these substrates originate (1) from readily decomposable organic matter (Burford and Bremner, 1975) and (2) from root exudates and sloughing cells (Woldendorp, 1963; Garcia, 1975).

5.3.3. Nitrate

Since nitrate is the electron acceptor in the denitrification process, nitrate is by definition a limiting factor. Control of denitrification by using nitrogen fertilizers containing no nitrate, such as ammonium nitrogen or urea, is actually ineffective because ammonium-nitrogen or urea is readily nitrified in most croplands. As noted previously (Section 5.2), nitrification may proceed in relatively acid soils, which often harbor neutral microsites, and even in waterlogged soils since ammonium-nitrogen diffuses from the anaerobic layer to the aerobic surface layer where nitrification is active (Reddy *et al.*, 1976). Controlling denitrification must then be achieved by preventing the accumulation of nitrate originating from the nitrification process. Different methods have been proposed involving the use of slow-release nitrogen fertilizers, such as urea-formaldehyde compounds, or the use of specific inhibitors such as 2-chloro-6-(trichloromethyl)pyridine (Goring, 1962).

5.4. Sulfate Reduction

Basically sulfate reduction is carried out by bacteria that use specific substrates (e.g., lactate or pyruvate) as electron and energy sources, with sulfate used as the terminal electron acceptor. Since the physiology of sulfate-reducing bacteria has broad analogies with that of denitrifying bacteria (Le Gall and Postgate, 1973), one may predict that factors limiting sulfate reduction are similar to factors limiting denitrification. In fact, such similitude exists. The three major limiting factors for both processes are presence of O_2, lack of energy and electron-yielding substrate, and lack of appropriate terminal electron acceptor.

5.4.1. Anaerobiosis

Sulfate-reducing bacteria, like the denitrifiers, are anaerobes, but they are more exacting in that sulfate reduction occurs at lower redox potentials than those allowing denitrification (Connell and Patrick, 1968). Since in most soils the number and volume of microsites characterized suitable for sulfate reduction are fewer than those favoring denitrification, evidence of sulfate reduction is not apparent. In soils where air diffusion is hampered by waterlogging or compaction, or where the rate of O_2 consumption by organisms is high, the number and volume of microsites favoring sulfate reduction may increase dramatically, causing damage to crops.

5.4.2. Energy and Electron-Yielding Substrates

Sources of energy and electrons used by sulfate-reducing bacteria are provided by decomposition products of plant origin, essentially crop residues and root exudates. This explains why sulfate reduction was shown to be restricted to the rhizosphere or the spermosphere of lucerne, maize, broad-bean, and rice grown in saline irrigated soils (Dommergues et al., 1969; Jacq and Dommergues, 1971). In sites devoid of proper carbonaceous substrates, sulfate reduction cannot occur.

5.4.3. Sulfate

Soil sulfate may originate from the parent material or from water (water table or irrigation water). Sulfate reduction is apparent only when soil sulfate content is higher than a certain threshold, which is commonly exceeded, especially in saline soils.

Damage to different crops caused by sulfate reduction has been reported frequently (cf. reviews of Le Gall and Postgate, 1973; Jacq, 1975). Such undesirable effects can be alleviated by manipulating one of the three limiting factors already mentioned. Owing to the high sensitivity of sulfate-reducing bacteria to O_2, manipulation of the O_2 factor appears to be most effective. The O_2 level in soil can be increased practically by draining and improving the soil structure; the latter objective can be achieved by using either proper tillage practices or proper rotations including the use of grasses with powerful root systems.

5.5. Formation of Peats and Mor Horizons

No unified explanation has been advanced for the accumulation of organic matter which leads to peat formation or to the development of mor horizons in forest soils. Actually such an accumulation may result simultaneously or separately from (1) the inherent resistance of some plant residues, (2) the development of unfavorable edaphic conditions induced by the plant itself or by the mineral soil substratum, and (3) unfavorable climatic conditions. Unfavorable edaphic and climatic conditions act as limiting factors preventing or delaying the decomposition of plant residues by the soil animals and microorganisms.

Sphagnum, one of the major peat-forming plants, is well known to decompose very slowly in comparison with many other plants. This tendency to resist microbial degradation was supposedly attributed to chemical characteristics of the moss, such as the unusual nature of its cellulose and lignin (Frankland, 1974). More probable is the view that edaphic and climatic factors are responsible for any limitation on the activity of decomposing organisms. Among the major edaphic limiting factors which are involved, one may cite the following: low-base status, acidity, low nitrogen or phosphorus content, waterlogging, and low temperature. Temperature is obviously a key factor in the zonal peats which are formed in cold and wet situations, but other factors are involved, especially in the case of intrazonal peats. Using the multiple-regression-analysis approach, Heal (1971) found that 40% of the variability in decomposition rate of sphagnum could be attributed to nitrogen content.

Factors governing the accumulation of mor in forest soils are similar to those involved in peat formation. Decomposition of some specific litter (especially conifers) was attributed to needle anatomy together with nutrient status of the trees and to the possible production of substances toxic to microorganisms (Melin and Wiken, 1946; Tsuru, 1967; Beck et al., 1969; Harrison, 1971). The influence of soil factors, particularly nitrogen and eventually phosphorus availability, is decisive in many situations. Garrett (1970) attributed the limitation of cellulose to the low nitrogen content of the soil. Since cellulose-degrading enzymes are adaptive, cellulolytic microorganisms require substrate contact before induction. In order to remain in contact with cellulose, the mycelium must maintain active growth and thus requires nitrogen in large amounts for the synthesis of its own protoplasm. The limitation of litter decomposition in a *Pinus jeffreyi* forest was reported to be spontaneously and transiently overcome by the pollen rain (Stark, 1972).

Reclamation of peat soil or improvement of decomposition of

coniferous leaf litter rests upon methods with the object of eliminating limiting factors. These include the use of artificial fertilization (e.g., Roberge and Knowles, 1966), drainage, liming, and other management practices.

6. Modeling Microbial Growth and Activity in Soil

Models which have incorporated limiting factors in microbial ecology fit into the subsystem of ecosystem models depicting mineral cycling and organic matter decomposition. It is the main purpose of such models to predict the rates at which specific microbiologically mediated transformations take place under various environmental conditions. It is hoped also that through modeling a better understanding can be attained of complicated ecosystems where various biological processes interact with one another and with the surrounding abiotic environment.

To achieve these goals, models should be constructed so as to have a physical or biological basis. Despite the rather obvious desirability for such an approach, in some models, often used for management purposes, mechanisms are neglected and algorithms are developed to best fit the data (e.g., Endelman *et al.*, 1973, 1974; Haith, 1973).

6.1. Components of the Model

Bunnell (1973a,b), Reichle *et al.* (1973), and Russell (1975) have outlined development criteria and philosophical aspects of modeling. Three points are prominent in their discussions: (1) Modeling should be held in proper perspective as one of several tools used to analyze an ecosystem. (2) Models should extract "certain essential features of the system and express the resulting abstraction as a system of mathematical functions capable of imitating some subset of the original systems behavior" (Reichle *et al.*, 1973). (3) The validity of the model should be testable.

In setting up a model, it is first required that the boundaries of the system to be modeled be defined. While for laboratory studies this may be straightforward, it is often a difficult task for natural systems. Laboratory systems are generally less complex than natural systems because they are physically confined and thus have only limited material exchange with the environment outside the confines of the system. Here the exchanges are mainly gases for batch-type systems, but for continuous-culture or flow systems there are, in addition to gases, well defined (usually constant) point source flows in and out. In natural systems, the flows in and out of the boundaries of the system are usually diffuse and vary with time. Systems with no exchange are known as "closed systems," while systems with exchange are known as "open systems."

6.2. Operation of the Model

After defining the boundaries of the ecosystem, in principle one should start by examining the entire system with its associated processes before trying to model one of its subsystems such as organic-matter decomposition or mineral cycling in soil environments. The initial process of setting up a model is to divide the ecosystem into compartments (usually represented as pools of biomass and pools of various inorganic and organic entities) and then to define the processes allowing the interaction of one compartment with another. This is basically a systems-analysis approach (Reichle *et al.*, 1973; Russell, 1975).

Diagrammatically the compartments are often presented as boxes with arrows representing the processes of conversion and/or interchange of matter between compartments. The subsystem to be modeled will consist of selected compartments and associated processes.

Generally some sort of material-balance equations are needed to account for material flowing into and out of the subsystem. Such equations also keep account of transformation material from one compartment to another and the spatial distribution of material within the system. Fluxes and rates are important features of such models, fluxes being the amount of material (or biomass) flowing across a unit boundary surface area of the system per unit time. Mathematically the rate of change of the total amount of each particular material substance within a system is equal to the sum of the fluxes of this substance across all the surfaces (total surface area) defining the physical limits of the system, plus or minus the net rate of production or destruction of the substance within the boundaries. A typical model would consist of a series of differential equations, one for each compartment.

It is important to note the relation between concentrations of substances and reaction rates and fluxes. The concentration of substance (components) alone tells one little about its transformation or production rates or the past history of the transformation process. Even in systems that have a high production rate (i.e., high activity for production), no accumulation would occur if there were a high flux of material out of the system or rapid metabolism within the system. For example, it is common to think of the flux of anionic nutrients such as nitrate in soil systems as being equal to the concentration of the nutrient times the flow rate of moisture through the system. If the moisture flow is high, nitrate could remain low in concentration even though active nitrification was taking place.

6.3. Treatment of Limiting Factors

Some models of differing complexity which involve limiting factors will be considered. The work of Bunnell and collaborators (Bunnell *et al.*, 1977a, b; Bunnell and Tait, 1974) on the decomposition of organic material involves essentially "closed system" laboratory respiration models. Their models treated soil moisture, O_2 availability, temperature, and the organic substrates as the most important factors limiting respiration in soil. They assumed that soil organic matter can be separated into several organic substrate fractions, with the rate at which each fraction decomposed being directly proportional to the amount of substrate (i.e., first-order kinetics). The total decomposition rate is the sum of the individual rates. Each rate was also dependent on moisture, according to Michaelis-Menten kinetics with respect to moisture content, and on temperature, according to functions similar to those known for enzymatic reactions. Their models contain no microbial biomass term.

The decomposition model of Reichle *et al.* (1973) is a simple "open system" model, with no spatial variation within the system. The actual algorithm for organic matter decomposition is less complex than algorithms proposed by Bunnell and Tait (1974), but the overall model is more complex because it considers the interaction between six compartments and includes the periodic flux of three kinds of litter (rapid, intermediate, and slow decomposing), which in turn are converted to another form of litter, which is subsequently decomposed to organic matter. Also incorporated into this model is the influence of earthworm biomass. No microbial biomass terms are included.

The model of Beek and Frissel (1973) represents perhaps the most complex model of a soil system. It incorporates not only organic matter decomposition processes but also nitrogen cycle transformations and allowances for spatial differences as a function of depth in a soil profile. Their model includes four categories: soil organic matter added to soil as protein, sugars, cellulose, and lignin, all of which are converted to humus, which is assumed to remain unaltered. Ammonification, nitrification, and leaching of nitrate are included also, along with some biomass terms. Of most interest here is the effect of limiting factors on growth of microorganisms. Only Beek and Frissel's model among those discussed so far has microbial biomass terms included or has any modeling of microbial growth in soil.

Growth rates in soil could be limited by substrate, energy source, temperature, pH, and moisture, among other things. Few models have tried to deal specifically with the way in which these factors may affect growth rates. McLaren (1969, 1971) has proposed a mechanistic model for nitrifier growth in soil columns constantly leached with inorganic nutrient solution. This model, which allows for spatial variation as a function of depth, expresses reaction kinetics in a Michaelis-Menten form. However, growth was characterized by a modified Verhulst-Pearl logistic equation primarily for those situations where growth is independent of substrate concentration (zero order with respect to substrate). Saunders and Bazin (1973) extended McLaren's type of model to include substrate-limited growth, assuming that growth was proportional to substrate concentration (first order with respect to substrate). Both zero-order and first-order growth with respect to substrate concentration are, of course, simplifications of the general Monod growth expression cited in Section 1.1 (zero order when $S >> K$ and first order when $S << K$). Such simplifications were made because it is generally impossible to get analytical solutions to differential equations which contain Monod-type expressions for growth.

Solutions can be obtained for differential equations containing Monod-type expressions through use of digital computing techniques. Laudelout *et al.* (1974, 1975, 1976) studied mixed cultures of nitrifiers. These models incorporated, not only substrate limitation on growth, but the effects of other factors, such as temperature, pH, and pO_2. Their most complex studies have been confined to aqueous systems. Beek and Frissel (1973) used similar computer techniques in their model of the terrestrial nitrogen cycle. Although the Beek and Frissel model included nitrifying populations, the growth was not limited by nitrogen substrate availability (zero-order growth). Limiting factors in their model included temperature and moisture.

6.4. The Model as an Investigative Tool

The models reviewed here represent the various approaches used for modeling microbial processes in soil. Although limiting factors were incorporated, the prime objective of the model was other than an examination as to how limiting factors regulate microbial growth and activity in soil. Thus, in the organic matter decomposition models, no populations of microorganisms were associated with these processes, and hence little advantage could be taken of one of the most powerful attributes of the model—the potential to resolve the effects of major limiting factors as they interact.

Only some of the models associated with nitrification have been designed to incorporate biomass and associated growth relations. Again, these models were not designed specifically to investigate growth as affected by limiting factors.

For the specific investigation of limiting factors, the modeling approach should be different from that emphasized for soil ecosystem analysis. First, models should be designed to study specific measurable microbial populations. Second, it is desirable that the population have a definable measurable activity or behavior. This second feature would allow independent checking of the validity of the model. Third, the modeled system should be as simple as possible, so that complicated exchange processes of natural open systems may be avoided. Fourth, simplifying assumptions that decrease the generality of the modeling should be avoided, with the model based on known or probable biological and physical mechanisms. The last requirement generally necessitates the use of computer techniques to solve the equations.

The closest approach to a model based on the foregoing considerations is that initiated by Laudelout *et al.* (1975) on the simulation of nitrification. Michaelis-Menten oxidation kinetics and Monod-type growth were combined in a general way, along with temperature effects. Important features of such an approach are the determination of an activity for nitrification on an organismal basis and the use of yield coefficients. The validity of the Laudelout model apparently has not been tested in soils. This is an important point since it is not clear that it is valid to take activity and yield coefficients from batch culture studies and apply them to soil models. This point was discussed earlier (Section 5.2.3), where it was noted that the growth rates and yield for nitrifiers were significantly lower in soil incubation than in comparable pure culture studies. One of the goals of an appropriate model would be to explain such discrepancies in a mechanistic manner and not merely determine constants that apply to soils with no relationship to pure culture kinetics.

Models of the type just described have the capability of being thoroughly tested for validity. Substrate oxidation and end-product generation can be followed, and populations can be enumerated simultaneously. In addition, since the various kinetic parameters can be determined independently, no adjustments to these parameters should be required if the model truly depicts the growth process in soil.

The nitrification system model appears to have excellent potential for the study of limiting factors such as O_2, moisture, and temperature. Information gained with such a model may have more general application to other microbial populations in the soil.

References

Ausmus, B. S., 1973, The use of the ATP assay in terrestrial decomposition studies, *Bull. Ecol. Res. Comm. (Stockholm)* 17:223.

Ayanaba, A., and Lawson, T. L., 1977, Diurnal changes in acetylene reduction in field grown cowpeas and soybeans, *Soil Biol. Biochem.* 9:125.

Babiuk, L. A., and Paul, E. A., 1970, The use of fluorescein isothiocyante in the determination of the bacterial biomass of grassland soil, *Can. J. Microbiol.* 16:57.

Balandreau, J. P., Millier, C. R., and Dommergues, Y. R., 1974, Diurnal variations of nitrogenase activity in the field, *Appl. Microbiol.* 27:662.

Balderston, W. L., Sherr, B., and Payne, W. J., 1976, Blockage by acetylene of nitrous oxide reduction in *Pseudomonas perfectomarinus*, *Appl. Environ. Microbiol.* 31:504.

Beck, G., Dommergues, Y., and Van der Driessche, R., 1969, L'éffet litière. II. Etude experimentale du pouvoir inhibiteur de composés hydrosolubles des feuilles et des litières foristières vis à vis de la microflore tellurique, *Oecol. Plant.* 4:237.

Beek, J., and Frissel, M. J., 1973, *Simulation of Nitrogen Behavior in Soils*, Simulation Monographs, Center for Agricultural Publishing and Documentation, Wageningen, The Netherlands.

Belser, L. W., 1977, Nitrate reduction to nitrite, a possible source of nitrite for growth of nitrite oxidizing bacteria, *Appl. Environ. Microbiol.* 34:403.

Belser, L. W., and Schmidt, E. L., 1978, Nitrification in soil, in: *Microbiology 1978* (D. Schlessinger, ed.), in press, American Society for Microbiology, Washington, D. C.

Beringer, J. E., and Hopwood, D. A., 1976, Chromosomal recombination and mapping in *Rhizobium leguminosarum*, *Nature (London)* 264:291.

Bock, E., 1976, Growth of Nitrobacter in the presence of organic matter. II. Chemoorganotrophic growth of *Nitrobacter agilis*, *Arch. Microbiol.* 108:305.

Bohlool, B. B., and Schmidt, E. L., 1973a, Persistence and competition aspects of *Rhizobium japonicum* observed in soil by immunofluorescence microscopy, *Soil Sci. Soc. Am. Proc.* 37:561.

Bohlool, B. B., and Schmidt, E. L., 1973b, A fluorescent antibody technique for determination of growth rates of bacteria in soil, *Bull. Ecol. Res. Comm. (Stockholm)* 17:336.

Bohlool, B. B., and Schmidt, E. L., 1974, Lectins: a possible basis for specificity in the *Rhizobium*–legume root nodule symbiosis, *Science* 185:269.

Bohlool, B. B., and Schmidt, E. L., 1976, Immunofluorescent polar tips of *Rhizobium japonicum*: possible site of attachment or lectin binding, *J. Bacteriol.* 125:1188.

Bohlool, B. B., and Schmidt, E. L., 1977, Nitrification in the intertidal zone: influence of effluent type and effect of tannin on nitrifiers, *Appl. Environ. Microbiol.* 34:523.

Brar, S. S., and Giddens, J., 1968, Inhibition of nitrification in Bladen grassland soil, *Soil Sci. Soc. Am. Proc.* 32:821.

Bremner, J. M., and Bundy, L. G., 1974, Inhibition of nitrification in soils by volatile sulfur compounds, *Soil Biol. Biochem.* 6:161.

Brian, P. W., 1957, The ecological significance of antibiotic production, in: *Microbial Ecology* (R.E.D. Williams and C. C. Spicer, eds.), pp. 168–188, Cambridge University Press, Cambridge.

Brockwell, J., 1977, Application of seed inoculants, in: *A Treatise on Dinitrogen Fixation: Section IV* (R.W.F. Hardy and A. H. Gibson, eds.), pp. 277–309, Wiley, New York.

Brown, M. E., 1976, Role of *Azotobacter paspali* in association with *Paspalum notatum*, *J. Appl. Bact.* 40:341.

Bunnell, F. L., 1973a, Decomposition models and the real world, in: *Modern Methods in the*

Study of Microbial Ecology (T. Rosswall, ed.), *Bull. Ecol. Res. Comm.* (*Stockholm*) **17**:407.

Bunnell, F. L., 1973b, Theological ecology or models and the real world, *Forestry Chronicle* **49**:167.

Bunnell, F. L., and Tait, D. E. N., 1974, Mathematical simulation models of decomposition processes, in: *Soil Organisms and Decomposition in Tundra* (A. J. Holding, O. W. Heal, S. F. Maclean, and P. W. Flanagan, eds.), pp. 207–227, Tundra Biome Steering Committee, Stockholm.

Bunnell, F. L., Tait, D. E. N., Flanagan, P. W., and Van Cleve, K., 1977a, Microbial respiration and substrate weight loss I. A general model of the influences of abiotic variables, *Soil Biol. Biochem.* **9**:33.

Bunnell, F. L., Tait, D. E. N., and Flanagan, P. W., 1977b, Microbial respiration and substrate weight loss. II. A model of the influences of chemical composition, *Soil Biol. Biochem.* **9**:41.

Burford, J. R., and Bremner, J. M., 1975, Relationships between the denitrification capacities of soils and total, water soluble and readily decomposable soil organic matter, *Soil Biol. Biochem.* **7**:389.

Burton, J. C., 1976, Pragmatic aspects of the *Rhizobium*–leguminous plant association, in: *Proceedings of the 1st International Symposium on Nitrogen Fixation* (W. E. Newton and C. J. Nyman, eds.), pp. 429–446, Washington State University Press, Pullman.

Carlucci, A. F., and Stickland, J. D. H., 1968, The isolation, purification and some kinetic studies of marine nitrifying bacteria, *J. Exp. Mar. Biol. Ecol.* **2**:156.

Chase, F. E., Corke, C. T., and Robinson, J. B., 1968, Nitrifying bacteria in soil, in: *The Ecology of Soil Bacteria* (T.R.G. Gray and D. Parkinson, eds.), pp. 593–611, Liverpool University Press, Liverpool.

Clark, C., and Schmidt, E. L., 1967, Uptake and utilization of amino acids by resting cells of *Nitrosomonas europaea*, *J. Bacteriol.* **93**:1309.

Conn, H. J., 1932, A microscopic study of the changes in the microflora of soil, *Tech. Bull. N.Y. State Agric. Exp. Stn.*, 204.

Connell, W. E., and Patrick, W. H., Jr., 1968, Sulfate reduction in soil: effects of redox potential and pH, *Science* **159**:86.

Crawford, D. L., Crawford, R. L., and Pometto, A. L., III, 1977, Preparation of specifically labeled ^{14}C-(lignin)- and ^{14}C-(cellulose)-lignocelluloses and their decomposition by the microflora of soil, *Appl. Environ. Microbiol.* **33**:1247.

Danso, S. K. A., and Alexander, M., 1975, Regulation of predation by prey density: the protozoan–*Rhizobium* relationship, *Appl. Microbiol.* **29**:515.

Dart, P. J., 1974, The infection process, in: *The Biology of Nitrogen Fixation* (A. Quispel, ed.), pp. 381–429, North Holland, Amsterdam.

Day, J. M., and Dobereiner, J., 1976, Physiological aspects of N_2-fixation by a *Spirillum* from *Digitaria* roots, *Soil Biol. Biochem.* **8**:45.

Deverall, B. J., 1977, *Defense Mechanisms of Plants*, Cambridge University Press, Cambridge.

Dixon, R. O. D., 1969, Rhizobia, *Annu. Rev. Microbiol.* **23**:137.

Dommergues, Y., 1962, Contribution à l'étude de la dynamique microbienne des sols en zone semi-aride et sèche, *Anns. Agron.*, Paris **13**:265, 391.

Dommergues, Y., 1977, *Biologie du Sol*, Presses Universitaire de France, Paris.

Dommergues, Y., Combremont, R., Beck, G., and Ollat, C., 1969, Note preliminaire concernant la sulfate-réduction rhizosphérique dans un sol salin tunisien, *Rev. Ecol. Biol. Sol.* **6**:115.

Dunican, L. K., and Rosswall, T., 1974, Taxonomy and physiology of tundra bacteria in relation to site characteristics, in: *Soil Organisms and Decomposition in the Tundra* (A. J. Holding, O. W. Heal, S. F. Maclean, Jr., and P. W. Flanagan, eds.), pp. 79–92, Tundra Biome Steering Committee, Stockholm.

Endelman, F. J., Northup, M. L., Hughes, R. R., Keeney, D. R., and Boyle, J. R., 1973, Mathematical modeling of soil nitrogen transformations. I, *Chem. Eng. Prog. Sym. Series* **70**:83.

Endelman, F. J., Box, G. E. P., Boyle, J. R., Hughes, R. R., Keeney, D. R., Northup, M. L., and Saffigna, P. G., 1974, The mathematical modeling of soil-water-nitrogen phenomena, EDFB-1BP-74-8 Oak Ridge National Laboratory, Oak Ridge, Tenn.

Evans, H. J., 1975, *Enhancing Biological Nitrogen Fixation*, National Science Foundation, Washington, D. C.

Evans, H. J., and Barber, L. E., 1977, Biological nitrogen fixation for food and fiber production, *Science* **197**:332.

Fliermans, C. F., Bohlool, B. B., and Schmidt, E. L., 1974, Autecological study of the chemoautotroph *Nitrobacter* by immunofluorescence, *Appl. Microbiol.* **27**:124.

Fliermans, C. B., and Schmidt, E. L., 1975a, Autoradiography and immunofluorescence combined for autecological study of single cell activity with *Nitrobacter* as a model system, *Appl. Microbiol.* **30**:676.

Fliermans, C. B., and Schmidt, E. L., 1975b, Fluorescence microscopy: direct detection, enumeration and spatial distribution of bacteria in aquatic systems, *Arch. Hydrobiol.* **76**:33.

Flühler, H., Ardakani, M. S., Szuszkieuricz, T. E., and Stolzy, L. H., 1976, Field measured nitrous oxide concentrations, redox potentials, oxygen diffusion rates, and oxygen partial pressures in relation to denitrification, *Soil Sci.* **122**:107.

Fogg, G. E., Stewart, W. D. P., Fay, P., and Walsby, A. E., 1973, *The Blue-green Algae*, Academic Press, London.

Foster, R. C., and Rovira, A. D., 1976, Ultrastructure of wheat rhizosphere, *New Phytol.* **76**:343.

Frankland, J. C., 1974, Decomposition of lower plants, in: *Biology of Plant Litter Decomposition* (C. H. Dickinson and G. J. F. Pugh, eds.), pp. 3–36, Academic Press, London.

Fred, E. B., Wilson, P. W., and Wyss, O., 1938, Light intensity and the nitrogen hunger period in the Manchu soybean, *Proc. Nat. Acad. Sci., U.S.A.* **24**:45.

Fry, J. C., and Staples, D. G., 1976, Distribution of *Bdellovibrio bacteriovorus* in sewage works, river water, and sediments, *Appl. Environ. Microbiol.* **31**:469.

Garcia, J. L., 1975, La dénitrification dans les sols, *Bull. Inst. Pasteur* **73**:167.

Garrett, S. D., 1970, *Pathogenic Root-Infecting Fungi*, Cambridge University Press, Cambridge.

Gibson, A. H., 1976, Limitation to dinitrogen fixation by legumes, in: *Proceedings of the 1st International Symposium on Nitrogen Fixation* (W. E. Newton and C. J. Nyman, eds.), pp. 400–428, Washington State University Press, Pullman.

Gibson, A. H., 1977, The influence of the environment and managerial practices on the legume–*Rhizobium* symbiosis, in: *A Treatise on Dinitrogen Fixation, Section IV* (R. W. F. Hardy and A. H. Gibson, eds.), pp. 393–450, Wiley, New York.

Goring, C.I.A., 1962, Control of nitrification by 2-chloro-6-(trichloromethyl)pyridine, *Soil Sci.* **93**:211.

Gunderson, K., Carlucci, A. F., and Bostrom, K., 1966, Growth of some chemoautotrophic bacterium at different oxygen tensions, *Experientia* **22**:229.

Hackett, W. F., Connors, W. J., Kirk, T. K., and Zeikus, J. G., 1977, Microbial decomposition of synthetic ^{14}C-labeled lignins in nature: lignin biodegradation in a variety of natural materials, *Appl. Environ. Microbiol.* **33**:43.

Haith, D. A., 1973, Optimal control of nitrogen losses from land disposal areas, *J. Environ. Engr. Div. EE6*, Am. Soc. Civil Engr. **Dec**:923.

Halliday, J., and Pate, J. S., 1976, The acetylene reduction assay as a means of studying nitrogen fixation in white clover under sward and laboratory conditions, *J. Br. Grassland Soc.* **31**:29.

Ham, G. E., Frederick, L. R., and Anderson, I. C., 1971, Serogroups of *Rhizobium japonicum* in soybean nodules sampled in Iowa, *Agron. J.* **63**:69.

Ham, G. E., 1977, The acetylene-ethylene assay and other measures of N_2 fixation in field experiments, in: *Conference Proceedings in Biological Nitrogen Fixation in Farming Systems of the Tropics* (A. Ayanaba and P. J. Dart, eds.), Wiley, London, in press.

Hardy, R. W. F., and Holsten, R. D., 1977, Methods for measurements of dinitrogen fixation, in: *A Treatise on Dinitrogen Fixation*, section IV (R. W. F. Hardy and A. H. Gibson, eds.), pp. 451–486, Wiley, New York.

Hardy, R. W. F., Holsten, R. D. Jackson, E. K., and Burns, R. C., 1968, The acetylene-ethylene assay for N_2 fixation: laboratory and field evaluation, *Plant Physiol.* **43**:1185.

Harley, J. L., 1969, *The Biology of Mycorrhiza*, Leonard Hill, London.

Harrison, A. F., 1971, The inhibitory effect of oak leaf litter tannins on the growth of fungi, in relation to litter decomposition, *Soil Biol. Biochem.* **3**:167.

Hattori, T., and Hattori, R., 1976, The physical environment in soil microbiology: an attempt to extend principles in microbiology to soil microorganisms, *Crit. Rev. Microbiol.* **May**:423.

Hauck, R. D., 1971, Quantitative estimates of nitrogen cycle processes, in: *Nitrogen-15 in Soil Plant Studies* (Proceedings, Panel Sofia, 1969), International Atomic Energy Agency, Vienna.

Heal, O. W., 1971, Decomposition, in: *IBP Tundra Biome Working Meeting on Analysis of Ecosystems* (O. W. Heal, ed.), pp. 262–278, Tundra Biome Steering Committee, Stockholm.

Hegazi, N. A., and Niemelä, S. I., 1976, A note on the estimation of *Azotobacter* densities by membrane filter technique, *J. Appl. Bacteriol.* **41**:311.

Hobbie, J. E., Daley, R. V., and Jasper, S., 1977, Use of Nucleopore filters for counting bacteria by fluorescence microscopy, *Appl. Environ. Microbiol.* **33**:1225.

Jacq, V. 1975, La sulfato-réduction en relation avec l'excretion racinaire, *Bull. Soc. Bot. Fr., La Rhizosphere* **122**:169.

Jacq, V., and Dommergues, Y., 1971, Sulfate-reduction spermospherique, *Ann. Inst. Pasteur* **121**:199.

Keya, S. O., and Alexander, M., 1975, Regulation of parasitism by host density: the *Bdellovibrio-Rhizobium* interrelationship, *Soil Biol. Biochem.* **7**:231.

King, J. D., and White, D. C., 1977, Muramic acid as a measure of microbial biomass in estuarine and marine samples, *Appl. Environ. Microbiol.* **33**:777.

Knowles, R., 1976, Factors affecting dinitrogen fixation by bacteria in natural and agricultural systems in: *Proceedings of the 1st International Symposium on Nitrogen Fixation* (W. E. Newton and C. J. Nyman, eds.), pp. 539–555, Washington State University Press, Pullman.

Knowles, R., 1977, The significance of asymbiotic dinitrogen fixation by bacteria, in: *A Treatise on Dinitrogen Fixation*, section IV (R. W. F. Hardy and A. H. Gibson, eds.), pp. 33–83, Wiley, New York.

Kong, K. C., and Dommergues, Y., 1970, Limitation de la cellulolyse dans les sols organiques, I. Etude respirometrique, *Rev. Ecol. Biol. Sol* **7**:442.

Kong, K. C., and Dommergues, Y., 1972, Limitation de la cellulolyse dans les sols organiques. II. Etude des enzymes du sol, *Rev. Ecol. Biol. Sol.* **9**:629.

Kowalski, M., Ham, G. E., Frederick, L. R., and Anderson, I. C., 1974, Relationship between strains of *Rhizobium japonicum* and their bacteriophages from soil and nodules of field-grown soybeans, *Soil Sci.* **118**:221.

Kouyeas, V., 1964, An approach to the study of moisture relations of soil fungi, *Plant Soil* **20**:351.

Langford, A. N., and Buell, M. F., 1969, Integration, identity and stability in plant associations, *Adv. Ecol. Res.* **6**:84.

Laudelout, H., Lambert, R., Fripiat, J. L., and Pham, M. L., 1974, Effet de la temperature sur la vitesse d'oxydation de l'ammonium en nitrate par des cultures mixtes de nitrifiants, *Ann. Microbiol. Inst. Pasteur* **125B**:75.

Laudelout, H., Frankart, R., Lambert, R., Mougenot, F., and Pham, M. L., 1975, Modeling of solute interactions with soils, in: *Modeling and Simulation of Water Resources Systems* (G. S. Vansteenkiste, ed.) pp. 361–366, North Holland, Amsterdam.

Laudelout, H., Lambert, R., and Pham, M. L., 1976, Influence du pH et de la pression partielle d'oxygene sur la nitrification, *Ann. Microbiol. Inst. Pasteur* **127A**:367.

Le Gall, J., and Postgate, J., 1973, The physiology of sulfate-reducing bacteria, *Adv. Microbial Physiol.* **10**:81.

Lie, T. A., 1974, Environmental effects on nodulation and symbiotic nitrogen fixation, in: *The Biology of Nitrogen Fixation* (A. Quispel, ed.), pp. 555–582, North Holland, Amsterdam.

Line, M. A., and Loutit, M. W., 1973, Nitrogen-fixation by mixed cultures of aerobic and anaerobic micro-organisms in an aerobic environment, *J. Gen. Microbiol.* **74**:179.

Mague, T. H., and Burris, R. H., 1972, Reduction of acetylene and nitrogen by field-grown soybeans, *New Phytol.* **71**:275.

Marks, G. C., and Kozlowski, T. T., 1973, *Ectomycorrhizae*, Academic Press, New York.

Marshall, K. C., 1975, Clay mineralogy in relation to survival of soil bacteria, *Annu. Rev. Phytopathol.* **13**:357.

Masterson, C. L., and Murphy, P. M., 1975, Application of the acetylene reduction technique to the study of nitrogen fixation by white clover in the field, in: *Symbiotic Nitrogen Fixation in Plants* (P. S. Nutman ed.), pp. 299–316, Cambridge University Press, Cambridge.

Mayaudon, J., 1971, Use of radiorespirometry in soil microbiology and biochemistry, in: *Soil Biochemistry*, vol. 2 (A. D. McLaren and J. Skujins, eds.), pp. 202–256, Marcel Dekker, New York.

McLaren, A. D., 1969, Nitrification in soil: systems approaching the steady state, *Soil Sci. Soc. Am. Proc.* **33**:551.

McLaren, A. D., 1971, Kinetics of nitrification in soil: growth of the nitrifiers, *Soil Sci. Soc. Am. Proc.* **35**:91.

Melin, E., and Wiken, T., 1946, Antibacterial substances in water extracts of pure forest litter, *Nature (London)* **158**:200.

Mikola, P., 1973, Application of mycorrhizal symbiosis in forest practice, in: *Endomycorrhizae* (G. C. Marx and T. T. Kozlowski, eds.), pp. 383–411, Academic Press, London.

Millar, W. N., and Casida, L. E., 1970, Evidence for muramic acid in soil, *Can. J. Microbiol.* **18**:299.

Moore-Landecker, E., and Stotzky, G., 1973, Morphological abnormalities of fungi induced by volatile microbial metabolites, *Mycologia* **65**:519.

Munns, D. N., 1977, Mineral nutrition and the legume symbiosis, in: *A Treatise on Dinitrogen Fixation*, section IV (R. W. F. Hardy and A. H. Gibson, eds.), pp. 353–392, Wiley, New York.

Nakos, G., 1975, Absence of nitrifying microorganisms from a Greek forest soil, *Soil Biol. Biochem.* **7**:335.

New, P. B., and Kerr, A., 1971, A selective medium for *Agrobacterium radiobacter* biotype 2, *J. Appl. Bacteriol.* **34**:1.

Obaton, M., 1971, Utilisation de mutants spontanes resistants aux antibiotiques pour l'etude ecologique des *Rhizobium*, *C. R. Acad. Sci, Paris*, Ser. D **272**:2630.

Odeyemi, O., and Alexander, M., 1977, Resistance of *Rhizobium* strains to phygon, spergon, and thiram, *Appl. Environ. Microbiol.* **33**:784.

Odum, E. P., 1971, *Fundamentals of Ecology*, Saunders, Philadelphia.

Old, K. M., and Nicholson, T. H., 1975, Electron microscopical studies of the microflora of roots of sand dune grasses, *New Phytol.* **74**:51.

Parker, C. A., and Grove, P. L., 1970, *Bdellovibrio bacteriovorus* parasitizing rhizobia in western Australia. *J. Appl. Bacteriol.* **33**:253.

Pattison, A. C., and Skinner, F. A., 1973, The effects of antimicrobial substances on *Rhizobium* spp. and their use in selective media, *J. Appl. Bacteriol.* **37**:239.

Pirt, S. S., 1965, The maintenance energy of bacteria in growing cultures, *Proc. R. Soc. London Ser. B* **163**:224.

Postgate, J., 1972, *Biological Nitrogen Fixation*, Merrow Publishing, Watford, Hertfordshire.

Ramsey, A., 1977, Direct counts by modified acridine orange method and activity of bacteria from marine and fresh waters, *Oecologia* in press.

Reddy, K. R., Patrick, W. H., Jr., and Phillips, R. E., 1976, Ammonium diffusion as a factor in nitrogen loss from flooded soils, *Soil Sci. Soc. Am. J.* **40**:528.

Reichle, D. E., O'Neill, R. V., Kay, S. V., Sollins, P., and Booth, R. S., 1973, Systems analysis as applied to modeling ecological processes, *Oikos* **24**:337.

Rennie, R. J., and Schmidt, E. L., 1974, Fluorescent antibody techniques to study the autecology of *Nitrobacter* in soils, *Abstracts of the Annual Meeting of the American Society of Microbiologists*, Chicago, p. 3.

Rennie, R. J., and Schmidt E. L., 1977, Autecological and kinetic analysis of competition between strains of *Nitrobacter* in soils. *Ecol. Bull.* (*Stockholm*) **25**:in press.

Reynaud, R., and Roger, R., 1976, N_2-fixing algal biomass in Senegal rice fields, *Ecol. Bull.* (*Stockholm*) **26**:in press.

Rice, E. L., and Pancholy, S. K., 1973, Inhibition of nitrification by climax ecosystems. II. Additional evidence and possible role of tannins, *Am. J. Bot.* **60**:691.

Roberge, M. R., and Knowles, R., 1966, Ureolysis, immobilization, and nitrification in black spruce (*Picea mariana*, mill.) humus, *Proc. Soil Sci. Soc. Am.* **30**:201.

Rovira, A. D., and Campbell, R., 1974, Scanning electron microscopy of microorganisms on the roots of wheat, *Microbial Ecol.* **1**:15.

Russell, J. S., 1975, Systems analysis of soil ecosystems, in: *Soil Biochemistry*, Vol. 3 (E. A. Paul and A. D. McLaren, eds.), pp. 37–82, Marcel Dekker, New York.

Salle, A. J., 1973, *Fundamental Principles of Bacteriology*, McGraw-Hill, New York.

Saunders, P. T., and Bazin, M. J., 1973, Non-steady state studies of nitrification in soil: Theoretical considerations, *Soil Biol. Biochem.* **5**:545.

Schmidt, E. L., 1973a, Fluorescent antibody techniques for the study of microbial ecology, *Bull. Ecol. Res. Comm.* (*Stockholm*) **17**:67.

Schmidt, E. L., 1973b, The traditional plate count technique among modern methods, *Bull. Ecol. Res. Comm.* (*Stockholm*) **17**:453.

Schmidt, E. L., 1974, Quantitative autecological study of microorganisms in soil by immunofluorescence, *Soil Sci.* **118**:141.

Schubert, K. R., and Evans, H. J., 1976, Hydrogen evolution: a major factor affecting the efficiency of nitrogen fixation in nodulated symbionts, *Proc. Nat. Acad. Sci. U.S.A.* **73**:1207.

Shields, V. A., Paul, E. A., Lowe, W. E., and Parkinson, D., 1973, Turnover of microbial tissue in soil under field conditions, *Soil Biol. Biochem.* **5**:753.

Simon, A., and Ridge, E. H., 1974, The use of ampicillin in a simplified selective medium for the isolation of fluorescent pseudomonads, *J. Appl. Bacteriol.* **37**:459.

Skinner, F. A., and Walker, N., 1961, Growth of *Nitrosomonas europaea* in batch and continuous culture, *Arch. Mikrobiol.* **38**:339.

Sloger, C., Bezicek, D., Milberg, R., and Boonkerd, N., 1975, Seasonal and diurnal variations in N_2 (C_2H_2)-fixing activity in field soybeans, in: *Nitrogen Fixation by Free-Living Microorganisms* (W. D. P. Stewart, ed.), pp. 271–284, Cambridge University Press, Cambridge.

Stark, N., 1972, Nutrient cycling pathways and litter fungi, *Bioscience,* **22**:355.

Starkey, R. L., 1938, Some influences of the development of higher plants upon the microorganisms in soil: VI. Microscopic examination of the rhizosphere, *Soil Sci.* **45**:207.

Stotzky, G., 1965, Microbial respiration, in: *Methods of Soil Analysis,* II (C. A. Black, D. D. Evans, J. L. White, L. E. Ensminger, and F. E. Clark, eds.), pp. 1550–1570, American Society of Agronomy, Madison.

Stotzky, G., 1972, Activity, ecology, and population dynamics of microbes in soil, *Crit. Rev. Microbiol.* **2**:59.

Stotzky, G., and Norman, A. G., 1961, Factors limiting microbial activities in soil, *Arch. Mikrobiol.* **40**:341.

Todd, R. L., Cromack, K., Jr., and Knutson, R. M., 1973, Scanning electron microscopy in the study of terrestrial microbial ecology, *Bull. Ecol. Res. Comm.* (*Stockholm*) **17**:109.

Tsien, H. C., and Schmidt, E. L., 1977, Polarity in the exponential phase *Rhizobium japonicum* cell, *Can. J. Microbiol.* **12**:1274.

Tsuru, S., 1967, On studies on the microbial decomposition of various litters and humus formation in volcanic soils, in: *Progress in Soil Biology* (O. Graff and J. E. Satchell, eds.), pp. 455–463, Vieweg, Braunschweig, Germany.

Watson, S. W., Novitsky, T. J., Quinby, H. L., and Valois, F. W., 1977, Determination of bacterial number and biomass in the marine environment, *Appl. Environ. Microbiol.* **33**:940.

Wheeler, C. T., 1971, The causation of the diurnal changes in nitrogen fixation in the nodules of *Alnus glutinosa,* *New Phytol.* **70**:487.

Wiebe, W. J., 1971, Perspectives in microbial ecology, in: *Fundamentals of Ecology* (E. P. Odum, ed.), pp. 487–497, Saunders, Philadelphia.

Wright, R. T., 1973, Some difficulties in using ^{14}C-organic solutes to measure heterotrophic bacterial activity, in: *Estuarine Microbial Ecology* (L. H. Stevenson and R. R. Colwell, eds.), pp. 199–217, University of South Carolina Press, Columbia.

Wilson, D. O., 1977, Nitrification in three soils and amended with zinc sulfate, *Soil Biol. Biochem.* **9**:277.

Woldendorp, T. W., 1963, L'influence des plantes vivantes sur la denitrification, *Ann. Inst. Pasteur* **105**:426.

Wong, P. T. W., and Griffin, D. M., 1974, Effect of osomtic potential on streptomycete growth, antibiotic production and antagonism to fungi, *Soil Biol. Biochem.* **6**:319.

Yoshinari, T., and Knowles, R., 1976, Acetylene inhibition of nitrous oxide reduction by denitrifying bacteria, *Biochem. Biophys. Res. Commun.* **69**:705.

Yoshinari, T., Hynes, R., and Knowles, R., 1977, Acetylene inhibition of nitrous oxide reduction and measurement of denitrification and nitrogen fixation in soil, *Soil Biol. Biochem.* **9**:177.

3

Eco-Physiological Aspects of Microbial Growth in Aerobic Nutrient-Limited Environments

D. W. TEMPEST AND O. M. NEIJSSEL

1. Introduction

Although the growth of microbial populations in natural environments may be limited from time to time (and from place to place) by extremes of temperature, pH, and pO_2, as well as by the presence of noxious substances, predators, and parasites, the one factor that will most consistently constrain the rate at which these populations grow must be the availability of essential nutrient substances. Natural environments rarely will contain all those substances that are essential for cell synthesis in concentrations sufficient to allow growth to proceed at its potentially maximum rate. This conclusion derives not so much from a detailed chemical analysis of natural ecosystems as from a basic appreciation of three facts: (1) that many microorganisms (e.g., most bacteria) grow and multiply by a process of binary fission, which is, in effect, an autocatalytic exponential function; (2) that the rate of substrate assimilation is proportional to, and invariably greater than, the rate of cell synthesis; and (3) that microbes are capable of growing at phenomenally fast rates when all nutrients are present in excess of the growth requirements and other conditions are propitious. For example, one *Escherichia coli* organism (weighing, say 0.2×10^{-12} g), and its progeny, dividing every 20 min, possesses the potential to generate a mass of organisms equivalent to 1000

D. W. TEMPEST and O. M. NEIJSSEL • Laboratorium voor Microbiologie, Universiteit van Amsterdam, Plantage Muidergracht 14, Amsterdam, The Netherlands.

times the mass of the earth in less than 3 days. That such a catastrophe could never occur is assured by the fact that organisms must consume substantially more than their own weight in nutrients in order to double in mass. Consequently, irrespective of other considerations, the availability of essential nutrient substances always would ultimately limit the extent to which any population could grow—be it in the laboratory culture or in some natural ecosystem. This being so, it then becomes important from an ecological point of view (if from none other) to understand how organisms accommodate to nutrient-limiting conditions—in other words, to appreciate how they adapt to circumstances in which one or the other essential nutrient is available only in low, enzyme-subsaturating concentration. This, of course, is a condition that is commonly realized in a chemostat culture, and the enormous number of physiological data that have now accumulated in the literature on the behavior of organisms in such culture systems enables one to discern, at least in outline, both the main physiological problems posed by nutrient-limited environments and some of the properties that organisms must have necessarily acquired in the course of evolution to accommodate to these ubiquitous conditions.

It is the purpose of this chapter to review and to critically assess those data relating to the structural and functional changes that are invoked in microbes as a consequence of exposure to low-nutrient (chemostat) environments and to place these in a putatively ecological context. However, since the authors are not ecologists—either by training or by inclination—they must leave it to the reader to judge whether the lines of argument advanced herein do or do not have direct relevance to real situations. Either way, it is hoped that the ideas and suggestions embodied in this article will provide ecologists with some food for thought.

Let us consider the sorts of problems that must face a small population of organisms located in some natural ecosystem in which one or more essential nutrient is present in low, enzyme-subsaturating concentration. Of course, the organisms will not be there alone but will be competing with other species for the limited supply of essential nutrient(s). How then might one expect them to react? What properties must they express in order to compete effectively? A detailed consideration of these questions leads to the inescapable conclusion that organisms would strive to modify their behavior in at least *four* respects: (1) It is reasonable to assume that they would induce (or derepress) the synthesis of some high-affinity uptake mechanism for the growth-limiting nutrient. Thus, the effectiveness with which they could compete with other organisms for the limited amount of nutrient present in the environment would depend critically upon the affinity (and activity) of their uptake system for that substrate (Tempest *et al.*, 1967a; Tempest, 1970; Harder and Veldkamp, 1971; Veldkamp and Jannasch, 1972; Neijssel, 1976). (2) At the same time, there would arise a need for organisms to modulate the rates of uptake of excess (nonlimiting) nutrients—particularly when one such nutrient is the carbon and energy source—so as to prevent intermediary metabolites from accumulating within the cell to "traumatic" levels (Calcott and Postgate, 1972, 1974). (3) Some rearrangement of metabolism presumably would be needed in order to circumvent (as far as was possible, physiologically) those "bottlenecks" imposed by the specific growth limitation (Ierusalimsky, 1967). This one would expect to be particularly evident when the growth-limiting nutrient was some component other than the carbon and energy source. (4) Finally, these organisms clearly would need to modulate coordinately the rates of synthesis of all their macromolecular components in order to allow "balanced" growth to proceed at a grossly submaximal rate.

Much evidence for the existence of each of these adaptive responses is to be found in the published literature, particularly in those papers dealing with the growth of microorganisms in chemostat culture. Though it is neither practicable nor desirable to review this body of literature exhaustively, it is instructive, we believe, to examine each of these aspects sequentially and in sufficient detail to establish the principle; that we do in the following four sections (2.1–2.4). But the regulation and modulation of substrate uptake, and of catabolism and anabolism, have important bioenergetic consequences which, in themselves, may impose additional limits on the extent to which organisms can accommodate to low-nutrient environments. Hence, relevant information on the energetics of microbial growth in chemostat cultures is considered in a further section; and this is followed by a section dealing with transient-state phenomena—a condition more likely to prevail in natural ecosystems than the steady-state conditions characteristic of chemostat cultures. Finally, in the context of microbial "eco-physiology," we have attempted a reappraisal of the conclusion (based mainly on batch culture studies) that regulatory processes function within the cell to ensure that microbial growth proceeds with a maximum of both efficiency and economy.

2. Mechanisms of Adaptation to Low-Nutrient Environments

2.1. Regulation and Modulation of Substrate-Uptake Systems

With those microorganisms that are incapable of forming "resting bodies," the most immediate problem which nutrient depletion imposes is a need to scavenge the residual nutrient from the environment more

Table I. Affinity Constants of Uptake Systems for Various Substrates of
Different Microorganisms and Some Examples of High- and Low-
Affinity Uptake Systems

Substrate	K_m (μM)	K_m (mM)	Organism	Reference
Alanine	10		*S. hydrogenans*	Alim and Ring (1976)
Amino acids	9–80		*B. subtilis*	Konings and Freese (1972)
Amino acids	3–40		*S. aureus*	Short *et al.* (1972)
Arginine	20		*S. hydrogenans*	Alim and Ring (1976)
Calcium	8.8		*B. megaterium*	Golub and Bronner (1974)
Citrate	240		*K. aerogenes*	Johnson *et al.* (1975)
Cytidine	0.3		*E. coli*	Mygind and Munch-Petersen (1975)
6-Deoxyglucose	210		*Chlorella* sp.	Tanner (1974)
Fumarate	7.5		*B. subtilis*	Bisschop *et al.* (1975)
D-Galactitol	3.3		*E. coli*	Lengeler (1975)
D-Glucitol	12		*E. coli*	Lengeler (1975)
Glutamate	23		*E. coli*	Kahane *et al.* (1976)
Homoserine	9.6		*E. coli*	Templeton and Savageau (1974)
D-Lactate	22		*B. subtilis*	Matin and Konings (1973)
D-Lactate	33		*E. coli*	Matin and Konings (1973)
L-Lactate	60		*B. subtilis*	Matin and Konings (1973)
L-Lactate	20		*E. coli*	Matin and Konings (1973)
L-Lactate	20		*Pseudomonas* sp.	Matin and Konings (1973)
Malate	13		*B. subtilis*	Bisschop *et al.* (1975)
Maltose	0.9		*E. coli*	Boos (1976)
Maltose	50		*P. fluorescens*	Guffanti and Corpe (1976)
Mannitol	0.37		*E. coli*	Lengeler (1975)
Methyl-α-D-glucoside	33.3		*E. coli*	Erlagaeva *et al.* (1977)
Succinate	4.5		*B. subtilis*	Matin and Konings (1973)
Succinate	5		*E. coli*	Matin and Konings (1973)
Succinate	300		*Pseudomonas* sp.	Matin and Konings (1973)
Sulfate	11.6		*B. licheniformis*	Unpublished result
Sulfate	8		*K aerogenes*	Unpublished result
Sulfate	10		*N. crassa*	Marzluf (1974)
Sulfate	50		*S. cerevisiae*	McCready and Din (1974)
Threonine	0.39		*E. coli*	Templeton and Savageau (1974)
Uridine	0.6		*E. coli*	Mygind and Munch-Petersen (1975)
Asparagine	3.5	0.08	*E. coli*	Willis and Woolfolk (1975)
Citrate	400	1.5	*B. subtilis*	Oehr and Willecke (1974)
Fructose	< 100	>1	*E. coli*	Jones-Mortimer and Kornberg (1976)
Gluconate	10	0.1	*E. coli*	Faik and Kornberg (1973)
Glucose	5–10	0.1	*E. coli*	Kundig (1974)

(Continued)

Table I. (*Cont.*)

Substrate	K_m (μM)	K_m (mM)	Organism	Reference
Glucose	10	8	*N. crassa*	Schulte and Scarborough (1975)
Iron	0.1	n.d.[a]	*E. coli*	Frost and Rosenberg (1973)
Lysine	12	0.134	*S. cerevisiae*	Morrison and Lichstein (1976)
Magnesium	n.d.[a]	n.d.[a]	*E. coli*	Nelson and Kennedy (1972)
Phosphate	0.4	0.033	*E. coli*	Willsky and Malamy (1974, 1976)
Phosphate	3	0.17	*N. crassa*	Lowendorf *et al.* (1974, 1975)
Potassium	2	1.5	*E. coli*	Rhoads and Epstein (1977)
Urea	14	0.5	*S. cerevisiae*	Cooper and Sumrada (1975)

[a]n.d., Not determined by the authors.

effectively than other, competitor organisms. This is especially the case with regard to depletion of carbon-containing substrates since, with heterotrophic microorganisms, these substances are needed to meet both the cell's maintenance demands and its basic growth requirements. Hence, one might expect microorganisms generally to possess high-affinity uptake mechanisms for all those nutrient substances (and particularly carbon-containing substrates) that are essential for growth and survival. The data contained in Table I show that the K_m values for different metabolites of various transport systems lie in the μM range—precisely as expected. But the possession of these high-affinity (low K_m) uptake systems does not, in itself, ensure that any particular species can compete effectively with other microbial species, since the uptake capacity (that is, the V_{max} of the uptake system) will exert a profound influence on the rate of penetration of substrate into the cell, even when that substrate is present only in low, enzyme-subsaturating concentration. This can be readily appreciated by reference to Fig. 1, which shows how changes in V_{max} affect both the concentration of limiting substrate s when the uptake rate is held constant, and the uptake rate v when s is held constant. Hence, one might expect organisms to respond to changing levels of nutrient in their environment by modulating (whenever possible) both the nature of the uptake system and the extent to which its synthesis is induced or derepressed. Many examples of enzyme hyperproduction

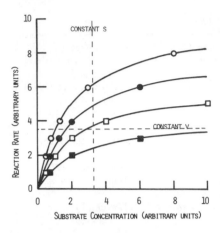

Figure 1. Theoretical relationship between reaction rate and substrate concentration following Michaelis-Menten kinetics and assuming a maximum velocity of (○) 10, (●) 8, (□) 6, and (■) 4 arbitrary units.

and of modulation between alternative high- and low-affinity uptake and assimilatory processes are to be found in the literature, but it is sufficient here to cite in detail just two of these.

Those microorganisms that can grow with ammonia providing the sole source of utilizable nitrogen generally possess the enzyme glutamate dehydrogenase (EC 1.4.1.4.), which is able to reductively aminate 2-oxoglutarate to form glutamate. This glutamate subsequently can act as an amino-group donor for transaminase reactions leading to the synthesis of all the other amino acids and hence to the synthesis of protein. Therefore, glutamate dehydrogenase must be considered a key enzyme in microbial nitrogen metabolism and clearly functions as the primary ammonia-assimilating reaction under conditions where ammonia is present in cell-saturating concentration. However, this assimilatory glutamate dehydrogenase (which most frequently is NADPH-specific) has only a low affinity for ammonia, the apparent K_m values being in the region of 1 to 5 mM, and therefore would not provide an effective mechanism for scavenging residual ammonia from an ammonia-depleted environment unless it was synthesized to very high levels within the cell. Not surprising, therefore, was the finding of Brown and Stanley (1972) that the yeast *Saccharomyces cerevisiae* synthesized glutamate dehydrogenase to extraordinarily high levels when grown in an ammonia-limited chemostat culture: when growing at a dilution rate of 0.1/hr (30°C, pH 5.5), the specific activity of this enzyme in cell-free extracts rose from 465 (nmol/min/mg protein) to 1730—that is, almost fourfold—when the culture was changed from being phosphate-limited to being ammonia-limited. Similar high levels of glutamate dehydrogenase activity were found with ammonia-limited cultures of other yeasts, such as *S. carlsbergensis* (1186 nmol/min/mg protein), *S. delbruckii* (1570 nmol/min/mg protein), and *Candida utilis* (2100 nmol/min/mg protein).

With a large number of bacterial species, however, as with a few yeast species (for example, *Schizosaccharomyces pombe*), exposure to low-ammonia environments did not evoke increased synthesis of glutamate dehydrogenase, and with several species (e.g., *Klebsiella aerogenes*), synthesis actually was suppressed totally. These organisms were nevertheless able to extract residual ammonia from their environments with remarkable effectiveness, which stemmed from their being able to induce, or derepress, the synthesis of two enzymes (glutamine synthetase and glutamate synthase), which together provided an alternative high-affinity mechanism for assimilating ammonia into glutamate (Fig. 2). The ammonia-scavenging potential of this alternative pathway resides, of course, in the first enzyme (glutamine synthetase—EC 6.3.1.2.), which has a very low K_m value for ammonia (Umbarger, 1969). Thus, this high

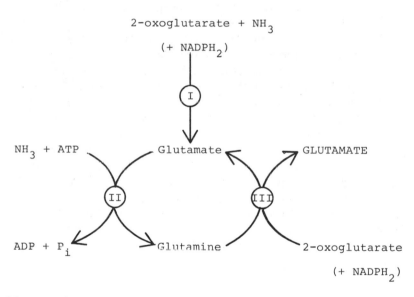

Overall reactions:

(I) Glutamate dehydrogenase:

NH_3 + 2-oxoglutarate + $NADPH_2 \longrightarrow$ Glutamate + NADP + H_2O

(II + III) Glutamine synthetase plus glutamate synthase:

NH_3 + 2-oxoglutarate + ATP + $NADPH_2 \longrightarrow$ Glutamate + ADP

+ P_i + NADP

Figure 2. The high-affinity pathway of ammonia assimilation.

affinity for ammonia allows for the efficient uptake of low concentrations of ammonia without the organisms having to commit themselves to synthesizing one particular enzyme to an exceptionally high level.

In just the same way that *K. aerogenes* possesses dual mechanisms for the uptake of ammonia, so too does this organism possess alternative assimilatory pathways for glycerol (Neijssel *et al.*, 1975). Both of these pathways (Fig. 3) operate in organisms growing aerobically, but not simultaneously. The glycerol kinase route is expressed in organisms exposed to carbon-limiting conditions, and the glycerol dehydrogenase route is induced, or derepressed, in organisms growing in glycerol-sufficient (ammonia-, sulfate-, or phosphate-limiting) environments. The essential difference between these two routes resides in the kinetic behavior of the first enzyme of each reaction sequence, and this also provides a clue to their functional significance. Thus, glycerol kinase has an exceedingly high affinity for glycerol (apparent $K_m = 1-5 \times 10^{-6}$ M) and therefore is effective in scavenging traces of glycerol from a glycerol-limited environment. On the other hand, glycerol dehydrogenase has an almost immeasurably low affinity for glycerol (apparent $K_m = 5 \times 10^{-2}$ M) and will therefore act to limit the extent to which excess glycerol is metabolized by glycerol-sufficient organisms. An additional feature of the high-affinity assimilatory pathway, evident in Fig. 3, is that the first dehydrogenase reaction (glycerol 3-phosphate dehydrogenase) is not NAD-linked but is linked to a flavoprotein, which then donates its electrons to the respiratory chain. Thus, this pathway could not function anaerobically unless some organic electron acceptor was provided (or generated) that could reoxidize either the reduced flavoprotein or some other reduced component of the respiratory chain. Actually, however, when the organisms were grown *anaerobically* in a glycerol-limited environment, it was found that they no longer synthesized glycerol kinase but, instead, formed glycerol dehydrogenase in massive amounts (Table II) (see also Lin *et al.*, 1960). This anaerobic glycerol dehydrogenase possessed kinetic properties that were closely similar to those of the

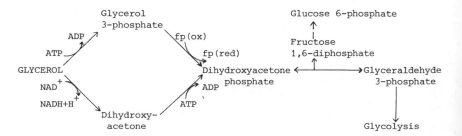

Figure 3. Pathways of glycerol assimilation in *Klebsiella aerogenes* NCTC 418.

Table II. Glycerol Kinase and Glycerol Dehydrogenase Contents of Variously
Limited *Klebsiella aerogenes* Organisms[a]

Growth limitation	Carbon source	Other conditions	Specific activity (nmol substrate used/ min/mg protein)	
			Glycerol kinase	Glycerol dehydrogenase
Carbon	Glycerol	Aerobic	122	<1
	Glucose	Aerobic	106	<1
	Glycerol	Anaerobic	<1	282
	Glucose	Anaerobic	<1	101
Sulfate	Glycerol	Aerobic	<1	11
	Glucose	Aerobic	<1	<1
	Glycerol	Anaerobic	<1	<50
	Glucose	Anaerobic	<1	22
Ammonia	Glycerol	Aerobic	<1	9

[a]Data from Neijssel *et al.* (1975)

enzyme produced aerobically by glycerol-sufficient organisms: they were both NAD-linked, but whether they were in fact identical enzymes was not determined.

The synthesis of glycerol kinase aerobically and high levels of glycerol dehydrogenase anaerobically by organisms exposed to carbon(glycerol)-limited environments is easy to rationalize in terms of the organisms' need to scavenge glycerol effectively from a substrate-depleted environment. But why they should suppress glycerol kinase synthesis when growing aerobically in a glycerol-sufficient medium is not immediately obvious. Put another way, one might argue that having acquired, in the course of evolution, a high-affinity uptake system for some substrate, it seemingly makes little sense to continue carrying the genetic burden of a parallel low-affinity uptake mechanism unless it confers upon the organism some additional advantage. So what might this be? In the case of ammonia assimilation, it can be argued that the low-affinity (glutamate dehydrogenase) pathway would be particularly advantageous to organisms when growing in environments that were carbon- and energy-limited, since this particular pathway is energetically less expensive than the glutamine synthetase (high-affinity) pathway. However, the same line of reasoning cannot be used to explain the existence of dual pathways of glycerol assimilation; here the physiological considerations are probably more subtle, as will be discussed later.

With the two substrates considered above (that is, ammonia and glycerol), their entry into the cell does not involve or require the participation of membrane-associated active transport components. Hence, the scavenging function is served by the first enzyme of the metabolic sequence effecting assimilation of the growth-limiting nutrient. However, the cell membrane (being a lipophilic environment) has almost no inherent permeability toward hydrophilic substances like glucose and amino acids, and for these to be taken into the cell, specific active transport systems are necessary. Thus, in the context of microbial adaptation to low-nutrient environments, one might expect either that high- and low-affinity active transport systems would be synthesized by organisms or else that they would respond to changes in the availability of essential nutrient substances by modulating the extent to which specific transport systems are elaborated. In this connection, there are to be found in bacteria at least three different types of active transport system: (1) the phosphoenolpyruvate phosphotransferase system (PTS), which transports sugars and simultaneously phosphorylates them (group translocation); (2) systems that involve specific substrate-binding proteins which are, in gram-negative bacteria, located in the periplasmic space from which they can be extracted by means of osmotic shock procedures; and (3) a system that involves only membrane-bound components. The last system is retained in membrane vesicles and transports solutes when the membrane is suitably energized. All three systems require the expenditure of energy in order to effect transport of the solute, and their affinities for specific substrates tend to be extremely high (Table I). Indeed, in some cases (as, for example, with the PT-system for glucose uptake by *E. coli*), the saturation constant of the uptake system is almost too low to be determined experimentally. Not surprisingly, although some of these uptake systems are formed constitutively, the synthesis of many, if not all, can be extensively modulated in response to changes in the growth environment (see, for example, Herbert and Kornberg, 1976).

The above statement implies that the high-affinity uptake systems may be, in general, energetically more expensive than the corresponding low-affinity systems, and there is much evidence to support this conclusion. Indeed, this seemingly also applies to the utilization of O_2 as the terminal electron acceptor in respiration. Thus, Meyer and Jones (1973) found that growth proceeded with a decreased efficiency when *E. coli* was exposed to O_2-limiting conditions, and they proposed that cytochrome a_1 (an oxidase with a high affinity for O_2) probably terminated a branch of the respiratory chain that had a lowered efficiency of energy conservation. They further proposed that when cell-saturating concentrations of O_2 were present, electron transfer to O_2 occurred via a branch of the respiratory chain, terminating in cytochrome o, that had an increased efficiency of energy conservation.

A lowered energetic efficiency may well be the "price to be paid" for the operation of a high-affinity substrate-uptake system, but the competitive advantage that such a system confers upon an organism is undeniable. The most graphic example that can be cited in support of this statement comes from the work of Hartley *et al.* (1972), who studied the competition between *K. aerogenes*, strain 2A, and an evolvant (with an improved xylitol dehydrogenase) when growing in a xylitol-limited chemostat culture (Fig. 4). In this experiment the initial population was of a strain of *K. aerogenes* which synthesized xylitol dehydrogenase (or, more properly, ribitol dehydrogenase) constitutively and which, when supplied with a medium containing 0.2% xylitol (w/v) as the growth-

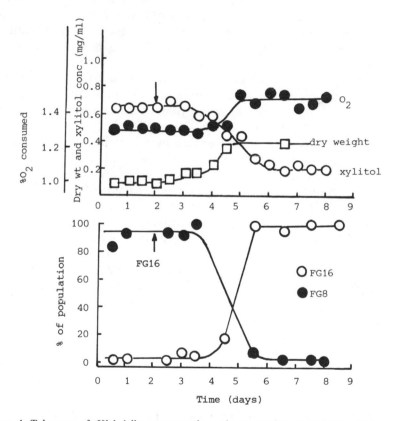

Figure 4. Takeover of *Klebsiella aerogenes* in a chemostat by an evolvant. At the point indicated by the arrow, approximately 20 cells of strain 1B (FG16, a guanine auxotroph of a mutant with improved xylitol dehydrogenase) were added to a steady-state culture containing approximately 10^{12} cells of strain 2A (FG8, an arginine auxotroph of the ancestral strain) grown on xylitol. The composition of the population was determined by plating on appropriate media. Data from Hartley *et al.* (1972).

limiting nutrient, established a population containing approximately 5×10^8 organisms/ml. At the point indicated by the arrow in Fig. 4, 20 cells of strain B (a mutant with a xylitol dehydrogenase that had a threefold improvement in the apparent K_m value) were added to the culture; after a period of a very few days, it was found that the evolvant had almost totally outgrown the initial population. It is clear, therefore, that a modest improvement in the kinetics of substrate uptake and assimilation confers upon the cell a powerful growth advantage in nutrient-limited environments. However, a single-step mutation leading to a detectable improvement in some enzymes' kinetic behavior is, in all probability, an exceedingly rare event, and, as suggested previously, a similar growth advantage accrues from cells synthesizing massive amounts of some less effective enzyme. In this connection, gene doubling provides a means whereby the rates of synthesis of constitutive enzymes can be increased substantially, effecting an enormous change in the ratio of these to the other proteins of the cell. Hartley (1974) cites the example of evolvants of *K. aerogenes* that synthesized ribitol dehydrogenase to a 20-fold higher level than found constitutively in the parent strain and in which 20% of the cells' total protein was this single enzyme. From a consideration of Michaelis-Menten kinetics [that is, $K_m/s = (V_{max}/v) - 1$], it is clear that when the ratio V_{max}/v is large, a further increase in V_{max} (effected by the cells' synthesizing an increased amount of enzyme) influences the growth-limiting substrate concentration s to an extent closely similar to that of a strictly proportional decrease in the affinity constant K_m. In other words, a threefold increase in the cells' content of some substrate-assimilating enzyme would be as effective (competitively speaking) as a threefold decrease in the value of K_m (or K_s). But again the price that has to be paid, so to speak, is a commitment of a substantial portion of the cells' biosynthetic potential toward synthesizing a single cellular component and, thereby, a commitment to metabolic specialization that, under other circumstances, would impose an intolerable burden. However, as a first step in enzyme evolution, gene doubling may play a crucial role, as has been argued by Hartley *et al.* (1972).

Apart from the mechanisms considered above, there are yet other discernible ways by which organisms scavenge for, and facilitate the uptake of, low (growth-limiting) concentrations of essential nutrients. For example, organisms may excrete substances that react extracellularly with specific nutrients (particularly cations) and thereby promote uptake by ancillary high-affinity transport processes. This is frequently observed with cultures that are exposed to environments containing only low concentrations of iron; here, iron-chelating compounds (siderochromes) are synthesized and secreted (Byers *et al.*, 1967; Frost and Rosenberg 1973). There is also good evidence for similar substances being secreted by Mg^{2+}-limited *Bacillus subtilis* organisms, which dramatically improve their ability to compete with other organisms for growth-limiting amounts of magnesium (Tempest *et al.*, 1967b; Meers and Tempest, 1968). Yet again, Ketchum and Owens (1975) described a molybdenum coordinating compound that was produced by *B. thuringiensis* and that may facilitate the uptake of this "trace" nutrient, and it has been further proposed that some peptide antibiotics, like bacitracin, may function in the uptake of divalent cations by the producer organism (Haavik, 1976).

The cell wall also can play a role in the uptake of "trace" nutrilites by providing binding sites at which these substances can be concentrated prior to translocation across the plasma membrane and into the cell. It has been shown, for instance, that intact cells and isolated walls of *B. subtilis* can bind magnesium and that, when these organisms are grown in a Mg^{2+}-limited environment, the walls are changed such as to possess an increased Mg^{2+}-binding affinity (Meers and Tempest, 1970). In contrast to this finding, Heckels *et al.* (1977) reported that the cell walls of *B. subtilis* W23 had almost the same affinity for magnesium ions whether the organisms had been cultured under magnesium, potassium, or phosphate limitation. Nevertheless, these authors showed that removal of teichoic acid from the walls of this organism, by periodic acid treatment, greatly diminished the affinity with which they could bind magnesium ions. It can therefore be concluded that teichoic acid (and teichuronic acid, which substitutes for teichoic acid in phosphate-limited organisms) plays a role in the binding of divalent cations at the cell surface, thereby facilitating their uptake.

Finally, organisms may respond to low-nutrient environments by secreting enzymes that are capable of mobilizing those substrates that otherwise are inaccessible to them. Thus, the synthesis and secretion of enzymes such as phosphatases, proteinases, and amylases may be, at least in part, considered to be properties that organisms have acquired in the course of evolution to cope primarily with conditions of nutrient depletion. It has been shown, for instance, that *B. licheniformis* 749/C produces a proteinase specifically when growing in an environment that contains growth-limiting concentrations of a utilizable nitrogen source and alkaline phosphatase and ribonuclease specifically when growing under conditions of phosphate limitation (Wouters and Buysman, 1977).

In conclusion, it can be stated that organisms react to conditions of nutrient insufficiency by synthesizing high-affinity uptake systems, trapper mechanisms, binding polymers, and chelating agents; they also may synthesize and excrete hydrolases. All these properties are of overwhelming importance to organisms in many natural ecosystems where their survival will depend critically upon their capacity to compete effectively with other organisms for the limited supply of essential nutrients that generally will be present. When the insufficiency is relieved, however, many of these properties are suppressed, and alternative uptake systems are elaborated—ones that have a lowered affinity for and activity toward the particular substrate. It is therefore appropriate to consider next these alternative uptake systems in the context of regulation of uptake and metabolism of nutrient substances that are present in excess of the growth requirement.

2.2. Regulation of Metabolism of Nonlimiting Nutrients

Clearly the possession of a high-affinity uptake mechanism for each of the potentially growth-limiting nutrients is of critical importance to organisms—particularly to those living in dilute aquatic environments, where the concentrations of most essential nutrients will be generally (if not invariably) low. But, as mentioned previously, it is not always obvious why, having acquired some high-affinity uptake mechanism, organisms should continue to carry the genetic burden of a parallel low-affinity mechanism; one can only suppose that such systems do confer upon the organisms some physiological advantage. However, in seeking an explanation for the presence of dual uptake mechanisms in general, it is perhaps more useful to consider the disadvantages that accrue from the possession of a high-affinity uptake system, rather than the advantages that arise from the possession of either one or the other alternative system. In particular, it is important to understand how organisms prevent (or at least circumvent) an excessive accumulation of nutrient substances and products of their metabolism within the cell when some specific growth-limitation is relieved. It is clear that an unrestricted accumulation of intermediary metabolites within the cell can have devastating consequences, as has been elegantly revealed by studies with both *E. coli* and *K. aerogenes.* In the first instance, a mutant of *E. coli* was obtained that was constitutive for glycerol kinase (the high-affinity glycerol-assimilating pathway) and that lacked feedback control of this enzyme (Zwaig *et al.*, 1970); in the second case, *K. aerogenes* organisms were induced to synthesize high levels of glycerol kinase by growing the organisms at a low dilution rate in a glycerol-limited aerobic chemostat culture (Postgate and Hunter, 1963). With both types of organisms, exposure to moderate concentrations of glycerol under conditions in which they could not grow (i.e., washed suspensions) led to a rapid loss of viability—a condition that was termed "substrate-accelerated death" (Postgate and Hunter, 1963, 1964). Although it was subsequently realized that, at least in the latter example, the cells were not actually dead but were suffering severe catabolite repression of protein synthesis (Calcott and Postgate, 1972, 1974), nevertheless it is abundantly obvious that a gross accumulation of products of glycerol metabolism within the cells had placed them in an uncompromising position from which they could not easily be extricated. Freedberg *et al.* (1971) have shown that the strain of *E. coli* mentioned above produced bactericidal amounts of methyloglyoxal because of the loss of feedback control of glycerol kinase by fructose 1,6-diphosphate. Yet other examples of accumulation of metabolites to concentrations that inhibit growth or interfere with metabolism have been reported: for example, growth inhibition of *Proteus mirabilis* by cyclic AMP (Schwarzhoff and Williams, 1976), growth inhibition of *Thiobacillus thiooxidans* as a consequence of the accumulation of pyruvate when the organism was grown on glucose (Borichewski and Umbreit, 1966), growth inhibition of *Shigella flexneri* by acetate (Baskett and Hentges, 1973), and intracellular accumulation of 3-phosphoglycerate and subsequent inhibition of growth and sporulation of *B. subtilis* because of manganese starvation (Oh and Freese, 1976). Hence, one may conclude that at least one advantage to organisms of the possession of a low-affinity uptake system for carbon substrate resides in their being able to coarsely "trim" the rate of penetration of nutrilite into the cell and thereby prevent intermediary metabolites from accumulating to potentially traumatic levels.

The above conclusion leads directly to a consideration of those substrates, like glucose, for which there is thought to be but a single uptake system (the PEP-phosphotransferase system, in the case of *E. coli* and *K. aerogenes*). Do these organisms in fact possess a finely tuned control mechanism capable of regulating the rate of penetration of glucose into the cell such as to just meet its bioenergetic and biosynthetic demands? And if not, how does the organism conspire to prevent the accumulation of intermediary metabolites to traumatic concentrations? With regard to *E. coli*, there is evidence that, at least with some strains, synthesis of some components of the phosphotransferase system is modulated so as to effect an approximate balance between the rate of glucose uptake and the rate of glucose utilization in growth-associated processes (Herbert and Kornberg, 1976). But with other strains of *E. coli* and of *K. aerogenes*, such a modulation, if it occurs at all, does not concomitantly modulate extensively the rate of penetration of glucose into the cell (Neijssel and Tempest, 1976b; Neijssel *et al.*, 1977). This conclusion can also be drawn from a consideration of the data shown in Table III, which compares the rates of glucose utilization associated with a rate of cell synthesis (at a dilution rate of 0.17/hr) equivalent to 20 milli-atoms cell carbon/hr when the cultures were glucose-limited and then, successively, ammonia-, sulfate-, phosphate-, magnesium-, and potassium-limited (with glucose being maintained in excess of the growth requirement). For further comparison, data obtained with variously limited chemostat cultures of *K. aerogenes* growing on glycerol, mannitol, or lactate (as alternative carbon and energy sources) are included.

It is clear from these data that whenever the carbon source was

Table III. Substrate Utilization Rates and Rates of Product Formation in Variously Limited Chemostat Cultures of Klebsiella aerogenes ($D = 0.17 \pm 0.01$/hr, pH = 6.8, 35°C)[a,b]

Carbon source	Limitation	Substrate used	Cells	CO_2	Pyr	2-OG	Acet	GA	2-KGA	Succ	Protein	CHO	Carbon recovery (%)
Glucose	Carbon	36.8	20	15.6	—	—	—	—	—	—	—	—	97
	Sulfate	98.7	20	20.8	21.9	1.8	9.6	—	—	2.3	1.8	7.0	91
	Ammonia	107.4	20	20.2	5.2	22.5	4.0	—	—	—	—	36.0	102
	Phosphate	112.8	20	20.4	—	9.6	4.5	9.5	39.9	—	—	15.1	97
	Magnesium	124.6	20	31.4	9.6	9.6	17.7	0.1	11.9	—	2.5	—	83
	Potassium	175.0	20	56.3	10.1	3.0	13.3	31.9	20.5	—	8.0	—	93
Glycerol	Carbon	35.9	20	15.3	—	—	—	—	—	—	—	—	98
	Sulfate	54.9	20	25.4	—	—	7.4	—	—	—	—	—	96
	Ammonia	45.1	20	20.2	—	5.7	2.0	—	—	—	—	—	106
	Phosphate	53.8	20	31.1	—	—	—	—	—	—	—	—	95
Mannitol	Carbon	33.1	20	13.1	—	—	—	—	—	—	—	—	100
	Sulfate	61.1	20	27.5	—	—	15.2	—	—	—	—	—	103
	Ammonia	64.0	20	25.0	—	15.8	3.3	—	—	—	—	—	100
	Phosphate	41.2	20	19.8	—	—	1.6	0.1	0.1	—	—	—	101
Lactate	Carbon	40.2	20	19.3	—	—	—	—	—	—	—	—	98
	Sulfate	60.4	20	36.5	—	—	8.5	—	—	—	—	—	108
	Ammonia	92.1	20	60.3	1.6	4.3	6.3	—	—	—	—	—	100
	Phosphate	62.0	20	39.2	—	—	2.5	—	—	—	—	—	100

[a]All values obtained were converted into their carbon equivalents and adjusted to a cell production rate of 20 milli-atoms carbon per hour. In this way the specific substrate-utilization rates and the rates of specific product formation can be compared directly. The following abbreviations have been used: Pyr, pyruvate; 2-OG, 2-oxoglutarate; Acet, acetate; GA, gluconate; 2-KGA, 2-ketogluconate; Succ, succinate; Protein, extracellular protein; and CHO, extracellular polysaccharide.
[b]Data from Neijssel and Tempest (1975) and unpublished results.

present in excess of the basic growth requirement and it was some other component of the medium that limited growth, then the carbon source was utilized more extensively than when its availability was growth-rate limiting. But the extent to which any particular carbon source was overutilized varied markedly both with the nature of the growth limitation and with the identity of the particular carbon substrate. In this latter connection, it may be significant that glycerol, with its low-affinity assimilation pathway, was overutilized less extensively than was glucose and that it was more completely oxidized. Thus, whereas many products of catabolism accumulated in the culture extracellular fluids when glucose was the carbon substrate, when glycerol was being used the principle products generally were cells and carbon dioxide (Table III).

It is also apparent from these data that, of the four carbon substrates studied, excess glucose was overutilized more extensively than each of the other three compounds—even though, just as with glucose uptake, no evidence could be found to indicate that *K. aerogenes* organisms possess more than one uptake system for either mannitol or lactate. But, clearly, in all of these cases the accumulation of intermediary metabolites to traumatic levels was prevented by the organisms' selectively excreting into the culture fluids a small number of intermediary metabolites. Significantly, these were all nonphosphorylated compounds which, presumably, could be more readily exported from the cell. Further, the spectrum of compounds that were excreted differed (both qualitatively and quantitatively) with the nature of the growth limitation. Thus, phosphate-limited cells as well as potassium-limited cells excreted large amounts of gluconate and 2-ketogluconate when growing on glucose, whereas the main overflow product of ammonia-limited cells was 2-oxoglutarate, and of sulfate-limited cells, pyruvate and acetate.

Apart from the synthesis of extracellular polysaccharide by glucose-sufficient cultures, the overflow products were all typical intermediates of oxidative metabolism, and it is obvious that any excess of reducing equivalents, released as a consequence of the excess utilization of substrate, must have been oxidized through to water. Thus, there was no significant accumulation of, say, ethanol or 2,3-butanediol, in any of these cultures, such as occurs under conditions of O_2 limitation (Pirt, 1957; Harrison and Pirt, 1967) or with Crabtree-positive yeasts growing in the presence of excess glucose.

From the data contained in Table III and from similar data obtained by us with a number of other organisms (including *E. coli, Pseudomonas aeruginosa, B. subtilis, B. stearothermophilus*), it is abundantly obvious that, when growing under conditions of carbon-substrate excess, organisms frequently do not modulate extensively the rates of substrate uptake sufficient just to meet the organisms' biosynthetic and bioenergetic demands and such as to achieve an optimization in the conversion of substrate carbon into cell carbon. Hence, two important questions require to be considered: (1) Why don't organisms regulate their uptake processes more effectively (in other words, what is the significance of overflow metabolism)? (2) How do organisms contrive to dispose of the excess reducing equivalents which follow from not fully modulating the rate of substrate uptake, without generating a severe energy imbalance?

With regard to the first question, it is reasonable to assume that when the culture is, say, ammonia-limited, it is advantageous to maintain an elevated "pool" of intermediary metabolites like 2-oxoglutarate and glutamate since these are key reactants in the assimilation of the growth-limiting nutrient (see Fig. 2). Under ammonia-limiting conditions, the efficient uptake and assimilation of ammonia will depend critically on the activity of two key enzymes (glutamine synthetase and glutamate synthase), each of which has a triple substrate requirement (ammonia, ATP, and glutamate in the first instance and glutamine, 2-oxoglutarate, and NADPH in the second). Of these six substrates, two (ammonia and glutamine) will be present in low concentration; thus, the other four must be maintained at enzyme-saturating concentrations for the total assimilation of ammonia into glutamate to proceed efficiently. Of these four substrates, one (glutamate) is in fact the product of ammonia assimilation; it is also deeply implicated in osmoregulatory processes, and homeostatic mechanisms have been found to exist within the cell to ensure that the "pool" free glutamate concentration is maintained at a high level (Tempest *et al.*, 1970). Therefore, ancillary processes must ensure that ammonia-limited organisms do not simultaneously experience some limitation in the rates of synthesis of 2-oxoglutarate, NADPH, and ATP, and this requires that stringent control of carbon-substrate catabolism be circumvented, which clearly is the case. Thus, in order to guarantee that the efficiency with which the growth-limiting substrate can be taken into the cell is not impeded, it is far better *not* to modulate extensively the rate of carbon-substrate uptake but to generate key intermediary metabolites at a high rate and to excrete (or otherwise dispose of) the excess reactants that cannot be utilized. Hence, one can adequately account for the excretion of 2-oxoglutarate by organisms limited in their growth by the availability of ammonia, for the excretion of pyruvate by sulfate-limited organisms, and even for the excretion of gluconate and 2-ketogluconate by phosphate- or potassium-limited organisms. In the particular case of phosphate-limitation, it must be remembered that, when microorganisms are growing aerobically, the respiratory chain provides the major mechanism of phosphate assimilation, and for this to run effectively requires an unimpeded supply of reducing equivalents (primarily NADH) to be delivered up by processes that do not demand the excessive participation of phosphate esters. In this regard, the conversion of glucose to gluconate and 2-ketogluconate (though proceeding through the synthesis of glucose 6-phosphate) represents an optimal compromise. The efficient uptake of low available concentrations of potassium likewise requires, with aerobic organisms, intense respiratory activity, and there has been found to exist (at least with *Candida utilis*) a stoichiometric relationship between the cellular potassium content of potassium-limited organisms and the yield value on O_2 (Y_{O_2} = g organisms synthesized/mmol O_2 consumed) (Aiking and Tempest, 1976; Aiking *et al.*, 1977).

But an inevitable consequence of maintaining a high rate of carbon-substrate catabolism is that organisms must invoke processes for disposing of the excess reducing equivalents (mainly in the form of NADH) that are generated simultaneously. As mentioned previously, from the measured rates of O_2 uptake (Table IV) and from the absence of typical (reduced) fermentation products (Table III), it is clear that a redox balance is maintained within the cell by stimulating the terminal respiratory processes and oxidizing NADH at an increased rate. However, in solving this problem of redox imbalance in this particular way (that is, by increased electron-transport-chain activity), a further problem is created in that the transfer of electrons to O_2 is associated with the generation of ATP from ADP and inorganic phosphate. This increased respiration rate must effect a severe imbalance in the cells' energy charge (Atkinson, 1968) unless, that is, mechanisms can be invoked either to circumvent concomitant ATP synthesis or to turn over the ATP "pool" by growth-unassociated energy-spilling reactions. That such "slip" reactions can occur in most, if not all, organisms and that respiration is not firmly coupled to biosynthesis are evident from the fact that washed suspensions of organisms generally will oxidize suitable substrates at high rates under conditions where they clearly cannot grow. This common finding has profound implications in our assessment of the energetics of microbial growth, as will be detailed later. But in concluding this section it is necessary to consider the broader ecological implications of both "overflow metabolism" and "energy-spilling reactions" and to inquire into whether these occur at all in natural ecosystems and, if so, how they are manifest.

Reports in the literature both of overflow metabolism and of energy-spilling reactions occurring in natural environments are not widespread. The main reason for this, no doubt, is that the concentrations of the possible excretion products will be extremely low in most natural ecosystems because of the low population densities and the consumption of excreted products by other organisms. Nevertheless, it has been shown, for instance, that many algae excrete glycollate both in fresh water and in marine environments. This substance has been detected in

D. W. Tempest and O. M. Neijssel

Table IV. Specific Rates of O_2 Consumption and Yield Values for Carbon
Substrate and O_2 Found with Variously Limited Chemostat Cultures of *Klebsiella
aerogenes* NCTC 418 Growing at a Fixed Dilution Rate ($D = 0.17/hr$),
Temperature (35°C), and pH Value (6.8) [a]

Substrate	Limitation	Q_{O_2} value [b]	$Y_{carbon\ substrate}$	Y_{oxygen}
Glucose	Carbon	101	81.0	20.4
	Sulfate	179	34.2	11.5
	Ammonia	179	28.8	11.5
	Phosphate	236	27.0	8.7
	Magnesium	272 [c]	24.4	7.6 [c]
	Potassium	420	17.1	5.2
Glycerol	Carbon	114	42.3	18.1
	Sulfate	274	27.6	7.5
	Ammonia	248	33.1	8.3
	Phosphate	254	27.6	8.1
Mannitol	Carbon	127	91.0	16.2
	Sulfate	278	49.1	7.4
	Ammonia	224	47.3	9.2
	Phosphate	184	72.8	11.2
Lactate	Carbon	166	36.9	12.4
	Sulfate	410	18.0	5.0
	Ammonia	411	15.3	5.0
	Phosphate	303	24.3	6.8

[a]Data from Neijssel and Tempest (1975) and unpublished results.
[b]The Q_{O_2} values are expressed as $\mu l\ O_2$ consumed/mg equivalent dry weight organisms × hr. The yield
values ($Y_{carbon\ substrate}$ and Y_0) are expressed as g organisms synthesized/mol carbon substrate or g atom
O_2, respectively, consumed.
[c]Because of the inherent instability of the O_2 consumption rate of this type of culture, only an
approximate figure is given.

lake waters in concentrations of up to 0.06 mg/liter (Fogg, 1975). That this
excretion of glycollate is a phenomenon dependent on the environmental
conditions and not an intrinsic property of phytoplankton is evident from
the fact that the amount of glycollate excreted during photosynthesis can
vary from a few percent to nearly 100% of the total carbon dioxide fixed by
laboratory cultures (Tolbert, 1974). In this connection, the following
processes are believed to account for glycollate excretion and photorespiration in plants and algae: (1) a competition between CO_2 and O_2 for
ribulose diphosphate carboxylation or oxygenation (which implies an
imbalance between the reactant, CO_2, and the product, O_2, of photosynthesis) and (2) disposal of excess carbon and reducing equivalents by

wasteful respiration, thereby maintaining the cofactors in the electron-transport chains within some oxidized/reduced state (Tolbert, 1974).

Similarly, glycollate excretion has been reported for *Rhodospirillum rubrum* grown heterotrophically and photoautotrophically at low O_2 tensions (Codd and Smith, 1974) and *Alcaligenes eutrophus* grown autotrophically (Codd *et al.*, 1976). In the latter case, addition of O_2 to the H_2 that was used for gassing stimulated excretion, whereas the further addition of CO_2 prevented stimulation of the excretion product formation by O_2.

Another overflow product that has been reported for photosynthetic organisms is H_2. Thus, nitrogen-starved cultures of *Anabaena cylindrica* (Weissman and Benemann, 1977) and nitrogen-limited cultures of *Rhodopseudomonas capsulata* (Hillmer and Gest, 1977a,b) both produce H_2 in considerable quantities. In *R. capsulata*, the production occurred via nitrogenase and was inhibited by nitrogen or ammonium salts. This illustrates again the point made previously, that microorganisms strive to saturate the uptake system of the growth-limiting nutrient with the key reactants: in this case reducing equivalents, which, in the absence of nitrogen, are disposed of as H_2.

Luminescence may be considered to be yet another example of energy-spilling reactions occurring in nature. Watanabe *et al.* (1975) have shown that the luminescence *in vivo* changes during growth of *Photobacterium phosphoreum* in batch culture, while the contents of luciferase and NAD(P)-FMN reductase remained almost constant. Just after the start of logarithmic growth, luminescence, which was occurring throughout the lag phase, decreased but increased again in the mid-logarithmic phase. The authors concluded that at the onset of growth the oxidation of cellular reducing equivalents by the cytochrome system prevailed over the oxidation of these by the luciferase system. Activity of the latter system obviously is invoked when growth, but not the generation of reducing equivalents, is impeded. Moreover, cyanide inhibited O_2 uptake and stimulated luminescence, again indicating that the luciferase system serves for the maintenance of an optimal redox balance in the cell.

Many more examples are available of overflow metabolism in laboratory cultures: production of citrate by zinc-limited *Aspergillus niger* (Wold and Suzuki, 1976), acetone production from ethane by *Methylosinus trichosporium* (Thomson *et al.*, 1976), excretion of 2-oxoglutarate and pyruvate by nitrogen-limited *Desulfovibrio* (Lewis and Miller, 1975), hydrogen cyanide production by several strains of *Pseudomonas aeruginosa* (Castric, 1975, 1977), production of waxes by *Acinetobacter calcoaceticus* growing in a nitrogen-limited chemostat culture (Fixter and Fewson, 1974), production of oxaloacetate and pyruvate by *Thiobacillus thiooxidans* (Borichewski and Umbreit, 1966; Borichewski 1967), and lastly production of mycotoxins in food by fungi (Turner, 1975).

In each of these cases, a detailed analysis of the culture conditions that prescribe the production of the particular metabolites reveals that the function of overflow metabolism resides in the maintenance of a balanced pool of metabolic intermediates within the cell. This, in turn, is essential for the regulation of synthesis of the cells' polymers, as will be considered in the next section.

2.3. Regulation and Modulation of Polymer Synthesis

The ability of microbial cells to undergo substantial changes in their chemical composition in response to changes in the growth environment is well documented (Herbert, 1961; Neidhardt, 1963); in particular, it has been established that many organisms accumulate "storage" polymers (such as glycogen, poly-β-hydroxybutyrate, and polyphosphate) when limited in their growth by the availability of an essential nutrient other than the carbon and energy source (Dawes and Senior, 1973). However, it is not as widely appreciated that organisms may actually diminish their requirement for some essential compound or element by synthesizing less of those particular cellular components that are heavy consumers of the specific growth-limiting nutrient. In this way, the yield of organisms (with respect to both biomass and cell number) can be maximized.

A good example of the latter type of response is seen in the way that some gram-positive bacteria accommodate to conditions of phosphate insufficiency (Tempest *et al.*, 1968; Ellwood and Tempest, 1969). It was found that *Bacillus subtilis* var. *niger* organisms contained almost twice as much phosphorus when grown in the presence of excess phosphate as when grown at corresponding rates, and at a similar temperature and pH value, in a phosphate-limited chemostat culture. The "extra" phosphorus of the phosphate-sufficient cells was organically bound and present largely, if not entirely, in the cell walls (Table V). Thus, whereas the phosphate-sufficient organisms had walls containing up to 6% (w/w) phosphorus, the walls of phosphate-limited cells were practically devoid of this element. The phosphorus present in the walls of these gram-positive bacteria is contained almost exclusively in teichoic acids (Baddiley, 1964), polymers that clearly can be deleted under conditions of phosphate-limitation, thereby effecting a considerable sparing in the cells' requirement for phosphate (e.g., from 32 mg P/g cells to 17 mg P/g cells when the culture was changed from being potassium-limited to being phosphate-limited). This deletion of teichoic acid was accompanied by a simultaneous deposition within the wall of teichuronic acid (a non-phosphorus-containing anionic polymer) which presumably served to functionally replace the teichoic acid (Fig. 5). In this connection, the precise physiological functions that teichoic acid and teichuronic acid

Table V. Influence of Growth-Limiting Component of the Medium on the Composition of the Cell Walls of *Bacillus subtilis* var. *niger* [a,b]

Component	Content (g of component/100 g of dry cell walls)	
	Phosphate-limited organisms	Mg^{2+}-limited organisms
Phosphorus	0.2	6.0
Glucose	<1	28(27)[c]
Glucuronic acid	22	<2
Galactosamine	14	<2

[a]Data from Tempest *et al.* (1968).
[b]Organisms were grown in chemostats at a dilution rate of 0.2/hr (at 35°C and pH 7.0), collected into cooled receivers, and fractionated. Glucose was determined enzymically on wall preparations that had been hydrolyzed with 1.0 N H$_2$SO$_4$ (at 100°C for 3 hr in a sealed tube) and neutralized with NaOH. Galactosamine also was determined enzymically on samples that had been hydrolyzed with 4 N HCl (at 100°C for 4 hr in a sealed tube) and neutralized with NaOH. Glucuronic acid and phosphorus estimations were carried out on unhydrolyzed suspensions of bacterial cell walls (10 mg/ml, in distilled water).
[c]Value in parenthesis was determined with anthrone on an unhydrolyzed suspension of cell walls.

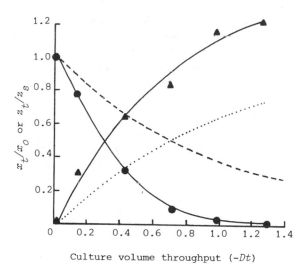

Culture volume throughput $(-Dt)$

Figure 5. Changes in the teichoic acid (●) and teichuronic acid (▲) contents of *Bacillus subtilis* var. *niger* cell walls after change-over from conditions of Mg^{2+}-limitation to those of phosphate-limitation in a chemostat culture. The broken line respresents the theoretical "washout" rate ($x_t/x_o = e^{-Dt}$), assuming that teichoic acid synthesis ceased immediately after the environment became phosphate-limited. The dotted line represents the theoretical rate of increase in teichuronic acid ($z_t/z_s = 1 - e^{-Dt}$), assuming that its synthesis started immediately after the environment became phosphate-limited and continued at a constant rate. Data from Ellwood and Tempest (1969).

serve are, as yet, uncertain; it remains a possibility, therefore, that besides substituting functionally for teichoic acid, teichuronic acid might act in some other way to facilitate the uptake of residual phosphate from the environment. Of relevance to this question is the finding that phosphate-limited *Bacillus cereus* T and *Pseudomonas diminuta* cells possess envelope structures in which certain of the phospholipids that usually are present have been replaced by glycolipids (Minnikin *et al.*, 1971, 1974); and these conditions were found (with some organisms at least) to also promote the secretion of alkaline phosphatase and the expression of a high-affinity phosphate-uptake system (Table I). But whether these particular structural and functional changes are directly interrelated to each other remains a matter of conjecture.

As with the above-mentioned changes in wall phosphorus content, so it was found that the sulfur content of the wall and soluble proteins of *K. aerogenes* decreased significantly when cultures were changed from being glucose-limited to sulfate-limited (Table VI). This was accompanied by a pronounced change in the spectrum of proteins present in the walls of these differently grown organisms (Fig. 6). A similar "sparing" of the cells' sulfur requirement also has been reported ot occur both with sulfate-limited cultures of the yeast *Candida utilis* (Light, 1972) and with *E. coli* K12 (Poole and Haddock, 1975). In these studies it was found that some of the iron-sulfur centers associated with site 1 energy conservation had been deleted and the loss of energetic efficiency compensated for by an increased respiration rate. In the latter case, however, it was necessary to omit manganese ions from the culture since these interfere with the

Table VI. Influence of Growth Rate and Growth Limitation on the Sulfur Content of Envelope, Ribosomal, and Soluble Proteins of *Klebsiella aerogenes*[a,b]

Growth limitation	Dilution rate (hr^{-1})	^{35}S content of total protein	^{35}S content (g/100 g protein)		
			In envelopes	In ribosomes	In soluble fraction
Glucose	0.2	0.74	0.79	0.68	0.74
Sulfate	0.2	0.61	0.64	0.66	0.60
	0.4	0.60	0.67	0.66	0.58
	0.8	0.60	0.62	0.60	0.58

[a]Organisms were grown in media containing ^{35}SO$_4^{2-}$ and disrupted in a Hughes press at $-20°C$. The disintegrated organisms were dispersed in 10 mM tris-HCl buffer (pH 7.6) containing 1 mM MgCl$_2$ and centrifuged at 20,000g for 1 hr to remove the envelope material. The supernatant fraction was further centrifuged (100,000g, 3 hr) to separate the ribosomal and soluble components. Each component was dispersed in tris-HCl buffer and analyzed for protein and ^{35}S contents.
[b]Data from Robinson and Tempest (1973).

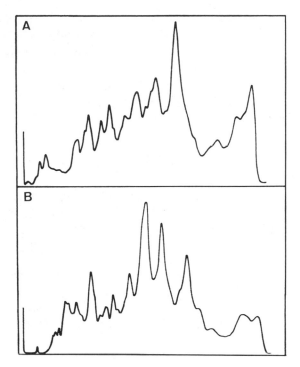

Figure 6. Scans of SDS-polyacrylamide gels of *Klebsiella aerogenes* envelope proteins: (A) sulfate-limited, $D = 0.2$/hr; (B) glucose-limited, $D = 0.2$/hr. Data from Robinson and Tempest (1973).

electron paramagnetic resonance (e.p.r.) signal of iron, and therefore interpretation of these data is not so straightforward.

But perhaps the most dramatic change in an organism's minimum requirement for some essential nutrient was observed with *C. utilis* cultures that were, respectively, glucose-limited (potassium-sufficient) and potassium-limited. In the former case, the cells were found to contain about 2% of their dry weight as potassium, irrespective of the growth rate, whereas in the latter cultures the potassium content varied markedly with growth rate from 0.15% (at $D = 0.1$/hr) to 1.3% (at $D = 0.5$/hr) (Aiking and Tempest, 1976). The precise nature of the metabolic compromises that allowed the minimum potassium requirement to be so greatly decreased were not elucidated, but again these seemed to involve substantial changes in the cells' energetic efficiency (Aiking *et al.*, 1977).

Clearly, then, microorganisms possess the ability to modulate extensively their structure and functioning so as to optimize and/or maximize their utilization of the growth-limiting nutrient. And it is equally clear that this optimization process may require compromises in the efficiency with which nonlimiting nutrients (particularly the carbon and energy source) are utilized. But irrespective of the extent to which metabolic compromises are invoked, the end product, it is reasonable to assume, must still be a viable functional cell. It is appropriate, therefore, to consider next those mechanisms by which organisms are able to coordinately regulate the rates of synthesis of all their macromolecular components so as to allow growth to proceed at a grossly submaximal rate.

2.4. Coordinated Regulation of Cell Synthesis

It follows from the fact that microbial populations can be made to grow at a variety of different rates in chemostat culture simply by varying the dilution rate that these creatures must possess effective mechanisms for coordinately modulating the rates of synthesis of all their macromolecular components so as to permit "balanced" growth to proceed at the imposed rate. With such cultures it has been found generally that changes in the growth rate are not accompanied by strictly proportional changes in the rates of synthesis of each of the cells' macromolecular components; thus, their proportions to each other may vary considerably (Fig. 7). This is particularly the case with respect to the cells' RNA content, which decreases sharply as the growth rate is progressively lowered, a phenomenon which relates to changes in the cellular content of ribosomes and to their quantitative involvement in protein synthesis (Schaechter *et al.*, 1958; Tempest and Hunter, 1965). These changes in cellular RNA and ribosome contents are such as to suggest that ultimately they may impose a lower limit on the extent to which growth can be diminished. Consequently, in seeking to understand the behavior of microbes in natural ecosystems, it is meaningful to establish whether there is for each organism a precise minimum rate at which growth can proceed and, if so, how it is manifest.

Of the few studies that have been undertaken and are reported in the literature, those of Postgate and Hunter (1964) and of Tempest *et al.* (1967a) are perhaps the most detailed. These studies were carried out with chemostat cultures of *K. aerogenes* which were either glycerol-limited or ammonia-limited and in which the dilution rate was progressively diminished from a value of about 1.0/hr (doubling time of 43 min) to a value of about 0.004/hr (doubling time of 173 hr). The most obvious effect on the population caused by this vast prolongation of the doubling time was to invoke the synthesis of cells possessing an unusual morphology (Fig. 8). At the same time, there was a marked fall in viability of the culture (Fig. 9), but nevertheless steady-state conditions could be maintained for many weeks at the lowest growth rate tested without the proportion of dead

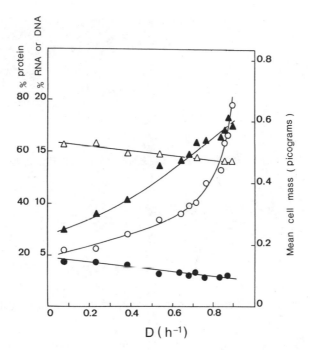

Figure 7. Growth of *Klebsiella aerogenes* in continuous culture; protein (△), DNA (●), and RNA (▲) contents and mean cell mass (○) as functions of the growth rate (μ = D). The organism was grown in a continuous culture apparatus of the chemostat type at a number of different flow rates and the cells analyzed after at least 2 days steady-state growth at each flow rate. Nucleic acid and protein contents expressed as percentage of cell dry weight; mean cell mass = dry weight/ml divided by total count/ml. Culture medium: glycerol-NH₃-salts with glycerol as limiting factor. Data from Herbert (1961).

cells in the population varying significantly. But since under these conditions the nonviable portion of the population could have been generated only from the viable portion, then clearly the viable cells in the population must have been growing at a rate that was substantially greater than the dilution rate. Plotting the actual rate of growth of the viable portion of the population at each growth rate (Fig. 9) revealed that this tended to a minimum value at about 0.01/hr (35°C, pH 6.8) and that this indeed correlated with the attainment of some minimum cellular RNA content. Moreover, this minimum growth rate value was not greatly different in ammonia-limited (glycerol-sufficient) cultures as compared with glycerol-limited cultures (Fig. 10), indicating that it was not related to the cells' maintenance energy requirements (see, by way of contrast, Schultze and Lipe, 1964). Finally it was found that the cellular RNA (and ribosome) content varied markedly with the growth temperature (Tem-

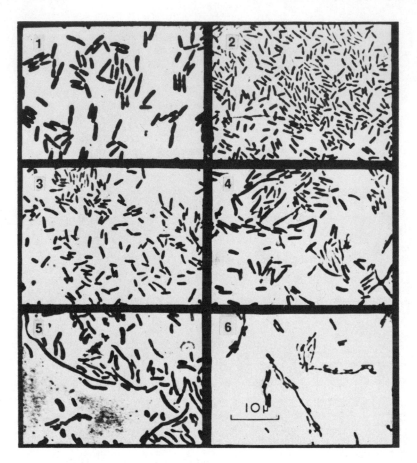

Figure 8. Photomicrographs of populations of glycerol-limited *Klebsiella aerogenes*, growing at "steady-state" dilution rates of: (1) 1.0/hr, (2) 0.24/hr, (3) 0.12/hr, (4) 0.015/hr, (5) 0.008/hr, and (6) 0.004/hr. Organisms were photographed using phase contrast optics. The magnification was uniform throughout. Data from Tempest *et al.* (1967a).

pest and Hunter, 1965) and that when the temperature of slowly growing glycerol-limited cultures of *K. aerogenes* was lowered from 35°C to 25°C, both the cellular RNA content and the viability increased markedly (Tempest *et al.*, 1967a).

Although slowly growing organisms were markedly different (both morphologically and physiologically) from organisms growing at higher rates (that is, at values in excess of $0.1\mu_{max}$), it would be false to conclude that they were in effect "abnormal," since this is precisely the condition in which they well might exist in natural ecosystems (Postgate, 1973). In this connection, a similar, though more dramatic, change in morphology has

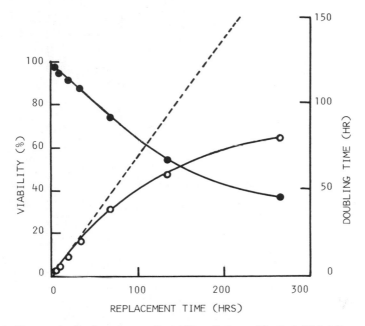

Figure 9. Changes in the "steady-state" viability of glycerol-limited *Klebsiella aerogenes* culture with growth rate. The viability (●) and doubling times of viable organisms (○) are plotted against the reciprocal of the dilution rate (termed the "replacement time"). The broken line represents the doubling time of the culture (($\ln 2$)/D). Data from Tempest *et al.* (1967a).

been observed with carbon-limited cultures of *Arthrobacter* species that were grown at different dilution rates (Luscombe and Gray, 1971). Here the morphological characteristics expressed at the low growth rates were closely similar to those observed in natural environments. Clearly much greater attention should be paid in the future (especially by microbial ecologists) to slowly growing chemostat cultures of microorganisms (see also Veldkamp, 1977).

3. Bioenergetic Considerations

Fundamental to any consideration of microbial behavior in ecologically relevant environments are the bioenergetic consequences of what Koch (1971) has called a "feast and famine existence." Thus, with chemoorganotrophic organisms, in which energy needed for biosynthesis is necessarily derived from the breakdown of some carbon substrate that simultaneously is being assimilated into cell substance, one might expect mechanisms to exist that allow the flow of intermediary metabolites to be

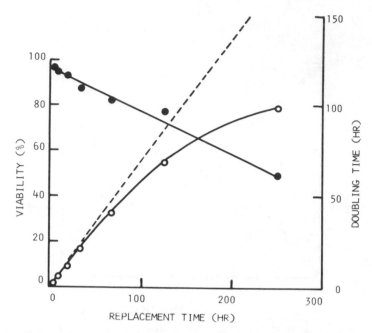

Figure 10. Changes in the "steady-state" viability of ammonia-limited *Klebsiella aerogenes* culture with growth rate. The viability (●) and doubling times of viable organisms (○) are plotted against the reciprocal of the dilution rate (the "replacement time"). The broken line indicates the culture doubling time [(ln 2)/D]. Data from Tempest *et al.* (1967a).

precisely partitioned between catabolic (energy-generating) reactions and the anabolic (energy-consuming) reactions of biosynthesis. This optimal partitioning would, of course, be of particular importance to organisms exposed to environments in which the carbonaceous substrate was present only in low, growth-rate-limiting concentration, that is, near-famine conditions. Thus, one could envisage control systems to exist within the cell that would act at specific branch points between catabolic and anabolic pathways of metabolism and which, further, would be "tuned" to the overall energy status of the cell, that is, to the "energy charge" (Atkinson, 1968, 1969). That such control systems do indeed exist within the microbial cell is abundantly obvious from the extensively reported involvement of adenine nucleotides as control elements in intermediary metabolism (for a recent review of this subject, see Chapman and Atkinson, 1977). The mode of action of these regulatory processes (i.e., allosteric effectors) leads not unreasonably to the concept of there being a "coupling" between ATP synthesis and growth, as

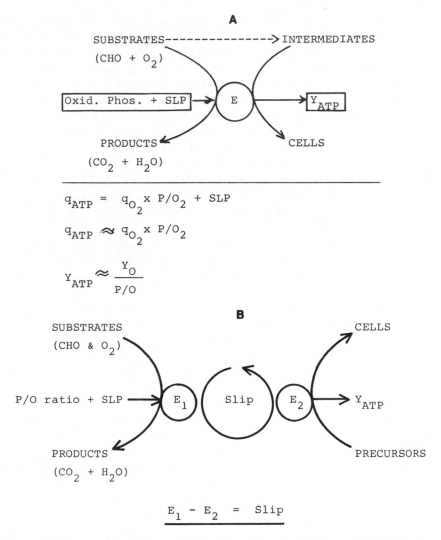

A

SUBSTRATES-------------->INTERMEDIATES
$(CHO + O_2)$

$\boxed{Oxid. \; Phos. \; + \; SLP} \rightarrow$ E $\rightarrow \boxed{Y_{ATP}}$

PRODUCTS   CELLS
$(CO_2 + H_2O)$

$q_{ATP} = q_{O_2} \times P/O_2 + SLP$

$q_{ATP} \approx q_{O_2} \times P/O_2$

$Y_{ATP} \approx \dfrac{Y_O}{P/O}$

B

SUBSTRATES   CELLS
$(CHO \; \& \; O_2)$

P/O ratio + SLP $\longrightarrow$ E_1 (Slip) E_2 $\longrightarrow Y_{ATP}$

PRODUCTS   PRECURSORS
$(CO_2 + H_2O)$

$E_1 - E_2 = Slip$

Figure 11. Two models of the energetics of aerobic microbial growth. (A) Obligatory coupling between energy-generating reactions and growth. (B) No direct coupling between energy-generating reactions and growth.

depicted schematically in Fig. 11A, an interrelationship that manifests itself as a Y_{ATP} value (i.e., g cells synthesized/mol ATP generated by catabolism) (Bauchop and Elsden, 1960; Senez, 1962; Belaich *et al.*, 1972; Stouthamer and Bettenhaussen, 1973). However, it is equally clear from data considered previously (Tables III and IV) that, when the growth of microorganisms is limited by the supply of some essential nutrient other than the carbon and energy source, reducing equivalents and ATP are generated at rates that are far in excess of those needed to meet the cells' biosynthetic demands. Clearly, at least under these conditions, ancillary growth-unassociated processes must operate to consume the excess of ATP and reducing equivalents that are generated and thereby to maintain the intracellular energy charge and redox state within their effective functional ranges. But if, as seems likely, such energy-spilling reactions can be readily invoked under these carbon-sufficient (feast) conditions, then it is reasonable to ask: How are they suppressed when the culture becomes carbon substrate-limited? Indeed, can they be totally suppressed, and, if not, what effect will they exert on biological productivity in severely carbon-limited environments? These are the principal questions considered below.

That the energy-consuming reactions of biosynthesis do not necessarily regulate closely the energy-generating reactions occurring within the microbial cell is evident not only from their behavior in carbon-sufficient chemostat environments but also from the fact mentioned previously that washed suspensions of organisms often will rapidly oxidize carbon substrates under conditions in which they simply cannot grow. Thus, it is clear that organisms possess a potential to oxidize substrates which they can largely, if not completely, express independent of the rate at which growth is able to proceed (see Belaich et al., 1972). The importance of this metabolic (or respiratory) potential to an assessment of the energetics of microbial growth in carbon-sufficient and carbon-limited chemostat culture is evident from the data contained in Fig. 12. This figure shows the relationship between the specific growth rate (μ = the dilution rate D, at steady state) and the specific rate of O_2 consumption (q_{O_2}) found with cultures of *Klebsiella aerogenes* that were carbon (glucose)-limited, ammonia-limited, and phosphate-limited. Clearly, at all comparable growth rates, the carbon-limited culture consumed O_2 at a lower rate than did either of the other two (carbon-sufficient) cultures. Since the specific rate of O_2 consumption is approximately proportional to the specific rate of ATP synthesis (that is, $q_{ATP} \simeq 2 \times q_{O_2} \times (P/O)$, where P/O is the ratio of phosphate assimilated into ATP to the O_2 concomitantly reduced by electron-transport-chain activity), then these data show that organisms grew with a maximum of energetic efficiency when they were carbon substrate growth-limited. This differential growth efficiency between carbon-sufficient and carbon-limited cultures was not so marked at high growth rates, but increased substantially as the growth rate was diminished toward zero, indicating, according to the Pirt (1965) hypothesis, that these three cultures had markedly different "maintenance energy" requirements. However, the essential

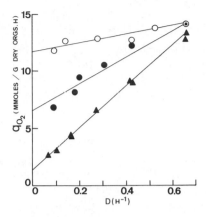

Figure 12. Relationship between the specific growth rate and the specific rate of O_2 consumption in variously limited chemostat cultures of *Klebsiella aerogenes* growing in a glucose-containing medium. Cultures were carbon-limited (▲), ammonia-limited (●), and phosphate-limited (○) (35°C; pH 6.8). Data from Neijssel and Tempest (1976a).

difference between these three cultures resided in the fact that only with the carbon-limited culture was the specific rate of carbon substrate, and O_2, consumption constrained by the limited availability of that substrate, the extracellular glucose concentration being vanishingly small at all growth rates below about 0.9 μ_{max}. Thus, particularly at low growth rates, the rate at which this substrate was being oxidized was far below that which the organisms were capable of achieving, had it been present in excess (as could be readily demonstrated by pulsing a cell-saturating concentration of glucose into the culture; Fig. 13). So what then constrains the rate at which carbon substrate is taken up and oxidized by carbon-sufficient cultures? In the case of phosphate-limited cells growing

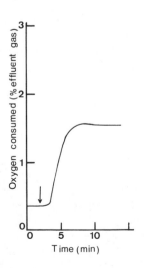

Figure 13. Comparison of the "potential" and "*in situ*" rates of O_2 consumption found with cultures of *Klebsiella aerogenes* growing at 35°C and pH 6.8. Recorder tracing of the change in O_2 uptake rate following the addition of glucose (final concentration 5 mg/ml) to a glucose-limited culture. The arrow marks the time of addition. Data from Neijssel and Tempest (1976b).

at a low rate (0.1/hr), the respiratory potential of the cells seemingly was almost fully expressed since, when the supply of fresh medium (and hence phosphate) was suspended for 20 min, the rate of glycerol oxidation was hardly affected (Fig. 14). Pulsing additional phosphate to the culture did, however, cause a transient rise in the specific rate of O_2 consumption, indicating that the transfer of electrons to O_2 was being impeded by the availability of inorganic phosphate (required for the phosphorylation of ADP). In other words, there did appear to be some measure of "respiratory control" extant in these glycerol-sufficient cells. Nevertheless, what is unequivocally shown by these data is (1) that the specific rate of carbon-substrate oxidation by carbon-sufficient cells is largely dependent on the metabolic potential of the cell extant at any particular growth rate and (2) that carbon-limited cells possess the potential to oxidize substrate at a vastly higher rate than they can achieve at most growth rates—a potential that can be fully and immediately expressed when the growth limitation is relieved (Fig. 13). It follows, therefore, that carbon-limited cells also must possess the capacity to spill excess energy, since the specific rate of ATP synthesis immediately following addition of a cell-saturating amount of glucose to a glucose-limited culture must be again far in excess of that which could be immediately consumed in biosynthesis.

One might go further and suggest that energy-spilling reactions are actually invoked by carbon-limited cultures, even when growing under steady-state conditions. In this connection, it should be emphasized that a

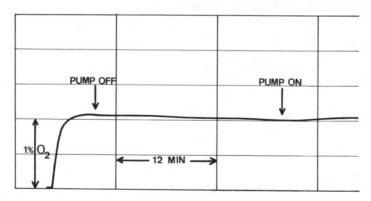

Figure 14. Effect of a sudden phosphate starvation on the O_2 consumption of *Klebsiella aerogenes* organisms growing in a phosphate-limited chemostat culture (carbon source: glycerol, $D = 0.11$/hr, 35°C, pH 6.8). As indicated by the arrows, the nutrient supply was interrupted for a period of ± 20 min. During that period, excess glycerol remained present in the culture extracellular fluids.

carbon-substrate limitation represents a complex situation in that the substrate provides not only a source of those intermediary metabolites required for polymer synthesis but also the necessary reducing equivalents (like NADH) and energy (as ATP) required for this process to proceed. Hence, depending on the nature of the carbon source and the redox state of the other essential nutrients, the culture may be either (1) carbon-limited (that is, the nutrient carbon supply is limiting growth and concomitant energy generation occurs in excess of the biosynthetic demand) or (2) energy-limited (including a limitation imposed by the organisms' requirement for reducing equivalents). A carbon *plus* energy dual limitation also could occur but, one might assume, would be rare. From the data obtained with one specific organism growing on one specific carbon-containing substrate, it would not be possible to decide whether the culture was carbon- or energy-limited, or both. But a comparison of data obtained from carbon-substrate-limited cultures growing on a range of closely similar compounds allows some conclusions to be drawn. In general, one might assume that when a culture was truly carbon-limited, the respiratory quotient would be relatively low since the q_{CO_2} value would be minimal and the excess reducing equivalents would be oxidized by respiratory processes; the concomitant spillage of excess energy would cause the Y_O value (g cells synthesized/g atom O consumed) to be relatively low, but the Y_s value (g cells synthesized/mol carbon-substrate assimilated) would be maximal. In contrast, with an energy limitation a greater proportion of the carbon substrate would be catabolized completely to CO_2; thus, the q_{CO2} value would be relatively high and the Y_s value relatively low. On the other hand, the Y_O value would be maximal since energy spillage would be minimal. A carbon *plus* energy dual limitation would represent the ideal situation for obtaining an optimal overall growth efficiency since, in this case, anabolism and catabolism would be perfectly matched.

From studies made of carbon-substrate-limited cultures of *K. aerogenes* growing on mannitol, glucose, and gluconic acid (Fig. 15), it is clear that neither mannitol limitation nor glucose limitation represents a true energy limitation since, at each growth rate value, the organisms consumed more O_2 (and therefore expressed a lower Y_O value) than they did when growing at corresponding rates in the gluconic acid-limited culture. But whether the latter represented a true energy limitation cannot be decided from the available evidence.

An inevitable consequence of all those findings contained in Figs. 12–15 and of the fact that washed suspensions of microorganisms can oxidize substrates at high rates under conditions in which they cannot grow is that the conceptual bioenergetic model depicted in Fig. 11A is

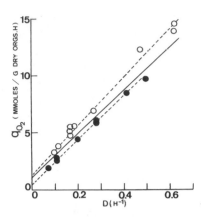

Figure 15. Relationship between the specific rate of O_2 consumption and dilution rate found with cultures of *Klebsiella aerogenes* that were mannitol-limited (○) and gluconate-limited (●). The center line represents the relationship routinely found with glucose-limited cultures (see Fig. 12). Data from Neijssel and Tempest (1976b).

totally inadequate. Clearly there is no *obligatory* coupling between the energy-generating reactions of the cell and biosynthesis; thus, it is necessary to interpose between these two processes some mechanism that allows for the fact that the organisms' metabolic machinery can, under appropriate conditions, deliver up energy at a high rate and maintain this rate independent of the energy demands of biosynthesis. Such a concept is embodied in Fig. 11B, in which is postulated the presence of so-called slip reactions that serve to turn over the ATP pool at a high rate whenever circumstances require, or lead to the generation of energy overplus (see also Dawes and Senior, 1973).

This lack of strict coupling (even under carbon-substrate-limiting conditions) between the specific rate of ATP synthesis and the specific rate of microbial cell synthesis makes a precise evaluation of the energetics of microbial growth difficult, if not impossible. But it leads automatically to a consideration of the more pragmatic question of why growth is not strictly coupled to energy generation or, to put it another way, to a consideration of those advantages that might accrue from growth not being so coupled. In order to appreciate fully the ecological implications inherent in this question, it is necessary to bear in mind the fact that, though chemostat environments have aspects in common with natural environments, they often differ in important respects. In particular, the prolonged steady-state growth conditions routinely imposed on organisms cultured in a chemostat may not be so readily achieved in natural ecosystems. Hence, before proceeding to a final detailed consideration of the eco-physiological significance of energy-spilling "slip reactions," it is appropirate to consider the ways in which microbes respond to transient changes in their growth environment.

4. Transient-State Phenomena: Microbial Reactivity

It was mentioned previously (Section 2.1) that one of the mechanisms by which organisms are able to accommodate effectively to nutrient-limited environments is by synthesizing in excessive amounts those cellular components that constitute the uptake system for the depleted nutrient. In this way the maximum rate at which the substrate can be taken into the cell (that is, its uptake potential) is maximized. The large differential thus established between the potential uptake rate (V_{max}) and the actual uptake rate that is expressed (v) allows the organisms to scavenge effectively those traces of growth-limiting nutrient that will be present in the environment. That such a large differential between the expressed and the potential uptake rates actually exists with nutrient-limited cultures can be concluded from the data contained in Fig. 16, which shows the respiration rate extant in a glucose-limited chemostat culture just prior to and immediately following the addition of a cell-saturating pulse of glucose to organisms that were growing at several different dilution rates. Clearly, although the actual uptake rate for glucose varied linearly with the dilution rate, the potential rate varied only slightly such that the differential was small at high growth rates and very large at low (severely nutrient-depleted) growth rates.

The establishment of this large differential between the actual and potential uptake rates, particularly at low dilution rates, confers upon the

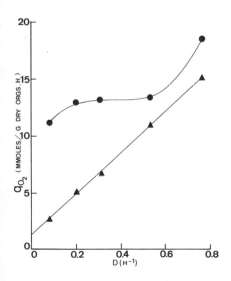

Figure 16. Comparison of the "potential" and "*in situ*" rates of O_2 consumption found with cultures of *Klebsiella aerogenes* growing at 35°C and pH 6.8. Plot of the changes in the "potential" rates of O_2 consumption (●) and "*in situ*" rates of O_2 consumption (▲) with changes in the dilution rate of a glucose-limited culture.

organisms additional benefits in that they can respond rapidly to environmental change, ultimately expressing this response as an increase in growth rate. The organisms may be said to be in a highly "reactive" or "responsive" state; and although this reactivity can, under some conditions, be detrimental, in that it can lead to the uncontrolled accumulation of traumatic metabolites within the cell (Section 2.2.), nevertheless, the competitive advantage that it otherwise confers upon the organisms must be considerable.

Of relevance to the question of "reactivity" is the extent to which organisms can coordinately step up or step down the activities of all those cellular processes involved in growth when the availability of the growth-limiting nutrient suddenly changes. That organisms cannot instantaneously express their maximum growth rate potential, when transferred from a severely nutrient-depleted environment to one that is nutrient-sufficient, has been clearly demonstrated by Koch (1971) and follows from the fact that the macromolecular composition and functional ability of fast-growing organisms is markedly different from that of organisms growing at slower rates. In particular, the cellular ribosome content is greatly decreased when the growth rate of organisms is nutritionally restricted (Maale and Kjeldgaard, 1966), thereby decreasing (though not proportionately) the maximum rate at which the cells can synthesize new protein. But clearly (Fig. 16) the capacity to oxidize substrate and to deliver up energy for biosynthetic processes is not extensively diminished at low growth rates so that when a carbon-substrate limitation is suddenly relieved, an energy overplus will be invoked, and the excess energy must be dissipated wastefully.

This form of behavior can be best rationalized in terms of the strategy and tactics needed to cope with the fiercely competitive situation posed by a nutrient-limited environment. Thus, as pointed out by Koch (1971), the advantage to some cell of being able to rapidly accelerate its rate of protein synthesis from a low value to one close to its potential maximum must be weighed against the cost of maintaining within the cell a ribosomal population far in excess of that required to effect protein synthesis at the restricted, submaximal rate. Of particular relevance to this question, of course, is the fact that ribosome synthesis is, energetically, exceedingly expensive. So clearly a marked advantage can be gained (in terms of the energetics of cell synthesis) from decreasing the cellular ribosome content, providing, that is, that the cell retains its capacity to rapidly synthesize extra ribosomes when required to do so. In this connection, the maintenance of a high respiratory potential (and hence the capacity to deliver up energy and intermediary metabolites at a high rate) is of fundamental importance. Indeed, the serious difficulties that would accrue from being energy-limited are such as to suggest that organisms would strive, wherever possible, to avoid this situation. It may well be primarily for this reason (i.e., to avoid an energy limitation) that adenylic acid and oxidized pyridine nucleotides are powerful promoters of catabolism. It is not unreasonable to suppose that this also may be the underlying consideration with respect to the "endogenous metabolism" expressed by organisms when they are totally starved of carbon substrate (see Dawes, 1976).

The tactical advantages gained from maintaining a high respiratory potential while modulating extensively the cells' biosynthetic potential raises the question of how organisms cope with continuously oscillating environmental conditions. In striving to cope with conditions of, alternately, feast and famine, do they in fact adopt some behavioral pattern relevant to a condition intermediate between the two extremes? The results of one such experiment that sought to investigate this situation is shown in Fig. 17. In this experiment the rate of growth of *K. aerogenes* was varied by altering the rate of supply of basal medium to a glucose-limited chemostat culture. The carbon substrate (glucose) was added discontinuously at approximately one drop of a 50% (w/v) aqueous glucose solution per 2 min, the rate being totally independent of the

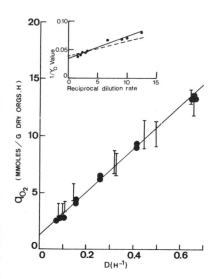

Figure 17. The relationships found between the specific rates of O_2 consumption and dilution rate with a glucose-limited chemostat culture of *Klebsiella aerogenes* in which the glucose was added either along with the bulk medium (●) or separately in regular small increments (1 drop every 2 min) as a concentrated solution (± 50%, w/v) (I). The upper and lower limits of the vertical bars show the maximum and minimum oxygen consumption rates, which oscillated with a periodicity of about 2 min. The insert shows the relationship between the reciprocal Y_0 value ($1/Y_0$) and reciprocal dilution rate ($1/D$) obtained with the pulsed culture (●) as compared with the nonpulsed culture (dashed line; symbols omitted for clarity). In this graph, the Y_0 values are the mean values derived from the mean of the O_2 uptake rate at each dilution rate. These data indicate that the pulsed culture not only expressed a higher maintenance rate (steeper slope) but also had an increased Y_0^{max} value (changed intercept on the ordinate). Data from Neijssel and Tempest (1976b).

overall dilution rate. The rate of glucose addition and of basal medium flow varied disproportionately and so too, therefore, did the population density in the culture. Nevertheless, steady transient-state conditions could be readily established and meaningfully analyzed. Although the O_2 consumption rate (and the rate of carbon dioxide production) oscillated with a regular 2-min periodicity, the maximum and minimum rates of O_2 consumption could be calculated and the average rate obtained. This could then be compared with the O_2 consumption rate extant in a minimally pulsed (ordinary) chemostat culture. As can be seen (Fig. 17), the results obtained showed clearly that when glucose was pulse-fed to a culture growing at a low dilution rate (that is, where the differential between the actual and potential respiration rates was large), the rate of O_2 consumption was clearly increased above the average value ordinarily found, showing that the growth efficiency was diminished. Surprisingly, at high growth rates the pulsed addition of glucose seemingly improved the efficiency with which growth occurred. However, overall, the net effect of a pulsed addition of glucose was such as to suggest that the maintenance energy requirement (expressed at low growth rates) was substantially increased. Hence, it remains a possibility that one component of that maintenance energy rate found generally with chemostat cultures (and perhaps a major one) is the cells' reaction to a discontinuous addition of substrate, which frequently, if not invariably, occurs at low dilution rates—particularly when using chemostats with a small culture volume (since then the flow rate, at low dilution rates, must of necessity be exceedingly slow).

Harrison and Loveless (1971a,b) studied the relationship between O_2 tension and respiratory activity of carbon-limited cultures of *K. aerogenes* and *E. coli*. The respiratory capacity of these organisms was always greater than the O_2 consumption rate expressed in the culture, irrespective of whether it was glucose-limited or glucose- and O_2-limited. When a transition of *K. aerogenes* organisms from anaerobic to aerobic growth conditions was followed closely, three transitional phases were observed: (1) a short period of increased acetate and pyruvate production and low carbon dioxide production immediately following reaeration; (2) a subsequent period of high respiration rate and low yield coefficients for glucose and O_2 utilization, and finally (3) the attainment of an aerobic steady state after about 8 hr. One can conclude from these yield data that, during the second phase of transition, a considerable amount of energy was spilled, and this could well play an essential role in the process of adaptation to the new environment. Thus, it should not be dismissed as being simply a weakness in the control system of the cells, as was argued by Harrison and Loveless (1971b).

In concluding this section it is necessary to point out that the vast majority of studies on the energetics of microbial growth have been performed with just a few bacterial species—particularly *E. coli* and *K. aerogenes*—and one must therefore avoid generalizing from the particular. Thus, whereas it has been found that the pulsed addition of methanol to a methanol-limited culture of a *Pseudomonas* species led to a marked decrease in the growth efficiency (Brooks and Meers, 1973) similar to that found with glucose-limited cultures of *K. aerogenes* (Neijssel and Tempest, 1976b), similar experiments performed with carbon-limited cultures of *Arthrobacter globiformis* failed to elicit any change in the growth yield (Luscombe, 1974). Whether this different behavior correlates with the different ecological niches generally occupied by these different species can be decided only after considerably more extensive investigations have been carried out.

5. Conclusions

Those persons who have read all of this chapter will realize that it is by no means a comprehensive and balanced literature survey, but rather a somewhat one-sided and chauvinistic assemblage of diverse observations that have been made over the past 20 years, largely by a few dedicated continuous culturists! However, though each of these different observations on the behavior of microorganisms in chemostat culture may have served as an end in itself (i.e., a publication), it is nevertheless reasonable to conclude that the properties so revealed do have broader relevance to the existence of these creatures in natural environments. One may safely assume, for instance, that the phenotypic changes elicited in chemostat culture do not represent an "abnormal" or "disordered" mode of behavior resulting from the exposure to some aberrant and artificial growth condition. Indeed, quite the opposite, for as argued in the Introduction there is every reason to suppose that in many natural ecosystems nutrient depletion is the rule rather than the exception. Hence, the ability of organisms to adapt to chemostat environments may reflect no more than their possession of a range of properties that routinely allow them to cope with the vicissitudes of life outside the laboratory. One might go even further and suggest that conditions of gross nutrient sufficiency, such as are more commonly provided by batch culture environments, represent the real "extreme" situations and that the properties which organisms express in these environments are similarly "extreme" (though still "normal"). But be that as it may, what seems abundantly obvious is that microorganisms do possess the ability to change *themselves*, structurally and functionally, to a remarkable degree, and that this physiological flexibility (requiring, as it does, a corresponding genetic complexity) must have some rational basis. This basis we have tried to distill from our own observations (and from those of many others) on the behavior of microbes (largely bacteria) when growing in chemostat culture.

The fact that, when growing in such cultures, organisms often express properties that are widely different from those expressed in batch culture has far-reaching consequences. This is particularly evident when one considers that much of the textbook information pertaining to the biochemistry and regulation of microbial growth is based largely, if not entirely, on batch culture studies. Thus, it has been concluded (Kornberg, 1973) that "coarse" and "fine" controls mediate the uptake of glucose by *E. coli* in such a way as to cause the rate of penetration of glucose into the cell to match precisely the rate at which intermediates of its metabolism can be assimilated into cell substance. Whereas such control systems unquestionably operate in organisms growing in nutrient-sufficient (batch) environments, they clearly do not operate in many nutrient-limited (chemostat) environments (Table III).

Again, it has been concluded (Mandelstam and McQuillen, 1968) that regulatory processes operate within the growing microbial cell in such a way as to "ensure that the cell is geared to get the maximum yield of protoplasm from its environment, and to do so in the minimum time." But the bioenergetic considerations underlying the growth of microorganisms in steady-state and transient-state chemostat culture, discussed in previous sections, suggest that other factors are of equal, if not greater, importance. For example, the capacity of organisms to respond rapidly to environmental change demands that the energy-generating reactions of respiration are not tightly coupled to, and regulated by, the energy-consuming reactions of biosynthesis. Consequently, much energy can be "spilled" in maintaining and expressing this reactivity—even under carbon-substrate-limiting growth conditions—and this inevitably results in a decrease in yield. Life, it would seem, is a compromise, even for the most humble of God's creatures!

Finally one must mention the intensive efforts that have been made by geneticists and molecular biologists to determine the mechanism by which synthesis of the bacterial genome is regulated and integrated with the cell division cycle (Cooper and Helmstetter, 1968; Donachie *et al.*, 1973; Matsushita and Kubitschek, 1975). Here it has been concluded that chromosome replication proceeds at a constant rate, irrespective of the growth rate, and that cell division follows termination of replication, again after a growth-rate-independent period of time. However, almost all this work has been done with organisms growing in batch culture with generation times from about 20 to 60 min. And since, in the human gut, *E. coli* may well not achieve a generation time less than 12 hr (Koch, 1971), it is pertinent to inquire whether the Cooper and Helmstetter (1968) model of chromosomal replication would apply under these more usual circumstances!

In this preliminary attempt to relate those properties of organisms expressed in chemostat environments to their physiological requirements when growing in natural ecosystems, a serious weakness can be discerned in that relatively few microbial species have been grown continuously. In fact, the vast majority of studies have been made with gram-negative species—principally on members of the Enterobacteriaceae. Consequently, as already mentioned, one must be cautious not to draw too broad a conclusion from so narrow a spectrum of studies. In this respect, we have been certainly less circumspect than befits a good reviewer, but our aim throughout has been to stress those ideas, generalizations, and principles which we believe may have real validity in an eco-physiological context. However, only time and a great deal more work will tell how near we have come to a proper understanding of the ecological significance of that behavior expressed by organisms in aerobic, nutrient-limited (chemostat) environments.

References

Aiking, H., and Tempest, D. W., 1976, Growth and physiology of *Candida utilis* NCYC 321 in potassium-limited chemostat culture, *Arch. Microbiol.* **108**:117.

Aiking, H., Sterkenburg, A., and Tempest, D. W., 1977, Influence of specific growth limitation and dilution rate on the phosphorylation efficiency and cytochrome content of mitochondria of *Candida utilis* NCYC 321, *Arch. Microbiol.* **113**:65.

Alim, S., and Ring, K., 1976, Regulation of amino acid transport in growing cells of *Streptomyces hydrogenans*. II. Correlation between transport capacity and growth rate in chemostat cultures, *Arch. Microbiol.* **111**:105.

Atkinson, D. E., 1968, The energy charge of the adenylate pool as a regulatory parameter. Interaction with feedback modifiers, *Biochemistry* **7**:4030.

Atkinson, D. E., 1969, Regulation of enzyme function, *Annu. Rev. Microbiol.* **23**:47.

Baddiley, J., 1964, Teichoic acids and the bacterial cell wall, *Endeavour* **23**:33

Baskett, R. C., and Hentges, D. J., 1973, *Shigella flexneri* inhibition by acetic acid, *Infec. Immun.* **8**:91.

Bauchop, T., and Elsden, S. R., 1960, The growth of microorganisms in relation to their energy supply, *J. Gen. Microbiol.* **23**:457.

Belaich, J. P., Belaich, A., and Simonpietri, P., 1972, Uncoupling in bacterial growth: effect of pantothenate starvation on growth of *Zymomonas mobilis*, *J. Gen. Microbiol.* **70**:179.

Bisschop, A., Doddema, H., and Konings, W. N., 1975, Dicarboxylic acid transport in membrane vesicles from *Bacillus subtilis*, *J. Bacteriol.* **124**:613.

Boos, W., 1976, Cell envelope proteins involved in the transport of maltose and *sn*-glycerol-3-phosphate in *Escherichia coli*, *J. Cell. Physiol.* **89**:529.

Borichewski, R. M., 1967, Ketoacids as growth limiting factors in autotrophic growth of *Thiobacillus thiooxidans*, *J. Bacteriol.* **93**:597.

Borichewski, R. M., and Umbreit, W. W., 1966, Growth of *Thiobacillus thiooxidans* on glucose, *Arch. Biochem. Biophys.* **116**:97.

Brooks, J. D., and Meers, J. L., 1973, The effect of discontinuous methanol addition on the growth of a carbon-limited culture of *Pseudomonas, J. Gen. Microbiol.* **77**:513.

Brown, C. M., and Stanley, S. O., 1972, Environment-mediated changes of the "pool" constituents and their associated changes in cell physiology, *J. Appl. Chem. Biotechnol.* **22**:363.

Byers, B. R., Powell, M. V., and Lankford, C. E., 1967, Iron-chelating hydroxamic acid (schizokinen) active in initiation of cell division in *Bacillus megaterium, J. Bacteriol.* **93**:286.

Calcott, P. H., and Postgate, J. R., 1972, On substrate accelerated death of *Klebsiella aerogenes, J. Gen. Microbiol.* **70**:115.

Calcott, P. H., and Postgate, J. R., 1974, The effects of β-galactosidase activity and cyclic AMP on lactose accelerated death, *J. Gen. Microbiol.* **85**:85.

Castric, P. A., 1975, Hydrogen cyanide, a secondary metabolite of *Pseudomonas aeruginosa, Can. J. Microbiol.* **21**:613.

Castric, P. A., 1977, Glycine metabolism by *Pseudomonas aeruginosa*: hydrogen cyanide biosynthesis, *J. Bacteriol.* **130**:826.

Chapman, A. G., and Atkinson, D. E., 1977, Adenine nucleotide concentrations and turnover rates. Their correlation with biological activity in bacteria and yeast, *Adv. Microbial Physiol.* **15**:253.

Codd, G. A., and Smith, B. M., 1974, Glycollate formation and excretion by the purple photosynthetic bacterium *Rhodospirillum rubrum, FEBS Letters* **48**:105.

Codd, G. A., Bowien, B., and Schlegel, H. G., 1976, Glycollate production and excretion by *Alcaligenes eutrophus, Arch. Microbiol.* **110**:167.

Cooper, S., and Helmstetter, C. E., 1968, Chromosome replication and the division cycle of *Escherichia coli* B/r, *J. Molec. Biol.* **31**:519.

Cooper, T. G., and Sumrada, R., 1975, Urea transport in *Saccharomyces cerevisiae, J. Bacteriol.* **121**:571.

Dawes, E. A., 1976, Endogenous metabolism and the survival of starved prokaryotes, in: *The Survival of Vegetative Microbes, 26th Symposium of the Society for General Microbiology* (T. R. G. Gray and J. R. Postgate, eds.), pp. 19–53, Cambridge University Press, Cambridge.

Dawes, E. A., and Senior, P. J., 1973, The role and regulation of energy reserve polymers in micro-organisms, *Adv. Microbial Physiol.* **10**:135.

Donachie, W. D., Jones, N. C., and Teather, R., 1973, The bacterial cell cycle, in: *Microbial Differentiation, 23rd Symposium of the Society for General Microbiology* (J. M. Ashworth and J. L. Smith, eds.), pp. 9–44, Cambridge University Press, Cambridge.

Ellwood, D. C., and Tempest, D. W., 1969, Control of teichoic acid and teichuronic acid biosynthesis in chemostat cultures of *Bacillus subtillis* var. *niger, Biochem. J.* **111**:1.

Erlagaeva, R. S., Bolshakova, T. N., Shulgina, M. V., Bourd, G. I., and Gershanovitch, V. N., 1977, Glucose effect in *tgl* mutant of *Escherichia coli* K12 defective in methyl-α-D-glucoside transport, *Eur. J. Biochem.* **72**:127.

Faik, P., and Kornberg, H. L., 1973, Isolation and properties of *E. coli* mutants affected in gluconate uptake, *FEBS Lett.* **32**:260.

Fixter, L. M., and Fewson, C. A., 1974, The accumulation of waxes by *Acinetobacter calcoaceticus* NCIB 8250, *Biochem. Soc. Trans.* **2**:944.

Fogg, G. E., 1975, *Algal Cultures and Phytoplankton Ecology*, 2nd ed., p. 87, The University of Wisconsin Press, Madison, Wisc.

Freedberg, W. B., Kistler, W. S., and Lin, E. C. C., 1971, Lethal synthesis of methylglyoxal by *Escherichia coli* during unregulated glycerol metabolism, *J. Bacteriol.* **108**:137.

Frost, G. E., and Rosenberg, H., 1973, The inducible citrate-dependent iron transport system in *Escherichia coli* K12, *Biochim. Biophys. Acta* **330**:90.

Golub, E. E., and Bronner, F., 1974, Bacterial calcium transport: energy dependent calcium uptake by membrane vesicles from *Bacillus megaterium, J. Bacteriol.* **119**:840.

Guffanti, A. A., and Corpe, W. A., 1976, Transport of maltose by *Pseudomonas fluorescens* W, *Arch. Microbiol.* **108**:75.

Haavik, H. I., 1976, On the role of bacitracin peptides in trace metal transport by *Bacillus licheniformis, J. Gen. Microbiol.* **96**:393.

Harder, W., and Veldkamp, H., 1971, Competition of marine psychrophilic bacteria at low temperatures, *Antonie van Leeuwenhoek J. Microbiol. Serol.* **37**:51.

Harrison, D. E. F., and Loveless, J. E., 1971a, The effect of growth conditions on respiratory activity and growth efficiency in facultative anaerobes grown in chemostat culture, *J. Gen. Microbiol.* **68**:35.

Harrison, D. E. F., and Loveless, J. E., 1971b, Transient responses of facultatively anaerobic bacteria growing in chemostat culture to a change from anaerobic to aerobic conditions, *J. Gen. Microbiol.* **68**:45.

Harrison, D. E. F., and Pirt, S. J., 1967, The influence of dissolved oxygen concentration on the respiration and glucose metabolism of *Klebsiella aerogenes* during growth, *J. Gen. Microbiol.* **46**:193.

Hartley, B. S., 1974, Enzyme families, in: *Evolution in the Microbial World, 24th Symposium of the Society for General Microbiology* (M. J. Carlile and J. J. Skehel, eds.), pp. 151–182, Cambridge University Press, Cambridge.

Hartley, B. S., Burleigh, B. D., Midwinter, G. G., Moore, C. H., Morris, H. R., Rigby, P. W. J., Smith, M. J., and Taylor, S. S., 1972, Where do new enzymes come from? in: *Enzymes: Structure and Function, 8th FEBS Meeting 1972,* vol. 29 (J. Drenth, R. A. Oosterbaan, and C. Veeger, eds.) pp. 151–176, North Holland, Amsterdam.

Heckels, J. E., Lambert, P. A., and Baddiley, J., Binding of magnesium ions to cell walls of *Bacillus subtilis* W23 containing teichoic acid or teichuronic acid, *Biochem. J.* **162**:359.

Herbert, D., 1961, The chemical composition of micro-organisms as a function of their environment, in: *Microbial Reaction to Environment, 11th Symposium of the Society for General Microbiology* (G. G. Meynell and H. Gooder, eds.), pp. 391–416, Cambridge University Press, Cambridge.

Herbert, D., and Kornberg, H. L., 1976, Glucose transport as rate-limiting step in the growth of *Escherichia coli* on glucose, *Biochem. J.* **156**:477.

Hillmer, P., and Gest, H., 1977a, H_2 metabolism in the photosynthetic bacterium *Rhodopseudomonas capsulata*: H_2 production by growing cultures, *J. Bacteriol.* **129**:724.

Hillmer, P., and Gest, H., 1977b, H_2 metabolism in the photosynthetic bacterium *Rhodopseudomonas capsulata*: production and utilization of H_2 by resting cells, *J. Bacteriol.* **129**:732.

Ierusalimsky, N. D., 1967, Bottle-necks in metabolism as growth rate controlling factors, in: *Microbial Physiology and Continuous Culture* (E. O. Powell, C. G. T. Evans, R. E. Strange, and D. W. Tempest, eds.), pp. 23–33, Her Majesty's Stationery Office, London.

Johnson, C. L., Cha, Y. A., and Stern, J. R., 1975, Citrate uptake in membrane vesicles of *Klebsiella aerogenes, J. Bacteriol.* **121**:682.

Jones-Mortimer, M. C., and Kornberg, H. L., 1976, Uptake of fructose by the sorbitol phosphotransferase of *Escherichia coli* K12, *J. Gen. Microbiol.* **96**:383.

Kahane, S., Marcus, M., Metzer, E., and Halpern, Y. S., 1976, Glutamate transport in membrane vesicles of the wild type strain and glutamate utilizing mutants of *Escherichia coli, J. Bacteriol.* **125**:770.

Ketchum, P. A., and Owens, M. S., 1975, Production of a molybdenum coordinating compound by *Bacillus thuringiensis, J Bacteriol.* **122**:412.

Koch, A. L., 1971, The adaptive responses of *Escherichia coli* to a feast and famine existence, *Adv. Microbial Physiol.* **6**:147.

Konings, W. N., and Freese, E., 1972, Amino acid transport in membrane vesicles of *Bacillus subtilis, J. Biol. Chem.* **247**:2408.

Kornberg, H. L., 1973, Fine control of sugar uptake by *Escherichia coli*, in: *Rate Control of Biological Processes, 17th Symposium of the Society for Experimental Biology*, pp. 175–193, Cambridge University Press, Cambridge.

Kundig, W., 1974, Molecular interactions in the bacterial phosphoenolpyruvate phospho-transferase system (PTS), *J. Supramol. Struct.* **2**:695.

Lengeler, J., 1975, Nature and properties of hexitol transport systems in *Escherichia coli, J. Bacteriol.* **124**:39.

Lewis, A. J., and Miller, D. A., 1975, Keto acid metabolism in *Desulfovibrio, J. Gen. Microbiol.* **90**:286.

Light, P. A., 1972, Influence of environment on mitochondrial function in yeast, *J. Appl. Chem. Biotechnol.* **22**:509.

Lin, E. C. C., Levin, A. P., and Magasanik, B., 1960, The effect of aerobic metabolism on the inducible glycerol dehydrogenase of *Aerobacter aerogenes, J. Biol. Chem.* **235**:1824.

Lowendorf, H. S., Slayman, C. L., and Slayman, C. W., 1974, Phosphate transport in *Neurospora.* Kinetic characterization of a constitutive, low affinity transport system, *Biochim. Biophys. Acta* **373**:369.

Lowendorf, H. S., Bazinet, G. F., Jr., and Slayman, C. W., 1975, Phosphate transport in *Neurospora.* Derepression of a high affinity transport system during phosphorus starvation, *Biochim. Biophys. Acta* **389**:541.

Luscombe, B. M., 1974, The effect of dropwise addition of medium on the yield of carbon limited cultures of *Arthrobacter globiformis, J. Gen. Microbiol.* **83**:197.

Luscombe, B. M., and Gray, T. R. G., 1971, Effect of varying growth rate on the morphology of *Arthrobacter, J. Gen. Microbiol.* **69**:433.

Maaløe, O., and Kjeldgaard, N. O., 1966, *Control of Macromolecular Synthesis*, W. A. Benjamin, New York.

Mandelstam, J., and McQuillen, K., 1968, *Biochemistry of Bacterial Growth*, p. 9, Blackwell, Oxford.

Marzluf, G. A., 1974, Uptake and efflux of sulfate in *Neurospora crassa, Biochim. Biophys. Acta* **339**:374.

Matin, A., and Konings, W. N., 1973, Transport of lactate and succinate by membrane vesicles of *Escherichia coli, Bacillus subtilis* and a *Pseudomonas* species, *Eur. J. Biochem.* **34**:58.

Matsushita, T., and Kubitschek, H. E., 1975, DNA replication in bacteria, *Adv. Microbial Physiol.* **12**:247.

McCready, R. G. L., and Din, G. A., 1974, Active sulfate transport in *Saccharomyces cerevisiae, FEBS Lett.* **38**:361.

Meers, J. L., and Tempest, D. W., 1968, The influence of extracellular products on the behavior of mixed microbial populations in magnesium-limited chemostat cultures, *J. Gen. Microbiol.* **52**:309.

Meers, J. L., and Tempest, D. W., 1970, The influence of growth-limiting substrate and medium NaCl concentration on the synthesis of magnesium-binding sites in the walls of *Bacillus subtilis* var. *niger, J. Gen. Microbiol.* **63**:325.

Meyer, D. J., and Jones, C. W., 1973, Oxidative phosphorylation in bacteria which contain different cytochrome oxidases, *Eur. J. Biochem.* **36**:144.

Minnikin, D. E., Abdolrahimzadeh, H., and Baddiley, J., 1971, The interrelation of polar lipids in bacterial membranes, *Biochim. Biophys. Acta* **249**:651.

Minnikin, D. E., Abdolrahimzadeh, H., and Baddiley, J., 1974, Replacement of acidic phospholipids by acidic glycolipids in *Pseudomonas diminuta, Nature (London)* **249**:268.

Morrison, C. E., and Lichstein, H. C., 1976, Regulation of lysine transport by feedback inhibition in *Saccharomyces cerevisiae, J. Bacteriol.* **125**:864.

Mygind, B., and Munch-Petersen, A., 1975, Transport of pyrimidine nucleosides in cells of *Escherichia coli* K12, *Eur. J. Biochem.* **59**:365.

Neidhardt, F. C., 1963, Effects of environment on the composition of bacterial cells, *Annu. Rev. Microbiol.* **17**:61.

Neijssel, O. M., 1976, The significance of overflow metabolism in the physiology and growth of *Klebsiella aerogenes*, Ph.D. thesis, University of Amsterdam.

Neijssel, O. M., and Tempest, D. W., 1975, The regulation of carbohydrate metabolism in *Klebsiella aerogenes* NCTC 418 organisms, growing in chemostat culture, *Arch. Microbiol.* **106**:251.

Neijssel, O. M., and Tempest, D. W., 1976a, Bioenergetic aspects of aerobic growth of *Klebsiella aerogenes* NCTC 418 in carbon-limited and carbon-sufficient chemostat culture, *Arch. Microbiol.* **107**:215.

Neijssel, O. M., and Tempest, D. W., 1976b, The role of energy-spilling reactions in the growth of *Klebsiella aerogenes* NCTC 418 in aerobic chemostat culture, *Arch. Microbiol.* **110**:305.

Neijssel, O. M., Hueting, S., Crabbendam, K. J., and Tempest, D. W., 1975, Dual pathways of glycerol assimilation in *Klebsiella aerogenes* NCIB 418. Their regulation and possible functional significance, *Arch. Microbiol.* **104**:83.

Neijssel, O. M., Hueting, S., and Tempest, D. W., 1977, Glucose transport capacity is not the rate limiting step in the growth of some wild type strains of *Escherichia coli* and *Klebsiella aerogenes* in chemostat culture, *FEMS Microbiol. Lett.* **2**:1.

Nelson, D. L., and Kennedy, E. P., 1972, Transport of magnesium by a repressible and a nonrepressible system in *Escherichia coli, Proc. Nat. Acad. Sci. U.S.A.* **69**:1091.

Oehr, P., and Willecke, K., 1974, Citrate Mg^{2+} transport in *Bacillus subtilis*: studies with 2-fluoro-L-*erythro*-citrate as a substrate, *J. Biol. Chem.* **249**:2037.

Oh, Y. K., and Freese, E., 1976, Manganese requirement of phosphoglycerate phosphomutase and its consequences for growth and sporulation of *Bacillus subtilis, J. Bacteriol.* **127**:739.

Pirt, S. J., 1957, The oxygen requirement of growing cultures of an *Aerobacter* species determined by means of the continuous culture technique, *J. Gen. Microbiol.* **16**:59.

Pirt, S. J., 1965, The maintenance energy of bacteria in growing cultures, *Proc. R. Soc. Lond. B Biol. Sci.* **163**:224.

Poole, R. K., and Haddock, B. A., 1975, Effects of sulphate limited growth in continuous culture on the electron transport chain and energy conservation in *Escherichia coli* K12, *Biochem. J.* **152**:537.

Postgate, J. R., 1973, The viability of very slow-growing populations: a model for the natural ecosystem, *Bull. Ecol. Res. Comm.* **17**:287.

Postgate, J. R., and Hunter, J. R., 1963, The survival of starved bacteria, *J. Appl. Bacteriol.* **26**:295.

Postgate, J. R., and Hunter, J. R., 1964, Accelerated death of *Aerobacter aerogenes* starved in the presence of growth-limiting substrates, *J. Gen. Microbiol.* **34**:459.

Rhoads, D. B., and Epstein, W., 1977, Energy coupling to net K$^+$ transport in *Escherichia coli, J. Biol. Chem.* **252**:1394.

Robinson, A., and Tempest, D. W., 1973, Phenotypic variability of the envelope proteins of *Klebsiella aerogenes, J. Gen. Microbiol.* **78**:361.

Schaechter, M., Maaløe, O., and Kjeldgaard, N. O., 1958, Dependency on medium and temperature of cell size and chemical composition during balanced growth of *Salmonella typhimurium, J. Gen. Microbiol.* **19**:592.

Schulte, T. H., and Scarborough, G. A., 1975, Characterization of the glucose transport systems in *Neurospora crassa* sl, *J. Bacteriol.* **122**:1076.

Schulze, K. L., and Lipe, R. S., 1964, Relationship between substrate concentration, growth rate, and respiration rate of *Escherichia coli* in continuous culture, *Arch. Microbiol.* **48**:1.

Schwarzhoff, R. H., and Williams, F. D., 1976, Growth inhibition of *Proteus mirabilis* by cyclic adenosine 3',5'-monophosphate, *J. Bacteriol.* **127**:1590.

Senez, J. C., 1962, Some considerations of the energetics of bacterial growth, *Bacteriol. Rev.* **26**:95.

Short, S. A., White, D. C., and Kaback, H. R., 1972, Mechanisms of active transport in isolated bacterial membrane vesicles. IX. The kinetics and specificity of amino acid transport in *Staphylococcus aureus* membrane vesicles, *J. Biol. Chem.* **247**:7452.

Stouthamer, A. H., and Bettenhaussen, C., 1973, Utilization of energy for growth and maintenance in continuous and batch culture of microorganisms. A reevaluation of the method for the determination of ATP production by measuring growth yields, *Biochim. Biophys. Acta* **301**:53.

Tanner, W., 1974, Energy coupled sugar transport in *Chlorella, Biochem. Soc. Trans.* **2**:793.

Tempest, D. W., 1970, The continuous cultivation of micro-organisms. I. Theory of the chemostat, in: *Methods in Microbiology*, vol. 2 (J. R. Norris and D. W. Ribbons, eds.), pp. 259–276, Academic Press, London.

Tempest, D. W., and Hunter, J. R., 1965, The influence of temperature and pH value on the macromolecular composition of magnesium-limited and glycerol-limited *Aerobacter aerogenes* growing in a chemostat, *J. Gen. Microbiol.* **41**:267.

Tempest, D. W., Herbert, D., and Phipps, P. J., 1967a, Studies on the growth of *Aerobacter aerogenes* at low dilution rates in a chemostat, in: *Microbial Physiology and Continuous Culture* (E. O. Powell, C. G. T. Evans, R. E. Strange, and D. W. Tempest, eds.), pp. 240–254, Her Majesty's Stationery Office, London.

Tempest, D. W., Dicks, J. W., and Meers, J. L., 1967b, Magnesium-limited growth of *Bacillus subtilis*, in pure and mixed cultures, in a chemostat, *J. Gen. Microbiol.* **49**:139.

Tempest, D. W., Dicks, J. W., and Ellwood, D. C., 1968, Influence of growth condition on the concentration of potassium in *Bacillus subtilis* var. *niger* and its possible relationship to cellular ribonucleic acid, teichoic acid and teichuronic acid, *Biochem. J.* **106**:237.

Tempest, D. W., Meers, J. L., and Brown, C. M., 1970, Influence of environment on the content and composition of microbial free amino acid pools, *J. Gen. Microbiol.* **64**:171.

Templeton, B. A., and Savageau, M. A., 1974, Transport of biosynthetic intermediates: homoserine and threonine uptake in *Escherichia coli, J. Bacteriol.* **117**:1002.

Thomson, A. W., O'Neill, J. G., and Wilkinson, J. F., 1976, Acetone production by methylobacteria, *Arch. Microbiol.* **109**:243.

Tolbert, N. E., 1974, Photorespiration, in: *Algal Physiology and Biochemistry* (W. D. P. Stewart, ed.), pp. 474–504, Blackwell, Oxford.

Turner, W. B., 1975, Biosynthetic origin of mycotoxins, *Int. J. Environ. Stud.* **8**:159.

Umbarger, H. E., 1969, Regulation of amino acid metabolism, *Annu. Rev. Microbiol.* **38**:323.

Veldkamp, H., 1977, Ecological studies with the chemostat, *Adv. Microbial Ecol.* **1**:59.

Veldkamp, H., and Jannasch, H. W., 1972, Mixed culture studies with the chemostat, in: *Environmental Control of Cell Synthesis and Function, Proceedings of the 5th International Symposium on the Continuous Culture of Microorganisms* (A. C. R. Dean, S. J. Pirt, and D. W. Tempest, eds.), pp. 105–123, Academic Press, London, New York.

Watanabe, H., Mimura, N., Takimoto, A., and Nakamura, T., 1975, Luminescence and respiratory activities of *Photobacterium phosphoreum*. Competition for cellular reducing power, *J. Biochem. (Tokyo)* **77**:1147.

Weissman, J. C., and Benemann, J. R., 1977, Hydrogen production by nitrogen starved cultures of *Anabaena cylindrica, Appl. Environ. Microbiol.* **33**:123.

Willis, R. C., and Woolfolk, C. A., 1975, L-asparagine uptake in *Escherichia coli*, *J. Bacteriol.* **123**:937.

Willsky, G. R., and Malamy, M. H., 1974, The loss of the phoS periplasmic protein leads to a change in the specificity of a constitutive inorganic phosphate transport system in *Escherichia coli*, *Biochem. Biophys. Res. Commun.* **60**:226.

Willsky, G. R., and Malamy, M. H., 1976, Control of the synthesis of alkaline phosphatase and the phosphate binding protein in *Escherichia coli*, *J. Bacteriol.* **127**:595.

Wold, W. S. M., and Suzuki, I., 1976, The citric acid fermentation by *Aspergillus niger*: regulation by zinc of growth and acidogenesis, *Can. J. Microbiol.* **22**:1083.

Wouters, J. T. M., and Buysman, P. J., 1977, Production of some exocellular enzymes by *Bacillus licheniformis* 749/C in chemostat cultures, *FEMS Microbiol. Lett.* **1**:109.

Zwaig, N., Kistler, W. S., and Lin, E. C. C., 1970, Glycerolkinase, the pacemaker for the dissimilation of glycerol in *Escherichia coli*, *J. Bacteriol.* **102**:753.

4

Role of Microorganisms in the Atmospheric Sulfur Cycle

JOHN M. BREMNER AND CHARLENE G. STEELE

1. Introduction

Microorganisms play key roles in the oxidation–reduction and assimilation–dissimilation steps of the sulfur cycle in nature. A substantial amount of information is available concerning the forms of sulfur in natural systems (ZoBell, 1963; Freney, 1967; Richmond, 1973; Williams, 1975) and the microbially mediated transformations of sulfur within the pedosphere and hydrosphere, the two major natural sites of microbial activity (Starkey, 1956; ZoBell, 1958, 1963, 1973; Postgate, 1959; Wood, 1965; Kelley, 1968; Roy and Trudinger, 1970; Rheinheimer, 1972, 1974; Freney and Swaby, 1975; Siegel, 1975; Trudinger, 1975; Weir, 1975). Very little is known, however, about the forms and amounts of volatile sulfur released to the atmosphere through microbial activity in the pedosphere or hydrosphere. The purpose of this chapter is to review current information relating to production of volatile sulfur compounds by terrestrial and aquatic microorganisms and the role of these microorganisms in the atmospheric sulfur cycle.

The global sulfur cycle has been the subject of a considerable number of reviews (e.g., Junge, 1960, 1963a,b, 1972; Eriksson, 1963; Kuznetsov *et al.*, 1963; Kellogg *et al.*, 1972; Deevey, 1973; Hill, 1973; Friend, 1973; LeGall, 1974; Rasmussen *et al.*, 1975; Trudinger, 1975; Eriksson and Rosswall, 1976; Granat *et al.*, 1976; Moss, 1976), and several models of the atmospheric sulfur cycle have been proposed (Eriksson,

JOHN M. BREMNER and CHARLENE G. STEELE • Department of Agronomy, Iowa State University, Ames, Iowa, U.S.A.

1963; Junge, 1963b; Robinson and Robbins, 1968, 1970a,b; Kellogg *et al.*, 1972; Rodhe, 1972; Friend, 1973; Granat *et al.*, 1976). Although these models differ in details, all indicate that a large proportion (usually about 50%) of the sulfur in the atmosphere is derived from biological transformations of sulfur in the pedosphere and hydrosphere and that most of the sulfur volatilized from natural systems through microbial activity is in the form of H_2S. One of the primary objectives of this chapter is to draw attention to the lack of evidence for these conclusions and to the need for reevaluation of current concepts of the atmospheric sulfur cycle.

Research on the role of microorganisms in the atmospheric sulfur cycle has been greatly hindered by the lack of sensitive and specific methods for detection and determination of volatile forms of sulfur. This problem has been reduced by the recent development of gas chromatographic techniques permitting identification and estimation of trace amounts of mercaptans, alkyl sulfides, and other volatile sulfur compounds identified as atmospheric constituents or as gaseous products of microbial activity. For example, techniques recently described by Bremner and Banwart (1974) permit detection and identification of nanogram amounts of the following 15 volatile sulfur compounds: sulfur dioxide, hydrogen sulfide, carbon disulfide, carbonyl sulfide, sulfur hexafluoride, methyl mercaptan, ethyl mercaptan, *n*-propyl mercaptan, *n*-butyl mercaptan, *iso*-butyl mercaptan, dimethyl sulfide, ethyl methyl sulfide, diethyl sulfide, dimethyl disulfide, and diethyl disulfide. These techniques are based on studies showing that a flame photometric detector fitted with a 394-nm optical filter exhibits high selectivity and sensitivity for sulfur (Brody and Chaney, 1966; Grice *et al.*, 1970; Stevens *et al.*, 1971; Ronkainen *et al.*, 1973; Banwart and Bremner, 1974; Bremner and Banwart, 1974) and on work by Stevens *et al.* (1971) showing that the sorption problems encountered in attempts to develop gas chromatographic methods for analysis of sulfur gases can be greatly reduced through use of fluorinated ethylene-propylene (FEP Teflon) tubing for chromatographic columns. Another recent development in research related to the role of microorganisms in the atmospheric sulfur cycle has been the use of mass spectrometry in combination with gas chromatography for identification of volatile sulfur compounds evolved from natural systems (Francis *et al.*, 1973, 1975; Smith, 1976; Smith *et al.*, 1977).

Despite recent advances in methods for analysis of air for sulfur compounds, very little is known about the forms and amounts of sulfur in the atmosphere. Recent work, however, indicates that unpolluted air contains parts-per-billion or parts-per-trillion amounts of SO_2, H_2S, CH_3SCH_3, COS, CS_2, and SF_6 (see Section 5). Significantly larger amounts of several of these gases, particularly SO_2, have been detected in polluted air, and work discussed in Section 3 suggests that improvements in analytical techniques will lead to identification of other sulfur gases in unpolluted air (e.g., CH_3SH and CH_3SSCH_3).

It is generally assumed that airborne microorganisms are dormant and play no part in the atmospheric sulfur cycle. Parker (1970) has suggested that some of these microorganisms may be physiologically active because they may be able to obtain water from clouds and to procure nutrients from airborne dust particles and by utilization of organic and inorganic volatiles in the atmosphere. However, there does not appear to be any basis for speculation that airborne microorganisms may play a significant role in the atmospheric sulfur cycle by metabolizing gaseous sulfur compounds.

2. Microbial Production of Volatile Sulfur Compounds

Much of the information currently available concerning production of volatile sulfur compounds by microorganisms is derived from culture studies reviewed by Kadota and Ishida (1972) and Alexander (1974). Although the methods used for identification of volatile sulfur compounds in some of these studies are open to criticism, recent work using more specific methods indicates that most of the identifications reported are reliable and that microorganisms can produce the volatile sulfur compounds listed in Table I.

Numerous studies have shown that microorganisms can produce H_2S both by reduction of sulfate and by degradation of organic sulfur compounds. Reduction of sulfate to H_2S is effected by *Desulfovibrio* and related bacteria. These bacteria are abundant in bogs, swamps, muds, and poorly drained soils because they can proliferate under anaerobic conditions by using sulfate as their terminal electron acceptor (see ZoBell, 1958, 1963; Postgate, 1959; Peck, 1962; Campbell and Postgate, 1965). As pointed out by Alexander (1974), they generally show little or no activity in natural environments containing dissolved O_2 or having a low pH because they are usually susceptible to acidity and to the presence of O_2. Many heterotrophic microorganisms can convert sulfur in organic substrates to H_2S under aerobic or anaerobic conditions (see Kadota and Ishida, 1972). The substrates for these cleavage reactions include proteins, polypeptides, cystine, cysteine, and glutathione.

Culture studies have demonstrated that many species of bacteria and fungi can produce volatile organic sulfur compounds, such as methyl mercaptan, dimethyl sulfide, and dimethyl disulfide. Since the microorganisms and substrates used in these culture studies are listed in recent reviews (Kadota and Ishida, 1972; Alexander, 1974), they will not be discussed here. It is noteworthy, however, that many of these culture studies have involved use of methionine or dimethyl-β-propiothetin as

Table I. Volatile Sulfur Compounds Produced by Microorganisms

Compound	Formula	References[a]
Hydrogen sulfide	H_2S	A
Sulfur dioxide	SO_2	B
Carbon disulfide	CS_2	C
Carbonyl sulfide (carbon oxysulfide)	COS	D
Methyl mercaptan (methanethiol)	CH_3SH	E
Ethyl mercaptan (ethanethiol)	CH_3CH_2SH	F
n-Propyl mercaptan (1-propanethiol)	$CH_3CH_2CH_2SH$	G
iso-Propyl mercaptan (1-methyl-1-ethanethiol)	$(CH_3)_2CHSH$	H
Allyl mercaptan (2-propene-1-thiol)	$CH_2{:}CHCH_2SH$	I
n-Butyl mercaptan (1-butanethiol)	$CH_3CH_2CH_2CH_2SH$	J
iso-Butyl mercaptan (2-methyl-1-propanethiol)	$(CH_3)_2CHCH_2SH$	H
Dimethyl sulfide (methylthiomethane)	CH_3SCH_3	K
Dimethyl disulfide (methyldithiomethane)	CH_3SSCH_3	L
Dimethyl trisulfide (methyltrithiomethane)	CH_3SSSCH_3	M
Ethyl methyl sulfide (methylthioethane)	$CH_3CH_2SCH_3$	N
Diethyl sulfide (ethylthioethane)	$CH_3CH_2SCH_2CH_3$	O
Diethyl disulfide (ethyldithioethane)	$CH_3CH_2SSCH_2CH_3$	P
Dipropyl disulfide (propyldithiopropane)	$CH_3CH_2CH_2SSCH_2CH_2CH_3$	Q
Diallyl sulfide (allyl sulfide)	$CH_2{:}CHCH_2SCH_2CH{:}CH_2$	Q
Diallyl disulfide (allyl disulfide)	$CH_2{:}CHCH_2SSCH_2CH{:}CH_2$	Q
Methyl allyl sulfide	$CH_3SCH_2CH{:}CH_2$	R
Allicin (S-oxodiallyl disulfide)	$CH_2{:}CHCH_2S({:}O)SCH_2CH{:}CH_2$	S

[a] A, Birkinshaw et al. (1942), Challenger and Charlton (1947), Clarke (1953), Labarre et al. (1966), Herbert et al. (1971), Kadota and Ishida (1972), Alexander (1974), Babich and Stotzky (1974), Bechard (1974), Bechard and Rayburn (1974), Herbert and Shewan (1976), McCready et al. (1976), Stotzky and Schenck (1976); B, Leinweber and Monty (1961, 1965), Soda et al. (1964), Rankine and Pocock (1969), Dittrich and Staudenmayer (1970), Minárik (1972), Kunert (1973), Dott and Trüper (1976); C, Banwart and Bremner (1974, 1975a,b, 1976a,b); D, Moje et al. (1964), Somers et al. (1967), Gobert et al. (1971), Elliot and Travis (1973), Banwart and Bremner (1975a,b, 1976a,b); E, Birkinshaw et al. (1942), Challenger and Charlton (1947), Frederick et al. (1957), Takai and Asami (1962), Asami and Takai (1963), Labarre et al. (1966), Jenkins et al. (1967), Ruiz-Herrera and Starkey (1969, 1970), Segal and Starkey (1969), Lewis and Papavizas (1970), Herbert et al. (1971), Kadota and Ishida (1972), Francis et al. (1973, 1975), Banwart and Bremner (1975a,b, 1976a,b), Laakso et al. (1976), Manning et al. (1976), Sharpe et al. (1976); F, Labarre et al. (1972), Banwart and Bremner (1975b); G, King and Coley-Smith (1969), Labarre et al. (1972); H, Jenkins et al. (1967); I, Challenger and Greenwood (1949), King and Coley-Smith (1969); J, Jenkins et al. (1967), Labarre et al. (1972), Francis et al. (1973); K, Challenger and Charlton (1947), Wagner and Stadtman (1962), Toan et al. (1965), Jenkins et al. (1967), Ishida (1968), Kadota and Ishida (1968), Ruiz-Herrera and Starkey (1969), Lewis and Papavizas (1970), Herbert et al. (1971), Kadota and Ishida (1972), Lovelock et al. (1972), Francis et al. (1973, 1975), Bechard (1974), Bechard and Rayburn (1974), Banwart and Bremner (1975a,b, 1976a,b), Herbert and Shewan (1976); L, Challenger and Charlton (1947), Kallio and Larson (1955), Frederick et al. (1957), Labarre et al. (1966), Jenkins et al. (1967), Segal and Starkey (1969), Lewis and Papavizas (1970), Ruiz-Herrera and Starkey (1970), Francis et al. (1973, 1975), Banwart and Bremner (1975a,b, 1976a,b); M, Miller et al. (1973); N, Banwart and Bremner (1975b); O, Kondo (1923), Challenger and North (1934), Drews et al. (1970), P, Drews et al. (1970), King and Coley-Smith (1969), Banwart and Bremner (1975b); Q, King and Coley-Smith (1969); R, Challenger and Greenwood (1949); S, Murakami (1960).

substrates and have resulted in detection of methyl mercaptan, dimethyl sulfide, or dimethyl disulfide. It should also be noted that very little is known about the mechanisms and enzymology of the reactions by which microorganisms produce volatile forms of sulfur from different substrates (for discussions of this topic, see Freney, 1967; Kadota and Ishida, 1972; Alexander, 1974).

Several heterotrophs have been found to produce SO_2 or sulfite (one of the two anionic forms of dissolved SO_2). This activity is important in the wine industry, and it has been demonstrated that several species of *Saccharomyces* (wine yeasts) can produce SO_2 (or sulfite) in cultures containing cysteine, methionine, or sulfate (Rankine and Pocock, 1969; Minárik, 1972; Dittrich and Staudenmayer, 1970). The production of sulfite by *Microsporum gypseum* (a dermatophyte) in cultures containing cystine has also been reported (Kunert, 1973). Alexander (1974) has suggested that cysteine sulfinic acid [$HOOSCH_2CH(NH_2)COOH$] may be an important intermediate in the biogenesis of SO_2 because there is evidence that this compound is cleaved enzymatically to yield SO_2 (or sulfite) and alanine by cultures of *Neurospora crassa, Escherichia coli,* and *Alcaligenes faecalis* (Leinweber and Monty, 1961, 1965; Soda *et al.,* 1964):

$$HOOSCH_2CH(NH_2)COOH \rightarrow SO_2 + CH_3CH(NH_2)COOH$$

$$SO_2 + H_2O \rightarrow H_2SO_3$$

Support for this suggestion has been provided by the detection of cysteine sulfinic acid as an intermediate in the conversion of cysteine to sulfate by soil microorganisms (Freney, 1960).

Moje *et al.* (1964) detected formation of COS through microbial decomposition in soil of the pesticide known as nabam (disodium ethylene-*bis*-dithiocarbamate), and Somers *et al.* (1967) found that COS was produced from captan (another pesticide) by axenic cultures of *N. crassa*. The production of COS by axenic cultures of *Trichomonas vaginalis* has also been reported (Gobert *et al.,* 1971), and this compound has been detected in the volatiles produced during anaerobic decomposition of animal manures (Elliott and Travis, 1973; Banwart and Bremner, 1975a) and in the gases evolved from unamended and amended soils under aerobic or waterlogged conditions (Banwart and Bremner, 1975b, 1976a,b).

Recent work has shown that CS_2 is evolved from animal manures under anaerobic conditions (Banwart and Bremner, 1975a) and is released from unamended and amended soils under aerobic or waterlogged conditions (Banwart and Bremner, 1975b, 1976a,b). Banwart and Bremner (1975b) found that CS_2 is produced by microbial decomposition of cysteine and cystine, and they have suggested that this process is responsible for the emission of CS_2 from soils and manures and may account for the recent detection of CS_2 in coastal and ocean waters (Lovelock, 1974). Munnecke et al. (1962) detected CS_2 and methyl-iso-thiocyanate in the volatiles produced by decomposition of several fungicides in soils. Carbon disulfide was produced in soils treated with nabam, thiram, and zineb, and methylisothiocyanate was produced in soils treated with vapam or mylone. Robbins and Kastelic (1961) observed production of CS_2 through decomposition of thiram by rumen microorganisms.

Iverson (1966) found that Desulfovibrio desulfuricans produced a volatile substance when grown on trypticase agar plus sulfate. He was unable to identify this gas by gas chromatographic techniques, but tentatively identified it as disulfur monoxide (S_2O, "Schenk's sulfur monoxide") by mass spectrometer analyses (Iverson, 1967). He found that this gas stimulated growth of D. desulfuricans on simple agar media (e.g., lactate plus inorganic salts) and could be used by this organism as an electron donor for the reduction of benzyl viologen. These observations led Iverson (1967) to suggest that S_2O may be an intermediate product of dissimilatory sulfate reduction. Roy and Trudinger (1970), however, have suggested that the gas detected by Iverson may have resulted from decomposition of an enzyme-bound intermediate with an oxidation state between that of sulfite and sulfide.

Several of the volatile sulfur compounds listed in Table I have been found to affect the growth of plants or microorganisms. For example, COS and CS_2 are toxic to fungi, and CS_2 inhibits oxidation of ammonium by nitrifying microorganisms in soils (Powlson and Jenkinson, 1971; Bremner and Bundy, 1974; Ashworth et al., 1975). Ethyl mercaptan, a product of Clostridium, inhibits ripening of the rice plant (Inoue et al., 1955), and several volatile organic sulfur compounds affect growth of pathogenic fungi (see Kadota and Ishida, 1972; Alexander, 1974; Stotzky and Schenck, 1976). For example, Lewis and Papavizas (1971) found that methyl mercaptan, dimethyl sulfide, and dimethyl disulfide retarded mycelial growth of Aphanomyces euteiches, a fungus causing root rot of peas, and prevented germination of spores of this fungus. In contrast, Coley-Smith and King (1969) and King and Coley-Smith (1969) found that alkyl sulfides and allyl sulfides stimulated germination of sclerotia of Sclerotium cepivorum, the fungus responsible for white rot of onions. Higher plants are remarkably sensitive to H_2S, and there is evidence that H_2S produced by microbial activity in waterlogged soils can severely damage the roots of rice and of fruit trees, particularly citrus and avocado (Vámos, 1959; Ford, 1973; Joshi and Hollis, 1977). Hydrogen sulfide is also highly toxic to nematodes and to many fishes, and it contributes significantly to the odor problems associated with sewage-treatment plants and animal manures.

3. Microbial Sources of Atmospheric Sulfur

Current concepts of the global sulfur cycle are based on the assumption that terrestrial and aquatic microorganisms contribute very substantially to atmospheric sulfur (see Granat *et al.*, 1976). This is illustrated by Table II, which shows estimates by Robinson and Robbins (1970a) of the relative contributions of various sulfur sources to atmospheric sulfur. According to these estimates, the amount of sulfur emitted as H_2S to the atmosphere through microbial activity on land and in water is more than 33 times the amount generated by industry and is appreciably greater than the total amount of sulfur emitted as H_2S or SO_2 by industrial and other nonbiological processes. It is difficult, however, to accept these estimates, because very little is known about the forms or amounts of volatile sulfur released to the atmosphere by terrestrial or aquatic microorganisms.

3.1. Terrestrial Microorganisms

3.1.1. Microbial Activity in Soils

Although it has been assumed that substantial amounts of H_2S are released from soils to the atmosphere through reduction of sulfate or cleavage of organic sulfur compounds by soil microorganisms, recent work has indicated that this assumption may be invalid and that most of the sulfur volatilized from soils through microbial activity is in the form of organic compounds such as dimethyl sulfide, dimethyl disulfide, and methyl mercaptan (Banwart and Bremner, 1975b, 1976a,b; Bremner, 1977).

Barjac (1952) and Greenwood and Lees (1956) detected evolution of mercaptan-like odors from soils treated with methionine, and Frederick *et al.* (1957) found that volatile compounds exhibiting some of the properties

Table II. Estimated Global Emissions of Sulfur Compounds[a]

Compound	Source	Estimated emission (tons S/yr, $\times 10^6$)
H_2S	Terrestrial emissions	70
	Marine emissions	30
	Industrial emissions	3
SO_2	Coal combustion	51
	Petroleum combustion and refining	14
	Smelting operations	8
Sulfate aerosols	Sea spray	44

[a]From Robinson and Robbins (1970a).

of methyl mercaptan and dimethyl disulfide were evolved from a sandy soil treated with methionine. Takai and Asami (1962) obtained evidence that methyl mercaptan is produced in paddy soils and that formation of this compound is promoted by high soil temperatures and by addition of manures (Asami and Takai, 1963). Lewis and Papavizas (1970) detected formation of methyl mercaptan, dimethyl sulfide, and dimethyl disulfide through decomposition of sulfur-rich plant materials in soil, and Frances *et al.* (1973) detected release of methyl mercaptan, dimethyl sulfide, dimethyl disulfide, and butyl mercaptan from a silt loam incubated under anaerobic conditions (argon atmosphere) after treatment with glucose and methionine (see also Francis *et al.*, 1975). Lovelock *et al.* (1972) detected emission of dimethyl sulfide from unamended soils, and Francis *et al.* (1975) observed release of this compound from a clay soil incubated anaerobically after treatment with glucose.

Besides confirming most of the observations by previous workers, recent gas chromatographic investigations by Banwart and Bremner (1975b, 1976a,b) have provided data concerning the amounts and forms of volatile sulfur released from both unamended and amended soils under aerobic and waterlogged conditions. The sulfur-containing materials used as amendments in these studies included organic and inorganic forms of sulfur commonly added to soils or known to be present in soils, e.g., sulfur-containing amino acids (Tables III and IV), plant proteins (Table V), plant materials (Tables VI and VII), animal manures (Tables VII and VIII), sewage sludges (Tables VII and VIII), and sulfate (Tables IX and X). The results of these studies indicate that volatilization of sulfur from unamended or amended soils occurs largely, if not entirely, through formation of CH_3SH, CH_3SSCH_3, CH_3SCH_3, COS, and CS_2 by microbial decomposition of organic sulfur compounds. They also indicate that most of the sulfur volatilized from soils under aerobic or waterlogged conditions is in the form of CH_3SH, CH_3SCH_3, and CH_3SSCH_3 derived from methionine, and that the remainder is in the form of CS_2 derived from cystine (or cysteine) and of COS, which may be derived from thiocyanates and isothiocyanates in plant residues (Table XI). They further indicate that very little, if any, of the sulfur volatilized from soils through microbial activity is in the form of H_2S.

Recent work discussed in Section 4 has shown that soils have the capacity to sorb the sulfur gases detected in the studies reported in Tables III–XI. This means that estimates of the amounts of volatile sulfur evolved from soils must be regarded as minimal estimates of the amounts of volatile sulfur actually produced in soils and that failure to detect evolution of H_2S or other sulfur gases from soils does not necessarily mean that these gases are not produced by soil microoorganisms. It should be noted, however, that although the sandy soil used in the experiments reported in Table XI had a relatively low capacity for

Table III. Volatile Sulfur Compounds Evolved from Soils Treated with Sulfur-Containing Amino Acids[a,b]

Amino acid	Incubation conditions[c]	Volatile S compounds detected[d]
Methionine	A	CH_3SSCH_3, CH_3SCH_3, CH_3SH
	W	CH_3SSCH_3, CH_3SCH_3, CH_3SH
Methionine sulfoxide	A	CH_3SSCH_3, CH_3SH, CH_3SCH_3
	W	CH_3SSCH_3, CH_3SH, CH_3SCH_3
Methionine sulfone	A	CH_3SSCH_3, CH_3SH, CH_3SCH_3
	W	CH_3SSCH_3, CH_3SH, CH_3SCH_3
S-Methyl methionine	A	None
	W	None
Ethionine	A	$CH_3CH_2SSCH_2CH_3$, $CH_3CH_2SCH_3$, CH_3CH_2SH
	W	$CH_3CH_2SSCH_2CH_3$, $CH_3CH_2SCH_3$, CH_3CH_2SH
Cysteine	A	CS_2
	W	CS_2
Cystine	A	CS_2
	W	CS_2
Cysteic acid	A	None
	W	None
S-Methyl cysteine	A	CH_3SSCH_3, CH_3SCH_3, CH_3SH
	W	CH_3SSCH_3, CH_3SCH_3, CH_3SH
S-Ethyl cysteine	A	$CH_3CH_2SSCH_2CH_3$, $CH_3CH_2SCH_3$, CH_3CH_2SH
	W	$CH_3CH_2SSCH_2CH_3$, $CH_3CH_2SCH_3$, CH_3CH_2SH
Lanthionine	A	CS_2, COS
	W	CS_2, COS
Djenkolic acid	A	CS_2, COS
	W	CS_2, COS
Homocystine	A	CH_3SCH_3, CS_2
	W	CH_3SCH_3, CS_2
Taurine	A	None
	W	None

[a]From Banwart and Bremner (1975b).
[b]Soil sample (5 g) treated with 2 mg of S as amino acid specified was incubated (30°C) under aerobic or waterlogged conditions for 40 days.
[c]A, aerobic conditions (60% of water-holding capacity); W, waterlogged conditions (10 ml of water).
[d]Where two or more compounds were detected, the first compound listed accounted for 60–99% of the total S evolved.

Table IV. Amounts of Sulfur Volatilized from Soils Treated with Sulfur-Containing Amino Acids[a,b]

Amino acid	S volatilized, calculated as % of S added as amino acid[c]	
	Aerobic conditions	Waterlogged conditions
Methionine	15.7–50.1	14.5–48.6
Methionine sulfoxide	1.6–7.8	1.8–13.0
Methionine sulfone	0.2–4.5	0.2–4.7
Ethionine	20.6–48.0	14.2–33.6
Cysteine	0.1–1.8	0.1–1.7
Cystine	0.1–1.4	0.1–1.6
S-Methyl cysteine	15.7–31.7	21.7–43.0
S-Ethyl cysteine	21.4–51.6	38.8–46.6
Lanthionine	0.03–0.2	0.03–0.2
Djenkolic acid	0.3–2.3	0.2–2.2
Homocystine	0.05–0.3	0.03–0.3

[a] From Banwart and Bremner (1975b).
[b] Soil sample (5 g) treated with 2 mg of S as amino acid specified was incubated (30°C) under aerobic (60% of water-holding capacity) or waterlogged conditions (10 ml of water) for 40 days.
[c] Range of values for three soils studied.

Table V. Volatile Sulfur Compounds Evolved from Soils Treated with Plant Proteins[a]

Protein	Soil	Conditions[b]	Amount of S volatilized (μg)[c]					
			As MM	As DMS	As DMDS	As CS	As CD	Total[d]
Zein	Dickinson	A	70.7	1.9	24.3	0	<0.1	97 (2.4)
	Dickinson	W	38.7	6.1	8.3	0	<0.1	53 (1.3)
	Webster	A	88.0	1.9	12.7	0	<0.1	103 (2.6)
	Webster	W	25.9	7.1	6.0	0	<0.1	39 (1.0)
Gluten	Dickinson	A	85.1	0.1	28.2	0	<0.1	113 (2.8)
	Dickinson	W	74.9	1.6	22.3	<0.1	<0.1	99 (2.5)
	Webster	A	59.9	5.4	13.3	0	<0.1	79 (2.0)
	Webster	W	54.4	5.1	16.3	0	<0.1	76 (1.9)
Gliadin	Dickinson	A	125	0.2	50.0	<0.1	<0.1	175 (4.4)
	Dickinson	W	82.6	2.9	62.1	<0.1	<0.1	148 (3.7)
	Webster	A	119	12.4	20.3	<0.1	<0.1	152 (3.8)
	Webster	W	83.4	2.0	37.2	<0.1	<0.1	123 (3.1)

[a] Banwart and Bremner (1976b).
[b] Soil (10 g) treated with 4 mg of S as plant protein was incubated (30°C) under aerobic (A) or waterlogged (W) conditions for 60 days.
[c] MM, methyl mercaptan; DMS, dimethyl sulfide; DMDS, dimethyl disulfide; CS, carbonyl sulfide; CD, carbon disulfide.
[d] Figures in parentheses indicate the total amount of S volatilized, calculated as a percentage of the S added as plant protein.

Table VI. Volatile Sulfur Compounds Evolved from Soils
Treated with Plant Materials[a]

Plant material[b]	Conditions[c]	Volatile sulfur compounds evolved[d]	
		Dickinson soil	Webster soil
Corn (F)	A	DMS	DMS
	W	<u>DMS</u>, CD	<u>DMS</u>, CD
Corn (D)	A	DMS	DMS
	W	DMS, CD	DMS, CD
Alfalfa (F)	A	MM, <u>DMS</u>, DMDS	MM, <u>DMS</u>, DMDS
	W	MM, DMS, DMDS	MM, <u>DMS</u>, DMDS
Alfalfa (D)	A	DMS	DMS
	W	MM, <u>DMS</u>, DMDS	MM, <u>DMS</u>, DMDS, CS
Orchardgrass (F)	A	<u>MM</u>, DMS, DMDS	MM, <u>DMS</u>, DMDS
	W	MM, DMS, DMDS	MM, DMS, DMDS
Orchardgrass (D)	A	DMS	DMS
	W	MM, <u>DMS</u>, DMDS	DMS
Cabbage (F)	A	<u>MM</u>, DMS, DMDS	<u>DMS</u>, DMDS
	W	<u>MM</u>, DMS, DMDS, CS, CD	<u>MM</u>, DMS, DMDS, CS
Cabbage (D)	A	<u>MM</u>, DMS, DMDS	MM, DMS, DMDS
	W	<u>MM</u>, DMS, DMDS, CS, CD	<u>MM</u>, DMS, DMDS
Brussels sprouts (F)	A	<u>MM</u>, DMS, DMDS, CS	MM, DMS, DMDS
	W	<u>MM</u>, DMS, DMDS, CS	<u>MM</u>, DMS, DMDS, CS
Brussels sprouts (D)	A	<u>MM</u>, DMS, DMDS	MM, <u>DMS</u>, DMDS
	W	<u>MM</u>, DMS, DMDS, CS	<u>MM</u>, DMS, DMDS, CS

[a]Banwart and Bremner (1976b).
[b](F), fresh material; (D), dried material.
[c]Soil (10 g) treated with 100 mg (oven-dry basis) of plant material was incubated (30°C) under aerobic (A) or waterlogged (W) conditions for 30 days.
[d]MM, methyl mercaptan; DMS, dimethyl sulfide; DMDS, dimethyl disulfide; CS, carbonyl sulfide; CD, carbon disulfide. Where two or more S compounds were detected, the compound underlined accounted for 52–98% of the total S evolved.

sorption of H_2S, no release of H_2S could be detected when this soil was incubated under aerobic or waterlogged conditions after treatment with inorganic or organic forms of sulfur (including sulfate and sulfide).

Nicolson (1970) observed significant losses of sulfur in sulfur-balance studies with an Australian sandy soil and concluded that substantial volatilization of sulfur can occur from unamended or sulfate-treated soils. This conclusion is not supported by a recent investigation of volatilization of sulfur from unamended and sulfate-treated samples of 25 Iowa soils (Banwart and Bremner, 1976a). No release of volatile sulfur was observed when 11 of these soils were incubated under aerobic or waterlogged conditions. Fourteen soils released volatile sulfur compounds when incubated under waterlogged conditions before or after addition of sulfate, but only four of these soils released volatile sulfur compounds when incubated under aerobic conditions (Table IX). Where volatilization of sulfur was observed, the amount of sulfur volatilized at 30°C in 60 days under aerobic or waterlogged conditions was very small and did not account for more than 0.05% of the sulfur in the unamended or sulfate-treated soils studied (Table X). The volatile sulfur detected was identified as dimethyl sulfide or as dimethyl sulfide associated with smaller amounts of carbonyl sulfide, carbon disulfide, methyl mercaptan, and/or dimethyl disulfide. No trace of H_2S could be detected.

The failure of Banwart and Bremner (1975b, 1976a,b) to detect release of H_2S from unamended soils or from soils treated with inorganic or organic forms of sulfur merits attention because, as noted previously, it has been assumed that volatilization of sulfur from soils as H_2S is a major step in the natural sulfur cycle. There is no doubt that soil microorganisms can produce H_2S by reduction of sulfate and by degradation of sulfur-containing organic substances, but there does not appear to be any

Table VII. Amounts of Sulfur Volatilized from Soils Treated with Different Materials[a]

Materials[b]	Conditions[c]	Amount of S volatilized (µg/g soil)		S volatilized, calculated as % of S in material added	
		Range	Average	Range	Average
Animal manures	A	0–0.26	0.06	0–0.49	0.10
Animal manures	W	0.05–0.92	0.27	0.05–1.77	0.48
Sewage sludges	A	0–0.25	0.09	0–0.11	0.04
Sewage sludges	W	0.05–0.98	0.38	0.02–0.44	0.16
Plant materials (F)	A	0.02–21.7	4.43	0.06–12.9	3.07
Plant materials (D)	A	0.01–16.5	3.39	<0.01–9.8	1.87
Plant materials (F)	W	0.07–26.2	5.53	0.21–15.6	3.34
Plant materials (D)	W	0.03–19.3	4.36	0.03–11.5	2.47

[a]Banwart and Bremner (1976b).
[b](F), fresh; (D), dried.
[c]Soil (10 g) treated with 100 mg (oven-dry basis) of material specified was incubated (30°C) under aerobic (A) or waterlogged (W) conditions for 30 days.

Material added to soil	Conditions[b]	Volatile sulfur compounds evolved[c]	
		Dickinson soil	Webster soil
Beef cattle manure	A	None	None
	W	MM, DMS	DMS, CS, CD
Dairy cattle manure	A	DMS	DMS
	W	MM, DMS, DMDS	MM, DMS, CS, CD
Poultry manure	A	DMS	None
	W	MM, DMS, DMDS	MM, DMS, CS
Sheep manure	A	DMS, DMDS	None
	W	DMS, DMDS	MM, DMS
Swine manure	A	MM, DMS, DMDS	DMS, DMDS
	W	MM, DMS, DMDS	MM, DMS, DMDS, CD
Anoka sludge	A	DMDS, CD	DMDS, CD
	W	MM, DMS, DMDS, CS, CD	DMS, DMDS, CS, CD
Hastings sludge	A	DMDS, CD	DMDS, CD
	W	MM, DMS, DMDS	DMS, DMDS, CS, CD
Orono sludge	A	DMS, DMDS, CD	DMS, DMDS
	W	MM, DMS, DMDS, CS, CD	MM, DMS, DMDS, CS
Metro sludge	A	DMS, DMDS	None
	W	MM, DMS, DMDS	MM, DMS, DMDS, CS, CD

[a] Banwart and Bremner (1976b).
[b] Soil (10 g) treated with 100 mg (oven-dry basis) of manure or sludge was incubated (30°C) under aerobic (A) or waterlogged (W) conditions for 30 days.
[c] MM, methyl mercaptan; DMS, dimethyl sulfide; DMDS, dimethyl disulfide; CS, carbonyl sulfide; CD, carbon disulfide. Where two or more sulfur compounds were detected, the compound underlined accounted for 55–98% of the total sulfur evolved.

Table IX. Volatile Sulfur Compounds Released from Unamended and Sulfate-Treated Soils Incubated under Aerobic and Waterlogged Conditions for 60 Days[a]

Soil		Conditions[d]	Volatile sulfur compounds released	
Type[b]	OM (%)[c]		Unamended soils	Sulfate-treated soils
Belinda sil	2.16	A	None	None
		W	CH_3SCH_3	CH_3SCH_3
Marshall sicl	3.01	A	None	None
		W	CH_3SCH_3, COS	CH_3SCH_3, COS
Kenyon l	3.10	A	None	None
		W	CH_3SCH_3	CH_3SCH_3
Tama sicl	3.58	A	None	None
		W	CH_3SCH_3	CH_3SCH_3
Sharpsburg sicl	3.91	A	None	None
		W	CH_3SCH_3, CH_3SSCH_3	CH_3SCH_3, CH_3SSCH_3
Muscatine sicl	4.05	A	CH_3SCH_3	CH_3SCH_3
		W	CH_3SCH_3	CH_3SCH_3
Shelby l	4.43	A	None	None
		W	CH_3SCH_3	CH_3SCH_3
Moody sicl	4.77	A	None	None
		W	CH_3SCH_3, CH_3SH	CH_3SCH_3, CH_3SH

Soil[b]	OM[c]	[d]		
Grundy sicl	4.81	A	None	None
		W	CH_3SCH_3, COS, CS_2	CH_3SCH_3, COS, CS_2
Hayden sal	5.78	A	CH_3SCH_3	CH_3SCH_3
		W	CH_3SCH_3	CH_3SCH_3
Weller sil	6.42	A	CH_3SCH_3	CH_3SCH_3
		W	CH_3SCH_3, CH_3SH	CH_3SCH_3, CH_3SH
Luton sic	7.88	A	CH_3SCH_3	CH_3SCH_3
		W	CH_3SCH_3	CH_3SCH_3
Glencoe c	10.6	A	None	None
		W	CH_3SCH_3, COS	CH_3SCH_3, COS
Okoboji sicl	12.1	A	None	None
		W	CH_3SCH_3	CH_3SCH_3

[a]From Banwart and Bremner (1976a).
[b]Sil, silt loam; sicl, silty clay loam; l, loam; sal, sandy loam; sic, silty clay; c, clay.
[c]Organic matter content.
[d]Soil (20 g) or soil (20 g) treated with 8 mg of S as K_2SO_4 was incubated (30°C) under aerobic (A) or waterlogged (W) conditions for 60 days.

Table X. Amounts of Sulfur Volatilized from Unamended and Sulfate-Treated
Soils Incubated under Aerobic and Waterlogged Conditions for 60
Days (25 Soils) [a]

		Amount of sulfur volatilized			
		As ng S/g soil		As % of S in unamended soil	
Soils	Conditions[b]	Range	Average[c]	Range	Average[c]
Unamended	Aerobic	0–19	1 (7)	0–0.0040	<0.0001 (0.0016)
Unamended	Waterlogged	0–84	17 (30)	0–0.0451	0.0003 (0.0098)
Sulfate-treated	Aerobic	0–19	1 (6)	0–0.0039	<0.0001 (0.0013)
Sulfate-treated	Waterlogged	0–71	15 (27)	0–0.0338	0.0002 (0.0083)

[a]Banwart and Bremner (1976a).
[b]Unamended soil or soil treated with sulfate (400 µg sulfate S/g soil) was incubated (30°C) under aerobic or waterlogged conditions for 60 days.
[c]Figures in parentheses indicate average for soils that released volatile S compounds.

evidence that significant amounts of H_2S are emitted from soils under natural conditions. This is not very surprising because soils have a substantial capacity for sorption of H_2S (Smith *et al.,* 1973), and there is good reason to believe that H_2S produced by reduction of sulfate and other microbial processes in soils is rapidly converted to metallic sulfides (chiefly FeS) and that very little, if any, of this gas escapes to the atmosphere (see Ogata and Bower, 1965; Ponnamperuma, 1972; Kittrick, 1976; Ayotade, 1977). Harter and McLean (1965) were unable to detect evolution of H_2S during incubation of a flooded Toledo soil that accumulated large amounts of sulfide (>2000 µg/g soil) when incubated under waterlogged conditions. Swaby and Fedel (1973) used lead acetate paper to detect evolution of H_2S from 56 Australian soils incubated under aerobic and waterlogged conditions after treatment with sulfate. They concluded that none of these soils released H_2S under aerobic conditions and that only four released H_2S under waterlogged conditions. Calculations from their estimates of the rate of emission of H_2S from these four soils show that, even if this rate did not decrease with time, the sulfur evolved as H_2S in one year would not account for more than 0.1% of the sulfate sulfur added. Lovelock *et al.* (1972) reported detection of H_2S release from unamended soils incubated in sealed flasks, but details of their work have not been published. Sachdev and Chhabra (1974) detected evolution of H_2S when an Indian alluvial soil was incubated under aerobic or waterlogged conditions after treatment with ^{35}S-labeled sulfate, but calculations from their data show that only 0.4–0.6% of the sulfate sulfur added was volatilized as H_2S in 4 months under aerobic or waterlogged conditions. Bloomfield (1969) detected evolution of H_2S when soils

Table XI. Volatile Sulfur Compounds Evolved from a Sandy Soil Incubated under Aerobic or Waterlogged Conditions after Treatment with Different Forms of Sulfur[a]

Form of S added	Volatile S compounds evolved[b]						S evolved, calculated as % of S added
	H_2S	CH_3SH	CH_3SCH_3	CH_3SSCH_3	COS	CS_2	
Inorganic forms[c]	−	−	−	−	−	−	0
Organic forms:							
Methionine	−	+	+	+	−	−	49–50
Cystine	−	−	−	−	−	+	1–2
Cysteine	−	−	−	−	−	+	1–2
Sulfate esters	−	−	−	−	−	−	0
Thiocyanates and isothiocyanates	−	−	−	−	+	−	0.1–0.3
Plant proteins	−	+	+	+	+	+	2–3[e]
Plant residues	−	+[d]	+	+[e]	+	+[d]	0.1–0.3[e]
Animal manures	−	+	+	+	−	−	0.1–0.5[e]
Sewage sludges	−	+[d]	+[d]	+	+[d]	+	0.1–0.4[e]

[a]From Bremner (1977).
[b]+, detected; −, not detected.
[c]Sulfate, sulfide, sulfite, thiosulfate, tetrathionate, and elemental S.
[d]Not detected under aerobic conditions.
[e]92–99% of the S evolved was in the form of CH_3SH, CH_3SCH_3, or CH_3SSCH_3.

treated with large amounts of sulfate and plant material were incubated under anaerobic conditions (N_2 atmosphere), but the conditions of his experiments differed greatly from natural field conditions and it has not been demonstrated that the technique he adopted for detection of H_2S is specific for this gas. The same comments apply to recent work by Simán and Jansson (1976a), who reported detection of H_2S release from a Swedish sandy soil. In summary, therefore, it is difficult to find any justification in the literature for the assumption that large amounts of H_2S are released to the atmosphere through microbial transformations of sulfur compounds in soils. There seems very little doubt, however, that H_2S and other volatile sulfur compounds are released to the atmosphere through decomposition of plant residues and other sulfur-containing materials commonly added to soils (e.g., animal manures and sewage sludges) if these materials are not thoroughly incorporated (i.e., if they are allowed to decompose on the surfaces of soils). Also, studies of the factors affecting reduction of sulfate to sulfide in soils (see Ogata and Bower, 1965; Chaudhry and Cornfield, 1967a,b; Bloomfield, 1969; Abed, 1976; Simán and Jansson, 1976a) suggest that some of the H_2S produced by sulfate-reducing microorganisms in soils could escape to the atmosphere under certain conditions, e.g., when soils containing high levels of both sulfate and readily available organic matter become waterlogged and lack the cations needed to precipitate the large amounts of H_2S produced under these conditions.

3.1.2. Microbial Activity in Manures

One of the most serious problems associated with the increased use of feedlots and other animal confinement systems for livestock production is that some of the gases released through decomposition of the manures accumulating in these systems have very offensive odors and can have toxic effects (Miner, 1973). Work reviewed by Miner (1973) has indicated that these gases are largely low-molecular-weight volatiles produced during anaerobic decomposition of manures, but attempts to identify these volatiles have been limited by the lack of sensitive and specific methods for direct analysis of complex mixtures of gaseous compounds. There is evidence, however, that volatile sulfur compounds, such as mercaptans, sulfides, and disulfides, and volatile nitrogen compounds, such as ammonia and amines, contribute substantially to the offensive odors of animal manures (Burnett, 1969; Merkel *et al.,* 1969; Miner and Hazen, 1969; White, 1969; Stephens, 1971; White *et al.,* 1971; Young *et al.,* 1971; Bethea and Narayan, 1972; Elliott and Travis, 1973; Miner, 1973; Banwart and Bremner, 1975a; Smith, 1976; Smith *et al.,* 1977).

The first attempt to use sensitive gas chromatographic techniques for direct identification of the sulfur gases evolved from animal manures was by Elliott and Travis (1973), who used a gas chromatograph fitted with a flame photometric detector and with Porapak Q and Chromosorb 101 columns to identify gases evolved from a beef cattle manure and a composted manure. They detected five gaseous compounds and identified four of them as H_2S, COS, CO_2, and CH_4. Banwart and Bremner (1975a) subsequently used a flame photometric detector and a variety of columns for gas chromatographic identification and estimation of the sulfur gases released when nine different manures, including samples of beef cattle, dairy cattle, poultry, sheep, and swine wastes, were incubated under anaerobic and aerobic conditions. They found that all nine of the manures studied released H_2S, CH_3SH, CH_3SCH_3, CO_2, and CH_4 when incubated under anaerobic conditions but that only four released COS (Table XII). They also found that several manures released CH_3SSCH_3 and/or CS_2 when incubated under anaerobic conditions and that only trace amounts of one sulfur gas (CH_3SCH_3) could be detected in the gaseous products of decomposition of manures under aerobic conditions. Their analyses using Porapak Q and Chromosorb 101 columns indicated that the unidentified compound detected by Elliott and Travis (1973) was CH_3SH. Their work also showed that the amount of sulfur volatilized on incubation of manures under anaerobic conditions is very small and does not represent a significant fraction of the total sulfur in manures. This is illustrated by Table XIII, which shows the amounts of sulfur volatilized as H_2S, CH_3SH, CH_3SCH_3, CH_3SSCH_3, COS, and CS_2 on incubation of different manures in sealed bottles for 30 days at 23°C. Calculations from the data reported show that 70–97% of the sulfur volatilized was in the form of H_2S and CH_3SH.

Comparison of Tables VIII and XIII shows that, whereas a large proportion of the sulfur volatilized from manures under anaerobic conditions is in the form of H_2S, no trace of this gas could be detected in the gases evolved from soils treated with the same manures. This is another indication that H_2S produced by microbial activity in soils is rapidly sorbed by soil constituents and does not escape to the atmosphere.

Combined gas chromatography–mass spectrometry has recently been used to identify volatile sulfur compounds and other odoriferous compounds released from chicken manure that were retained by Porapak QS-Carbosieve B traps (Smith et al., 1977). Dimethyl sulfide, dimethyl disulfide, and various alcohols, ketones, esters, and carboxylic acids were identified when fresh chicken manure was incubated anaerobically (argon atmosphere), and CH_3SCH_3, CH_3SSCH_3, and a few nonsulfur compounds were detected when this manure was incubated under air. Unlike Banwart

Table XII. Sulfur Gases Detected in Atmospheres of Manures Incubated in Sealed Bottles[a,b]

Manure	Conditions[c]	S gases[d]					
		H_2S	COS	CS_2	CH_3SH	CH_3SCH_3	CH_3SSCH_3
Beef cattle I	A	+	−	−	+	+	−
	B	+	−	−	+	+	−
	C	+	−	−	+	+	−
Beef cattle II	A	+	+	+	+	+	+
	B	+	+	+	+	+	+
	C	+	−	+	+	+	−
Beef cattle III	A	+	−	−	+	+	−
	B	+	−	−	+	+	−
	C	+	−	−	+	+	−
Beef cattle IV	A	+	+	+	+	+	+
	B	+	+	+	+	+	+
	C	+	−	+	+	+	−
Beef cattle V	A	+	−	−	+	+	+
	B	+	−	−	+	+	+
	C	+	−	−	+	+	−
	D	+	−	−	+	+	+
Dairy cattle	A	+	+	−	+	+	−
	B	+	+	−	+	+	−
	C	+	−	−	+	+	−
	D	+	+	−	+	+	−
Poultry	A	+	−	−	+	+	+
	B	+	−	−	+	+	+
	C	+	−	−	+	+	+
	D	+	−	−	+	+	+
Sheep	A	+	−	−	+	+	+
	B	+	−	−	+	+	+
	C	+	−	−	+	+	+
	D	+	−	−	+	+	+
Swine	A	+	+	−	+	+	+
	B	+	+	−	+	+	+
	C	+	−	−	+	+	−
	D	+	+	−	+	+	+

[a]Banwart and Bremner (1975a).
[b]Manure was incubated at 23°C in a sealed 65-ml bottle, and the gas phase was analyzed for S gases at 2-day intervals for 30 days.
[c]A, fresh manure (20 g); B and C, homogenized manure (20 g); D, air-dry manure (3 g) treated with 20 ml of water. In A, B and D, the initial gas phase was air. In C, the initial gas phase was helium.
[d]+, detected; −, not detected.

Table XIII. Amounts of Sulfur Volatilized on Incubation of Manures in Sealed Bottles [a,b]

Manure	Amount of S volatilized (μg/20 g of homogenized manure)						
	As H_2S	As CH_3SH	As CH_3SCH_3	As CH_3SSCH_3	As COS	As CS_2	Total [c]
Beef I	6.41	0.62	0.23	0	0	0	7.26 (0.09)
Beef II	1.17	0.42	0.31	0.17	0.13	0.08	2.28 (0.02)
Beef III	4.21	0.43	0.25	0	0	0	4.89 (0.09)
Beef IV	2.47	1.06	0.17	0.24	0.04	0.31	4.29 (0.05)
Beef V	5.17	2.37	0.10	0.22	0	0	7.86 (0.07)
Dairy	0.48	0.34	0.20	0	0.11	0	1.13 (0.02)
Poultry	22.8	51.5	1.74	1.70	0	0	77.7 (0.31)
Sheep	31.1	16.7	1.59	0.13	0	0	49.5 (0.53)
Swine	1.28	1.35	0.36	0.08	0.24	0	3.31 (0.03)

[a] From Banwart and Bremner (1975a).
[b] Homogenized manure (20 g) was incubated (25°C) in a sealed 65-ml bottle for 30 days.
[c] Figures in parentheses indicate the total amount of S volatilized, calculated as a percentage of the total S in the manure sample incubated.

and Bremner (1975a), Smith *et al.* (1977) found that the amount of volatile organic sulfur released from manure incubated anaerobically was smaller than the amount released from manure incubated under air. It is difficult to account for this finding because it is well established that aeration of manures greatly reduces release of odorous gases (see Miner, 1973), and Smith *et al.* (1977) found that, whereas the odor of chicken manure incubated under argon was extremely disagreeable, the aroma of this manure incubated under air was not particularly intense or unpleasant. The most likely explanation is that the techniques adopted by Smith *et al.* (1977) failed to detect important odor-causing compounds such as H_2S.

Rasmussen (1974) reported that the amount of sulfur volatilized as CH_3SH, CH_3SCH_3, CH_3SSCH_3, and $CH_3CH_2SCH_2CH_3$ from an anaerobic dairy manure lagoon into an aerobic atmosphere was about 100 times the amount volatilized as H_2S, but no details of this work have been published.

3.1.3. Microbial Activity in Phyllospheres and Rhizospheres

Although it is well established that a variety of saprophytic and parasitic microorganisms inhabit the leaf, stem, bark, and root surfaces of higher plants (see Last and Deighton, 1965; Pugh and Buckley, 1971; Last and Warren, 1972), the possibility that these microorganisms may be sources of atmospheric sulfur has received little attention.

Several workers have detected release of volatile sulfur compounds from intact plants or detached plant parts (e.g., Peterson, 1914; Oaks *et al.*, 1964; Asher and Grundon, 1970; Lovelock *et al.*, 1972; Richmond, 1973; Rasmussen, 1974; Grundon, 1975). Lovelock *et al.* (1972) found that living intact leaves of cotton, oak, pine, and spruce emitted dimethyl sulfide and reported CH_3SCH_3 emission rates ranging from 2×10^{-12} to 43×10^{-12} grams per gram oven-dried tissue per hour (see also Rasmussen, 1974). They also found that the rates of emission of CH_3SCH_3 from senescent leaves were from 10 to 100 times greater than the corresponding rates for leaves that were physiologically active. This suggests that microorganisms may contribute significantly to emission of volatile sulfur compounds from leaves. The role of epiphytic microorganisms in release of dimethyl sulfide from senescent leaves clearly merits attention because it has been demonstrated that senescent leaves have large microbial populations (Hudson, 1971; Jensen, 1971) and exude large quantities of metabolites (Tukey, 1971).

Very little work has been reported relating to production of volatile sulfur compounds by microorganisms in plant rhizospheres. Jacq and Dommergues (1970) found that high-intensity illumination of maize plants grown in a waterlogged saline soil led to a marked increase in the numbers of sulfate-reducing bacteria and in sulfide levels in the rhizosphere. They deduced that high light intensity increases exudation by plant roots and thereby promotes production of H_2S by sulfate-reducers in the rhizosphere. However, Joshi and Hollis (1977) have reported evidence for destruction of H_2S by *Beggiatoa* (a gliding filamentous bacterium) in the rhizosphere of rice.

Simán and Jansson (1976b) found that, when soils were amended with sulfur as $K_2{}^{35}SO_4$, larger quantities of ^{35}S were volatilized from cropped than from uncropped soils. This suggests that rhizosphere microorganisms may promote volatilization of sulfur from soils.

3.2. Aquatic Microorganisms

Although models of the global sulfur cycle indicate that a substantial proportion of atmospheric sulfur is derived from H_2S emitted to the atmosphere through microbial activity in aquatic systems and suggest that most of this H_2S is produced by microbial reduction of sulfate in anoxic waters and sediments, no measurements of H_2S emissions from aquatic systems have been reported. The hypothesis that large amounts of H_2S are released to the atmosphere through microbial reduction of sulfate in aquatic environments was advanced by Conway (1943) and has been accepted in numerous discussions of the global sulfur cycle (e.g., Altshuller, 1958; Junge, 1960, 1963b; Eriksson, 1963; Kellogg *et al.*, 1972; Deevey, 1973; Hill, 1973; Moss, 1976). This hypothesis has been criticized (Lovelock *et al.*, 1972; Rasmussen, 1974) and defended (Hitchcock, 1975, 1976a,b, 1977a,b; Hitchcock *et al.*, 1977) in recent articles. Its validity is clearly open to question.

Current knowledge indicates that the key factors affecting bacterial reduction of sulfate in aquatic systems are the redox potential (an E_n below $+100$ mV is required) and the supply of sulfate and of assimilable organic matter or molecular hydrogen (Kuznetsov *et al.*, 1963; Wood, 1965; Rheinheimer, 1974; Kuznetsov, 1975). Salt concentration does not appear to be an important factor because freshwater strains of *D. desulfuricans* can adapt to seawater and vice versa (Ochynski and Postgate, 1963). However, seawater has a higher sulfate content than freshwater, suggesting that production of H_2S by bacterial reduction of sulfate is most extensive in sediments and coastal waters containing significant amounts of assimilable organic matter. Quammen *et al.* (1973) found that addition of pulp mill wastes to a shallow river in Florida led to an increase in the number of sulfate-reducing bacteria and the concentration of dissolved sulfide and to a decrease in the concentration of dissolved oxygen, presumably because organic substances in these wastes promoted microbial reduction of sulfate. Jorgensen (1977) recently demonstrated by radiotracer techniques that sulfate reduction can occur in oxic as well as anoxic layers of marine sediments. He also showed that bacterial sulfate reduction occurs in reduced microsites within sediment particles ranging from 50 to 200 μm in diameter. Morita (1974) has suggested that sulfate-reducing bacteria in deep-sea sediments are generally inactive due to low temperature, but there is evidence that some *Desulfovibrio* strains can grow at or below 0°C (Campbell and Postgate, 1965). Moreover, despite the low temperature of the Great Salt Lake in Utah (ZoBell, 1973), Grey and Jensen (1972) concluded that bacteriogenic emission of H_2S through sulfate reduction in this lake may be an important source of atmospheric sulfur in and near Salt Lake City.

Besides being produced through microbial reduction of sulfate, H_2S is formed in aquatic systems through microbial degradation of organic sulfur compounds (Wood, 1965; Rheinheimer, 1974). Heterotrophic bacteria are widely distributed in marine and estuarine sediments (Litchfield, 1973; Stevenson et al., 1974), and Stevenson et al. (1974) found that microorganisms capable of degrading cysteine are abundant in estuarine sediments.

Although H_2S production in aquatic environments through reduction of sulfate is believed to be more important than H_2S production through degradation of organic sulfur compounds (see Gunkel and Oppenheimer, 1963; Kuznetsov, 1975), few attempts have been made to assess the relative importance of these processes. Gunkel and Oppenheimer (1963) estimated that about 80% of the H_2S formed in sediments from the German coast of the North Sea was produced by sulfate reduction and that about 50% of the H_2S formed in sediments from a shallow Texas bay was produced by this process. Deuser (1970) concluded from a combination of ^{13}C, H_2S, and total CO_2 analyses of water from the Black Sea that 3–5% of the H_2S in this water was produced through degradation of organic sulfur compounds and that this percentage increased with depth.

Although it seems to have been assumed that much of the H_2S produced in aquatic environments is oxidized chemically to polysulfide and thiosulfate by dissolved O_2, there is evidence that microorganisms contribute to oxidation of H_2S (Rheinheimer, 1974; Gourmelon et al., 1977). *Thiobacillus*, *Beggiatoa*, and *Thiothrix* are probably the most important sulfur oxidizers in aquatic habitats (Rheinheimer, 1974), but it seems likely that oxidation of H_2S diffusing into the photic zone is effected largely by the photoautotrophic green and purple sulfur bacteria (Wood, 1965; Pfennig, 1967; Rheinheimer, 1972, 1974; Schlegel, 1974; Matheron and Baulaigue, 1976).

Hydrogen sulfide and other gases in aquatic environments can be transported to the water surface by molecular and turbulent diffusion (Hill, 1973) and in gas bubbles (Deevey, 1973). Measurements of the rate of oxidation of H_2S in seawater by Östlund and Alexander (1963) indicated that this gas is rapidly oxidized in the oxygenated surface layer of ocean water, and it has been concluded from their work that shallow stagnant water and waterlogged soils probably represent the major sites of biogenic emissions of H_2S to the atmosphere. However, recent work by Chen and Morris (1972) indicates that H_2S may persist in aquatic systems for several hours, and it seems possible that at least some of the H_2S produced in aqueous environments escapes to the atmosphere in gas bubbles (Deevey, 1973). Turbulent mixing of water within lakes and ponds is produced by wind action and by convective flow resulting from seasonal cooling of surface water, and the H_2S concentration in surface water has been found to increase during periods of seasonal overturn (Orr and Gaines, 1974) or during cold-front conditions (Brinkmann and Santos, 1974).

The following volatile organic sulfur compounds have recently been detected in aquatic systems: dimethyl sulfide, dimethyl disulfide, methyl mercaptan, and carbon disulfide. It seems almost certain that all four of these compounds are produced by aquatic microorganisms.

Dimethyl sulfide appears to be the predominant volatile organic sulfur compound in freshwater, seawater, and algal cultures (Lovelock *et al.,* 1972; Bechard, 1974; Bechard and Rayburn, 1974; Rasmussen, 1974), and its concentration in pond water has been found to exceed that of H_2S (Bechard, 1974). It seems likely that the CH_3SCH_3 detected in aquatic systems is derived from dimethyl-β-propiothetin (DMPT) because this compound is widely distributed in marine and freshwater algae (Challenger, 1951; Ackman *et al.,* 1966; Tocher *et al.,* 1966; Richmond, 1973), which constitute a large part of the biomass in aquatic environments, and it is well established that DMPT is decomposed biologically with the formation of CH_3SCH_3. Ishida (1968) found that DMPT accounted for as much as 55% of the total sulfur in *Gyrodinium cohnii,* a unicellular marine alga. Although dimethyl sulfide is produced by enzymatic cleavage of DMPT in algal cells (Challenger, 1951; Challenger *et al.,* 1957; Ishida, 1968; Kadota and Ishida, 1968), recent work indicates that production of dimethyl sulfide in aquatic environments results largely from bacterial decomposition of DMPT leaked from aged algal cells. Maloney (1963) found that malodorous organic compounds are retained within algal cells and are released through autolysis or mechanical disruption of these cells, and Jenkins *et al.* (1967) noted that maximum odor release from eutrophic surface waters occurs during bacterial decay of algae following an algal bloom. Bechard (1974) and Bechard and Rayburn (1974) found that dimethyl sulfide was produced largely in old bacterized algal cultures (see also Jenkins *et al.,* 1967). They also found that volatile organic sulfides were not produced in axenic cultures of algae belonging to the divisions

Chlorophyta, Xanthophyta, and Bacillariophyta but that dimethyl sulfide was produced in axenic cultures of unicellular and nonheterocystous filamentous blue-green algae.

Bechard (1974) and Bechard and Rayburn (1974) measured the concentrations of volatile sulfur compounds in a shallow eutrophic pond at regular intervals for 14 months. They found that dimethyl sulfide was the predominant volatile sulfide in this pond and that its concentration following bacterial putrefaction of an algal bloom was as high as 75,000 ng/liter. The maximal concentrations of H_2S and dimethyl disulfide observed were 1150 and 249 ng/liter, respectively.

Recent work by Lovelock (1974) indicates that CS_2 is widely distributed in coastal and ocean waters. The highest concentrations of this compound were detected in relatively stagnant bays, and the lowest concentrations were observed in waters off the tip of the Beara Peninsula in Ireland. Since a very high CS_2 concentration (29.5 ng/liter) was observed in anaerobic mud, Lovelock (1974) suggested that CS_2 in seawater originates under anaerobic conditions on the sea floor.

Rasmussen (1974) has reported detection of methyl mercaptan as well as dimethyl sulfide and dimethyl disulfide in pond water.

Although Hidy and Brock (1971) have suggested that significant amounts of SO_2 may be produced biologically in the oceans, there is no evidence that SO_2 is produced in, or emitted from, aquatic environments. Several workers have concluded from gas exchange models and laboratory studies that oceans act as sinks for atmospheric SO_2 and cannot be direct sources of this gas (e.g., Beilke and Lamb, 1974; Broecker and Peng, 1974; Liss and Slater, 1974; Hicks, 1976).

Several workers have reported detection of unpleasant odors during the seasonal overturn or following an algal bloom in aquatic habitats (e.g., Jenkins *et al.,* 1967; Kendler and Donagi, 1970; Brinkmann and Santos, 1974). Although this has been interpreted as evidence for evolution of H_2S or other volatile sulfides, there does not appear to be any unequivocal evidence that volatile sulfur compounds are released from aquatic systems. However, recent work has provided indirect evidence that tidal marshes emit "volatile sulfides" (Hitchcock *et al.,* 1977) and that oceans are sources of atmospheric CH_3SCH_3 (Maroulis and Bandy, 1977). Hitchcock *et al.* (1977) analyzed air over a North Carolina tidal marsh by a method designed to determine H_2S and other reactive sulfides and found that sulfide values obtained depended on the tidal cycle and were highest during low tide. Maroulis and Bandy (1977) analyzed air over two coastal sites in Virginia for CH_3SCH_3 and deduced from their results that the oceans may represent an important source of atmospheric dimethyl sulfide. They found that the atmospheric concentration of dimethyl sulfide at one site exhibited a diurnal pattern and concluded from lapse measurements that this pattern was determined primarily by low-level radiation inversions. They also suggested that the lower concentrations of CH_3SCH_3 observed at their other site resulted from destruction of this compound by smog from nearby Norfolk.

To summarize, although there is evidence that H_2S, CH_3SCH_3, CH_3SH, CH_3SSCH_3, and CS_2 are produced in aquatic environments through microbial transformations of sulfur compounds, there is no evidence that substantial amounts of H_2S or other volatile forms of sulfur are released to the atmosphere through microbial activity in the hydrosphere. There is an obvious and urgent need for research to assess emissions of volatile sulfur compounds from aquatic systems.

4. Microbial Sinks of Atmospheric Sulfur

It has been known for many years that soils can sorb sulfur dioxide and ammonia from the atmosphere, and recent work has shown that they also can sorb other gases identified as atmospheric constituents, including nitrogen dioxide, ethylene, and carbon monoxide (for references, see Bohn, 1972; Hidy, 1973; Smith *et al.*, 1973; Bohn and Miyamoto, 1974; Rasmussen *et al.*, 1975). This has stimulated research on the ability of soils to sorb sulfur gases identified as atmospheric constituents and has generated interest in the possibility that soil microorganisms may act as sinks for atmospheric sulfur gases.

It is now well established that soils have the ability to sorb SO_2 (Alway *et al.*, 1937; Fuhr *et al.*, 1948; Roberts and Koehler, 1965; Terraglio and Manganelli, 1966; Faller and Herwig, 1969/1970; Seim, 1970; Abeles *et al.*, 1971; Smith *et al.*, 1973; Payrissat and Beilke, 1975; Yee *et al.*, 1975; Bremner and Banwart, 1976; Ghiorse and Alexander, 1976), H_2S (Fuhr *et al.*, 1948; Carlson and Gumerman, 1966; Carlson and Leiser, 1966; Gumerman, 1968; Carlson *et al.*, 1970; Kanivets, 1970; Smith *et al.*, 1973; Bremner and Banwart, 1976), and CH_3SH (Carlson and Gumerman, 1966; Smith *et al.*, 1973; Bremner and Banwart, 1976). Recent work has shown that soils also have the capacity to sorb CH_3SCH_3, CH_3SSCH_3, COS, and CS_2 (Bremner and Banwart, 1976). However, very little is known about the factors affecting uptake of sulfur gases by soils or the role of soil microorganisms in uptake of these gases.

Recent gas chromatographic studies by Smith *et al.* (1973) and by Bremner and Banwart (1976) have shown that both air-dry and moist soils sorb H_2S, SO_2, CH_3SH, CH_3SCH_3, CH_3SSCH_3, COS, and CS_2, but do not sorb SF_6 (Table XIV). Moist soils sorb larger amounts of CH_3SCH_3, CH_3SSCH_3, COS, and CS_2 than do dry soils (Table XV), but the capacity of moist or air-dry soils for sorption of these gases is much smaller than their capacity for sorption of H_2S, SO_2, or CH_3SH (Table XVI).

Bremner and Banwart (1976) found that the ability of moist soils to sorb COS is considerably greater than their ability to sorb CH_3SCH_3,

Table XIV. Sorption of Different Sulfur Gases by Air-Dry and Moist Soils from Air Initially Containing 100 Parts/10^6 (v/v) of Sulfur Gas [a,b]

Soil [c]	Percent sorption of gas in 10 min							
	SO_2	H_2S	CH_3SH	CH_3SCH_3	CH_3SSCH_3	CS_2	COS	SF_6
Weller AD	100[d]	100[d]	100[d]	81	78	34	22	0
Weller M	100[d]	100[e]	90	17	15	7	68	0
Okoboji AD	100[d]	100[e]	100[e]	94	92	57	19	0
Okoboji M	100[d]	100[e]	90	20	26	14	29	0
Regina AD	100[d]	100[e]	90	84	83	53	17	0
Regina M	100[d]	100[e]	87	12	25	9	36	0
Harps AD	100[d]	100[e]	100[e]	72	76	38	8	0
Harps M	100[d]	100[e]	85	39	48	16	24	0
Average AD	100	100	98	83	82	46	17	0
Average M	100	100	88	22	29	12	39	0

[a]From Bremner and Banwart (1976).
[b]Sulfur gas was injected into sealed 65-ml bottle containing 5 g air-dry soil or 5 g air-dry soil moistened to 50% of its water-holding capacity.
[c]AD, air-dry soil; M, moist soil (50% of water-holding capacity).
[d]Sorption was complete in <3 min.
[e]Sorption was complete in 3–8 min.

CH_3SSCH_3, or CS_2, and they concluded from experiments with sterilized soils that microorganisms are partly responsible for the sorption of CH_3SCH_3, CH_3SSCH_3, COS, and CS_2 by moist soils (Table XV). Support for this conclusion was provided by their finding that the rate of sorption of these sulfur gases by moist soils increases with time (Table XVII), suggesting that exposure of soils to CH_3SCH_3, CH_3SSCH_3, COS, or CS_2 promotes growth of microorganisms able to utilize these gases.

Abeles *et al.* (1971) found that autoclaved soil samples sorbed considerably less SO_2 than did similar samples that were not autoclaved, suggesting that soil microorganisms have the capacity to sorb SO_2. Smith *et al.* (1973), however, found that neither the rate nor the extent of sorption of SO_2 by samples of air-dry or moist soils was significantly affected by steam sterilization of these samples, and they concluded that microorganisms play little, if any, part in sorption of SO_2 by soils. Similarly, Ghiorse and Alexander (1976) found that sterilization of a Lima loam by autoclaving or exposure to gamma-irradiation did not affect uptake of SO_2 by this soil and concluded that viable microorganisms are not directly involved in removal of SO_2 from air by soils. They suggested, however, that microorganisms may participate passively in sorption of

Table XV. Amounts of Different Sulfur Gases Sorbed by Air-Dry and Moist Soils from Air Initially Containing 500 Parts/10^6 (v/v) of Sulfur Gas [a,b]

	Amount of gas sorbed in 15 days (μg/g soil) [d]				
Soil [c]	CH_3SCH_3	CH_3SSCH_3	CS_2	COS	SF_6
Weller AD	23 (21)	102 (98)	14	123	0 (0)
Weller M	316 (12)	155 (43)	71 (14)	2060 (21)	0 (0)
Okoboji AD	33 (31)	101	18 (20)	72	0 (0)
Okoboji M	64 (8)	130 (9)	35 (5)	1688 (13)	0 (0)
Regina AD	41	140	18 (16)	87	0
Regina M	409 (10)	172 (35)	57 (14)	1709 (25)	0 (0)
Harps AD	31 (28)	119	21 (23)	106	0 (0)
Harps M	442 (10)	306 (9)	104 (13)	2340 (77)	0 (0)
Average AD	32	116	18	97	0
Average M	308 (10)	191 (24)	67 (12)	1950 (34)	0 (0)

[a]From Bremner and Banwart (1976).
[b]Sulfur gas was injected into sealed 65-ml bottle containing 1 g of air-dry soil or 1 g of air-dry soil moistened to 50% of its water-holding capacity.
[c]AD, air-dry soil; M, moist soil (50% of water-holding capacity).
[d]Figures in parentheses indicate amount of gas sorbed by sterilized soil.

Table XVI. Capacities of Soils for Sorption of SO_2, H_2S, and CH_3SH [a]

	Sorption capacity (mg gas/g soil)					
	Air-dry soil			Moist soil [b]		
Soil	SO_2	H_2S	CH_3SH	SO_2	H_2S	CH_3SH
Astoria	8.9	62.9	9.8	37.1	51.6	8.1
Weller	15.3	61.0	28.5	31.9	58.1	21.4
Okoboji	10.2	52.9	21.2	31.4	44.5	19.7
Thurman	1.1	15.4	2.4	9.3	11.0	2.2
Regina	13.3	65.2	21.4	50.4	62.5	18.0
Harpster	10.2	46.6	32.1	66.8	40.6	19.7
Average	9.8	50.7	19.2	37.8	44.7	14.9

[a]From Smith et al. (1973).
[b]50% of water-holding capacity.

Table XVII. Effect of Previous Exposure of Moist Soils to CH_3SCH_3,
CH_3SSCH_3, CS_2, and COS on Rate of Sorption of These Gases [a,b]

Soil	Gas	Time (hr) required for complete sorption of gas by soil from air initially containing 100 parts/10^6 (v/v) of gas		
		1st exposure	2nd exposure	3rd exposure
Weller	CH_3SCH_3	150	100	95
	CH_3SSCH_3	68	60	55
	CS_2	340	310	300
	COS	0.8	0.7	0.6
Harps	CH_3SCH_3	45	24	19
	CH_3SSCH_3	53	43	21
	CS_2	100	95	90
	COS	1.8	1.6	1.5

[a]From Bremner and Banwart (1976).
[b]Sulfur gas was injected into sealed 65-ml bottle containing moist (50% of water-holding capacity) soil (5 g air-dry material) in an amount such that the initial concentration of this gas in the air in the bottle was 100 parts/10^6 (v/v). This injection was repeated when gas chromatographic analysis of the gas phase showed no trace of sulfur gas.

SO_2 by providing organic material to react with this gas and may be implicated in removal of the sorption products.

Craker and Manning (1974) used $^{35}SO_2$ and radiotracer techniques to study uptake of SO_2 by bacteria and fungi isolated from soil. They found that, of the microorganisms tested, only the fungi appeared capable of SO_2 uptake, an *Alternaria* isolate being the most efficient (Table XVIII). They concluded that soil fungi may function as biological sinks for atmospheric SO_2.

Carlson and Gumerman (1966) found that passage of air containing H_2S and CH_3SH through columns of soil led to removal of these gases and attributed this to microbial activity (see also Carlson and Leiser, 1966; Gumerman, 1968). But Smith *et al.* (1973) found that neither the rate nor the extent of sorption of H_2S or CH_3SH by samples of air-dry or moist soils was significantly affected by steam sterilization of these samples, and they concluded that microorganisms are not involved to any appreciable extent in sorption of H_2S or CH_3SH by soils.

Very little is known about the biological or nonbiological processes by which soils sorb sulfur gases because attempts to study their mechanisms have been greatly hindered by the lack of sensitive and specific methods for identification and estimation of different forms of sulfur in soils. It seems likely, however, that sorption of SO_2 by soils involves formation of sulfite and sulfate (Faller and Herwig, 1969/1970; Seim, 1970; Ghiorse and Alexander, 1976) and that sorption of H_2S involves formation of metallic sulfides and elemental sulfur (Gumerman, 1968).

Bremner and Banwart (1976) found that sorption of COS by moist soils was accompanied by release of small amounts of CS_2, but they could not detect any release of sulfur gases during sorption of H_2S, SO_2, CH_3SH, CH_3SCH, CH_3SSCH, or CS_2. Experiments with sterilized soils indicated that the release of CS_2 observed when moist unsterilized soils were exposed to air containing COS resulted from metabolism of COS by soil microorganisms.

In summary, there is good evidence that soils have the capacity to sorb SO_2, H_2S, CH_3SH, CH_3SCH_3, CH_3SSCH_3, COS, and CS_2 and that microorganisms are at least partly responsible for the ability of soils to sorb some of these gases. But the mechanisms by which soils sorb sulfur gases remain obscure, and it has not been demonstrated that soil microorganisms are important biological sinks for atmospheric sulfur compounds. It seems reasonable to conclude, however, that soils may represent important natural sinks for atmospheric SO_2, H_2S, and CH_3SH because both air-dry and moist soils sorb these gases very rapidly by processes that do not involve microorganisms (see Figs. 1 and 2) and have very substantial capacities for sorption of these gases (see Table XVI).

Although there is abundant evidence that plants have the capacity to take up SO_2 from the atmosphere (see Fried, 1948; Faller, 1971; Faller *et al.*, 1970; Hill, 1971; Babich and Stotzky, 1972; Hill, 1973; Simán and

Table XVIII. Uptake of SO_2 by Soil Fungi [a,b]

Species	Trial (cpm/g fresh weight)		Average
	1 [c]	2 [d]	
Alternaria sp.	17,440	29,547 [e]	21,476
Penicillium sp.	5,370	—	5,370
Fusarium oxysporum	2,560	9,589	4,903
Chaetomium sp.	4,032	—	4,032
Rhizoctonia solani	4,487	1,263	3,412
Penicillium sp.	1,539 [f]	6,283	3,120
Trichoderma sp.	2,249	726	1,921
Aspergillus sp.	—	1,388	1,388
Rhizopus sp.	1,242	816	1,100
Colletotrichum sp.	688	436	604
Yeast	402 [f]	68	235

[a]From Craker and Manning (1974).
[b]Colonies of fungi growing on agar plates were exposed to $^{35}SO_2$ for 3 hr. After exposure, the fungi were transferred from agar plates to planchets and cpm/g fresh weight determined using a radionuclide detector.
[c]Initial concentration of $^{35}SO_2$ = 4.0 μCi/liter, SO_2 = 1.09 mg/m³. Each number in the column represents the mean of samples taken from six agar plates, except where indicated.
[d]Initial concentration of $^{35}SO_2$ = 3.8 μCi/liter, SO_2 = 1.04 mg/m³. Each number in the column represents the mean of samples taken from three agar plates, except where indicated.
[e]Mean of two samples.
[f]Mean of three samples.

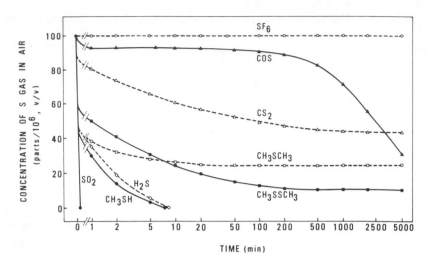

Figure 1. Sorption of different sulfur gases by air-dry soil from air initially containing 100 parts/10^6 (v/v) of sulfur gas (Bremner and Banwart, 1976). Gas was injected into sealed 65-ml bottle containing 5 g of air-dry Harps soil.

Jansson, 1976b), no attempts appear to have been made to determine whether microorganisms on the surfaces of plants can act as sinks for SO_2 or other atmospheric sulfur gases. Brinkmann and Santos (1974), however, have suggested that a substantial amount of the H_2S emitted from

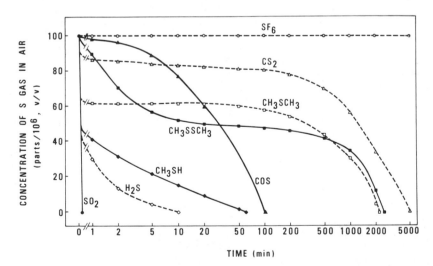

Figure 2. Sorption of different sulfur gases by moist soil from air initially containing 100 parts/10^6 (v/v) of sulfur gas (Bremner and Banwart, 1976). Gas was injected into sealed 65-ml bottle containing 5 g of air-dry Harps soil moistened to 50% of its water-holding capacity.

Amazonian flood plains is taken up by microbial epiphytes in the surrounding forests. Also, Rasmussen and Hutton (1972) have speculated that the microflora in the phylloplane and rhizosphere of some plant communities, particularly tropical forests, may be important global sinks for volatile organic substances in the atmosphere.

The possibility that microorganisms in aquatic systems have the capacity to remove sulfur gases from air does not appear to have been explored.

5. Conclusions

To evaluate present knowledge concerning the role of microorganisms in the atmospheric sulfur cycle, it is necessary to review current information concerning the forms and amounts of sulfur in the atmosphere.

Until very recently, it was generally assumed that atmospheric sulfur is largely, if not entirely, in the form of SO_2, H_2S, and sulfates. Considerable doubt now exists, however, concerning the reliability of most of the published data concerning the forms and amounts of sulfur in air (see Junge, 1972; Kellogg et al., 1972; Adams, 1976; Tanner and Newman, 1976; Urone, 1976), and Adams (1976) recently concluded that most of the methods for determination of sulfur compounds in air "must be considered in a state of evolution" and that "information on the accuracy, precision, and reliability of these methods is very limited." The methods that have been used for H_2S analysis of air are particularly suspect (see Kellogg et al., 1972; Urone, 1976), and most of the published information concerning the H_2S content of air is probably unreliable. According to Rasmussen (1974), there are no reports of successful attempts to detect H_2S in remote, unpopulated areas.

Table XIX lists the sulfur gases currently recognized as components of tropospheric air and reports estimates of the atmospheric concentrations of these gases. Although an attempt was made to find SO_2 and H_2S data for unpolluted air, it is very difficult to identify such data, and it seems unlikely that any of the atmospheric concentration values reported in Table XIX are representative background values.

It is noteworthy that, of the six sulfur gases listed in Table XIX, four (CH_3SCH_3, COS, CS_2, and SF_6) have only recently been detected. Crutzen (1976) has emphasized the need for an inventory of all possible releases of sulfur compounds to the atmosphere and has pointed out that small amounts of SF_6 and SO_2F_2 are released to the atmosphere from industrial sources. As shown in Table XIX, Singh (1977) has recently detected SF_6 in air, but SO_2F_2 has not thus far been detected.

Table XIX. Atmospheric Concentrations of Sulfur Gases

Gas	Concentration[a]	References
SO_2	0.2 ppb (average)	Robinson and Robbins (1970a)
	3.5 ppb (average)	Altshuller (1973)
	From less than 0.3 ppb to about 2 ppb	Junge (1972), Granat et al. (1976)
	Less than 5 ppb	Stevens et al. (1977)
H_2S	0.2 ppb (average)	Robinson and Robbins (1970a)
	From 0.06 ppb to several ppb	Junge (1972), Granat et al. (1976)
	20–60 ppt	Natusch et al. (1972)
	Less than 1 ppb	Stevens et al. (1977)
COS	0.20–0.24 ppb	Hanst et al. (1975)
	0.40–0.56 ppb	Sandalls and Penkett (1977)
CS_2	0.07–0.37 ppb	Sandalls and Penkett (1977)
CH_3SCH_3	58 ppt (average)	Maroulis and Bandy (1977)
SF_6	0.24 ppt (average)	Singh (1977)

[a]ppb, parts per billion (volume); ppt, parts per trillion (volume).

The recent detection of CH_3SCH_3, COS, and CS_2 in air is not surprising because work discussed in Section 3 has shown that these gases are released to the atmosphere through microbial transformations of sulfur compounds in soils and manures and that CH_3SCH_3 and CS_2 are produced by aquatic microorganisms. Although it seems likely that microorganisms are the major sources of these atmospheric gases, the possibility that significant amounts of CH_3SCH_3, COS, or CS_2 may be emitted to the atmosphere by industrial and other nonbiological processes has received little attention. This possibility cannot be ignored because COS has been detected in emissions from volcanoes and fumaroles (Stoiber et al., 1971) and in products emitted from coal-to-energy conversion plants (Gibson et al., 1974). Since recent work discussed in Section 3 has shown that CH_3SH and CH_3SSCH_3 are released to the atmosphere through microbial activity in soils and manures, there seems little doubt that these compounds will be detected in further research to identify atmospheric sulfur gases.

The recent finding that COS and CS_2 are emitted by microorganisms in soils and manures and can be detected in air is of considerable interest because Crutzen (1976) has suggested that these gases may contribute significantly to the nonvolcanic background sulfate layer in the stratosphere and may affect the earth's radiation balance.

Although literature discussed in Section 2 indicates that some microorganisms can produce SO_2, it has not been demonstrated that this gas is released to the atmosphere from natural ecosystems through the activity of terrestrial or aquatic microorganisms. It is generally assumed, however, that a significant amount of the SO_2 in air originates through atmospheric oxidation of H_2S produced by terrestrial and aquatic microorganisms (Junge, 1960; Robinson and Robbins, 1968; Kellogg et al., 1972). But recent work discussed in Section 3 indicates that this assumption may be invalid, and very little is known about the rate and mechanism of oxidation of H_2S in air (Cadle and Ledford, 1966; Kellogg et al., 1972; Penkett, 1972; Cox and Sandalls, 1974; Hales et al., 1974; Levy, 1974; Cadle, 1976).

Work reviewed in Section 3 suggests that most of the CH_3SCH_3 in air originates through microbial degradation of methionine or dimethyl-β-propiothetin and that most of the CS_2 in air originates through microbial degradation of cysteine or cystine. Also, it seems likely that at least some of the COS in air is produced by microbial decomposition of thiocyanates and isothiocyanates in plant materials.

Work discussed in Section 4 indicates that terrestrial microorganisms have no capacity for uptake of SF_6 but can act as sinks for other sulfur gases listed in Table XIX. It has not been demonstrated, however, that terrestrial or aquatic microorganisms have the capacity to remove SO_2, H_2S, CH_3SCH_3, COS, or CS_2 from air containing atmospheric concentrations of these gases, and current knowledge does not permit assessment of the global importance of microorganisms as sinks for atmospheric sulfur gases. But it is evident from recent work discussed in Sections 3 and 4 that there is good reason to question the current belief that soils are important sources of atmospheric H_2S and SO_2. It seems much more likely that soils are natural sinks for these gases because they sorb H_2S and SO_2 very rapidly by nonbiological mechanisms and appear to have very little, if any, capacity to emit these gases.

As noted previously, current concepts of the global sulfur cycle are based on the assumption that large amounts of sulfur are released to the atmosphere through microbial activity in the biosphere. This is illustrated by Table XX, which reports estimates by various workers of biogenic emissions of sulfur to the atmosphere. It is important to note that these estimates resulted from attempts to balance global sulfur budgets and have not been validated by research to assess biogenic emissions of sulfur compounds.

Until very recently, it was generally assumed that most, if not all, of the sulfur emitted to the atmosphere through microbial activity in the biosphere is in the form of H_2S. This assumption has been questioned by Lovelock et al. (1972), and they have suggested that most of the sulfur emitted to the atmosphere through biological processes is in the form of dimethyl sulfide (see also Rasmussen, 1974). This hypothesis has been criticized by Hitchcock (1975, 1976b, 1977a), and recent calculations have

Table XX. Estimated Biogenic Emissions of Sulfur

Reference	Estimated emissions (tons S/yr, × 10⁶)		
	Terrestrial	Marine	Total
Eriksson (1963)	110	170	380
Junge (1963b)	70	160	230
Robinson and Robbins (1968, 1970a)	70	30	100
Robinson and Robbins (1970b)	68	30	98
Kellogg *et al.* (1972)	—	—	89
Friend (1973)	58	48	106
Granat *et al.* (1976)	5	27	32

suggested that biogenic emissions of sulfur as CH_3SCH_3 represent considerably less than 10% of the estimated total biogenic emission of sulfur (Liss and Slater, 1974; Hitchcock, 1975, 1977a; Maroulis and Bandy, 1977). But these calculations are based on very limited data for biogenic emissions of CH_3SCH_3, and it is important to note that lack of H_2S emission data has prevented similar calculations of biogenic emissions of H_2S. Moreover, there is a clear need for consideration of the possibility that the total biogenic emission of sulfur has been grossly overestimated. In our opinion, the evidence for significant biogenic emissions of dimethyl sulfide and other volatile sulfur compounds is much stronger than the evidence for significant biogenic emissions of H_2S.

Sulfur isotope ratios have been used to identify and estimate atmospheric sulfur derived from biological sulfate reduction on the assumption that biogenic sulfur is depleted in the [34]S isotope (see Grey and Jensen, 1972; Kellogg *et al.*, 1972; Hitchcock, 1976b). According to Deevey (1973), however, this technique does not identify biogenic sulfur unambiguously because much industrial sulfur is [34]S-depleted and fertilizer sulfate is [34]S-enriched.

Although numerous models of the global sulfur cycle have been proposed, it is evident from this review that we lack the information needed to develop acceptable models. The most obvious need for realistic modeling and for assessment of the role of microorganisms in the global sulfur cycle is an extensive program of research to determine the forms and amounts of sulfur emitted from terrestrial and aquatic systems. The major hindrance to such research is the lack of highly sensitive and specific methods of determining volatile forms of sulfur in air. As recently noted by Granat *et al.* (1976), there is a clear need for research to assess emissions of H_2S and volatile organic sulfur compounds from soils and forests in different areas and from open oceans and coastal waters. In our view, there is a special need for studies of fluxes of volatile sulfur compounds at land–air and water–air interfaces because such studies are much more meaningful and much easier to interpret than studies of the concentrations of sulfur gases in air over land and oceans. And it is evident that, to develop a realistic model of the global sulfur cycle, we must acquire a more thorough understanding of the atmospheric chemistry of sulfur compounds.

Acknowledgments

We thank Pergamon Press, Oxford, for permission to reproduce material from *Soil Biology and Biochemistry* (Figs. 1 and 2, Tables III–X, XIV, XV, XVII); the American Society of Agronomy, Crop Science Society of America, and Soil Science Society of America for permission to reproduce material from *Journal of Environmental Quality* (Tables XII and XIII); and The Williams and Wilkins Co., Baltimore, for permission to reproduce material from *Soil Science* (Table XVI). Preparation of this article was supported in part by the U.S. Energy Research and Development Administration under contract E(11-1)-2530.

References

Abed, M. A. H. F., 1976, Sulfate reduction in poorly-drained soils as influenced by organic matter and soil texture, *Beitr. Trop. Subtrop. Landwirtsch. Veterinarmed.* **14**:89.

Abeles, F. B., Craker, L. E., Forrence, L. E., and Leather, G. R., 1971, Fate of air pollutants: removal of ethylene, sulfur dioxide, and nitrogen dioxide by soil, *Science* **173**:914.

Ackman, R. G., Tocher, C. S., and McLachlan, J., 1966, Occurrence of dimethyl-β-propiothetin in marine phytoplankton, *J. Fish. Res. Board Can.* **23**:357.

Adams, D. F., 1976, Sulfur compounds, in: *Air Pollution*, vol. III, 3rd ed. (A. C. Stern, ed.), pp. 213–257, Academic Press, New York.

Alexander, M., 1974, Microbial formation of environmental pollutants, *Adv. Appl. Microbiol.* **18**:1.

Altshuller, A. P., 1958, Natural sources of gaseous pollutants in the atmosphere, *Tellus* **10**:479.

Altshuller, A. P., 1973, Atmospheric sulfur dioxide and sulfate: distribution of concentration at urban and nonurban sites in the United States, *Environ. Sci. Technol.* **7**:709.

Alway, F. J., Marsh, A. W., and Methley, W. J., 1937, Sufficiency of atmospheric sulfur for maximum crop yields, *Soil Sci. Soc. Am. Proc.* **2**:229.

Asami, T., and Takai, Y., 1963, Formation of methyl mercaptan in paddy soils II., *Soil Sci. Plant Nutr. (Tokyo)* **9**:23.

Asher, C. J., and Grundon, N. J., 1970, Volatile losses of mineral constituents from forage plants, in: *Proceedings of the XI International Grassland Congress*, pp. 329–332, University of Queensland Press, Brisbane, Australia.

Ashworth, J., Briggs, G. G., and Evans, A. A., 1975, Field injection of carbon disulphide to inhibit nitrification of ammonia fertiliser, *Chem. Ind.* **1975**:749.

Ayotade, K. A., 1977, Kinetics and reactions of hydrogen sulfide in solution of flooded rice soils, *Plant Soil* **46**:381.

Babich, H., and Stotzky, G., 1972, Ecologic ramifications of air pollution, *Soc. Automot. Eng. Trans.* **81**:1955.

Babich, H., and Stotzky, G., 1974, Air pollution and microbial ecology, *CRC Crit. Rev. Environ. Control* **4**:353.

Banwart, W. L., and Bremner, J. M., 1974, Gas chromatographic identification of sulfur gases in soil atmospheres, *Soil Biol. Biochem.* **6**:113.

Banwart, W. L., and Bremner, J. M., 1975a, Identification of sulfur gases evolved from animal manures, *J. Environ. Qual.* **4**:363.

Banwart, W. L., and Bremner, J. M., 1975b, Formation of volatile sulfur compounds by microbial decomposition of sulfur-containing amino acids in soils, *Soil Biol. Biochem.* **7**:359.

Banwart, W. L., and Bremner, J. M., 1976a, Volatilization of sulfur from unamended and sulfate-treated soils, *Soil Biol. Biochem.* **8**:19.

Banwart, W. L., and Bremner, J. M., 1976b, Evolution of volatile sulfur compounds from soils treated with sulfur-containing organic materials, *Soil Biol. Biochem.* **8**:439.

Barjac, H. de., 1952, Contribution a l'étude du métabolisme des acides aminés soufrés, et spécialement de la méthionine dans le sol, *Ann. Inst. Pasteur* **82**:623.

Bechard, M. J., 1974, Emission of volatile organic sulfides by freshwater algae, M.S. Thesis, Washington State University, Pullman.

Bechard, M. J., and Rayburn, W. R., 1974, Emission of volatile sulfur compounds from freshwater algae, *J. Phycol. (Suppl.)* **10**:10.

Bethea, R. M., and Narayan, R. S., 1972, Identification of beef cattle feedlot odors, *Trans. A.S.A.E.* **15**:1135.

Beilke, S., and Lamb, D., 1974, On the absorption of SO_2 in ocean water, *Tellus* **26**:268.

Birkinshaw, J. H., Findlay, W. P. K., and Webb, R. A., 1942, Biochemistry of the wood-rotting fungi. 3. The production of methyl mercaptan by *Schizophyllum commune* Fr., *Biochem. J.* **36**:526.

Bloomfield, C., 1969, Sulphate reduction in waterlogged soils, *J. Soil Sci.* **20**:207.

Bohn, H. L., 1972, Soil absorption of air pollutants, *J. Environ. Qual.* **1**:372.

Bohn, H. L., and Miyamoto, S., 1974, Soil as a sorbent and filter of waste gases, in: *Proceedings of the International Conference on Land for Waste Management* (B. P. Warkentin, chairperson), pp. 104–114, Department of the Environment, National Research Council of Canada, Ottawa.

Bremner, J. M., 1977, Role of organic matter in volatilization of sulfur and nitrogen from soils, in: *Proceedings of Symposium on Soil Organic Matter Studies*, vol. II, Braunschweig, Federal Republic of Germany, Sept. 6–10, 1976, pp. 229–240, International Atomic Energy Agency, Vienna.

Bremner, J. M., and Banwart, W. L., 1974, Identifying volatile S compounds by gas chromatography, *Sulphur Inst. J.* **10**:6.

Bremner, J. M., and Banwart, W. L., 1976, Sorption of sulfur gases by soils, *Soil Biol. Biochem.* **8**:79.

Bremner, J. M., and Bundy, L. G., 1974, Inhibition of nitrification in soils by volatile sulfur compounds, *Soil Biol. Biochem.* **6**:161.

Brinkmann, W. L. F., and Santos, U. D. M., 1974, The emission of biogenic hydrogen sulfide from Amazonian floodplain lakes, *Tellus* **26**:261.

Brody, S. S., and Chaney, J. E., 1966, The application of a specific detector for phosphorus and for sulfur compounds—sensitive to subnanogram quantities, *J. Gas Chromatogr.* **4**:42.

Broecker, W. S., and Peng, T.-H., 1974, Gas exchange rates between air and sea, *Tellus* **26**:21.

Burnett, W. E., 1969, Air pollution from animal wastes, *Environ. Sci. Technol.* **3**:744.

Cadle, R. D., 1976, The photo-oxidation of hydrogen sulphide and dimethyl sulphide in air, *Atmos. Environ.* **10**:417.

Cadle, R. D., and Ledford, M., 1966, Reaction of ozone with hydrogen sulfide, *Air Water Pollut.* **10**:25.

Campbell, L. L., and Postgate, J. R., 1965, Classification of the sporeforming sulphate-reducing bacteria, *Bacteriol. Rev.* **29**:359.

Carlson, D. A., and Gumerman, R. C., 1966, Hydrogen sulfide and methyl mercaptan removal with soil columns, in: *Proceedings of the 21st Industrial Waste Conference,* pp. 177–191, Purdue University, Lafayette, Ind.

Carlson, D. A., and Leiser, C. P., 1966, Soil beds for the control of sewage odors, *J. Water Pollut. Control Fed.* **38**:829.

Carlson, D. A., Leiser, C. P., and Gumerman, R., 1970, *The Soil Filter: A Treatment Process for Removal of Odorous Gases,* University of Washington Final Report to Federal Water Pollution Control Administration WP 00883-03 (February 1970).

Challenger, F., 1951, Biological methylation, *Adv. Enzymol.* **12**:429.

Challenger, F., and Charlton, P. T., 1947, Studies on biological methylation. Part X. The fission of the mono- and di-sulphide links by moulds, *J. Chem. Soc.* **1947**:424.

Challenger, F., and Greenwood, D., 1949, Sulphur compounds of the genus *Allium:* detection of *n*-propylthiol in the onion. The fission and methylation of diallyl disulphide in cultures of *Scopulariopsis brevicaulis, Biochem. J.* **44**:87.

Challenger, F., and North, H. E., 1934, The production of organo-metalloidal compounds by microorganisms. Part II. Dimethyl selenide, *J. Chem. Soc.* **1934**:68.

Challenger, F., Bywood, R., Thomas, P., and Hayward, B. J., 1957, Studies on biological methylation. XVII. The natural occurrence and chemical reactions of some thetins, *Arch. Biochem. Biophys.* **69**:514.

Chaudhry, I. A., and Cornfield, A. H., 1967a, Effect of moisture content during incubation of soil treated with organic materials on changes in sulphate and sulphide levels, *J. Sci. Food Agric.* **18**:38.

Chaudhry, I. A., and Cornfield, A. H., 1967b, Effect of temperature of incubation on sulphate levels in aerobic soils and sulphide levels in anaerobic soils, *J. Sci. Food Agric.* **18**:82.

Chen, J. Y., and Morris, J. C., 1972, Kinetics of oxidation of aqueous sulfide by O_2, *Environ. Sci. Technol.* **6**:529.

Clarke, P. H., 1953, Hydrogen sulphide production by bacteria, *J. Gen. Microbiol.* **8**:397.

Coley-Smith, J. R., and King, J. E., 1969, The production by species of *Allium* of alkyl sulfides and their effect on germination of sclerotia of *Sclerotium cepivorum* Berk., *Ann. Appl. Biol.* **64**:289.

Conway, E. J., 1943, Mean geochemical data in relation to oceanic evolution, *Proc. R. Ir. Acad. Sect. A* **48**:119.

Cox, R. A., and Sandalls, F. J., 1974, The photo-oxidation of hydrogen sulphide and dimethyl sulphide in air, *Atmos. Environ.* **8**:1269.

Craker, L. E., and Manning, W. J., 1974, SO_2 uptake by soil fungi, *Environ. Pollut.* **6**:309.

Crutzen, P. J., 1976, The possible importance of CSO for the sulfate layer of the stratosphere, *Geophys. Res. Lett.* **3**:73.

Deevey, E. S., Jr., 1973, Sulfur, nitrogen, and carbon in the biosphere, in: *Carbon and the Biosphere* (G. M. Woodwell and E. V. Pecan, eds.), pp. 182–190, U.S. Atomic Energy Commission, Washington, D.C.

Deuser, W. G., 1970, Carbon-13 in Black Sea waters and implications for the origin of hydrogen sulfide, *Science* **168**:1575.

Dittrich, H. H., and Staudenmayer, T., 1970. Über die Zusammenhänge zwischen der Sulfit-Bildung und der Schwefelwasserstoff-Bildung bei *Saccharomyces cerevisiae, Zentralbl. Bakteriol. Parasitenk.* Abt. 2 **124**:113.

Dott, W., and Trüper, H. G., 1976, Sulfite formation by wine yeasts III. Properties of sulfite reductase, *Arch. Microbiol.* **108**:99.

Drews, B., Baerwald, G., and Niefind, H. J., 1970, Determination of some volatile sulfur compounds of primary and secondary fermentation by gas chromatography, *Eur. Brew. Conv., Proc. Congr. 1969* **12**:419.

Elliott, L. F., and Travis, T. A., 1973, Detection of carbonyl sulfide and other gases emanating from beef cattle manure, *Soil Sci. Soc. Am. Proc.* **37**:700.

Eriksson, E., 1963, The yearly circulation of sulfur in nature, *J. Geophys. Res.* **68**:4001.

Eriksson, E., and Rosswall, T., 1976, Man and biogeochemical cycles: impacts, problems and research needs, in: *Nitrogen, Phosphorus, and Sulphur—Global Cycles* (B. H. Svensson and R. Söderlund, eds.), SCOPE Report 7, Ecological Bulletins **22**:11.

Faller, N., 1971, Effects of atmospheric SO_2 on plants, *Sulphur Inst. J.* **7**:5.

Faller, N., and Herwig, K., 1969/1970, Untersuchungen über die SO_2-Oxydation in verschiedenen Böden, *Geoderma* **3**:45.

Faller, N., Herwig, K., and Kühn, H., 1970, Die Aufnahme von Scheweldioxid ($^{35}SO_2$) aus der Luft. I. Einfluss auf den pflanzlichen Ertrag, *Plant Soil* **33**:177.

Ford, H. W., 1973, Levels of hydrogen sulfide toxic to citrus roots, *J. Am. Soc. Hort. Sci.* **98**:66.

Francis, A. J., Adamson, J., Duxbury, J. M., and Alexander, M., 1973, Life detection by gas chromatography–mass spectrometry of microbial metabolites, *Bull. Ecol. Res. Commun.* **17**:485.

Francis, A. J., Duxbury, J. M., and Alexander, M., 1975, Formation of volatile organic products in soils under anaerobiosis—II. Metabolism of amino acids, *Soil Biol. Biochem.* **7**:51.

Frederick, L. R., Starkey, R. L., and Segal, W., 1957, Decomposability of some organic sulfur compounds in soil, *Soil Sci. Soc. Am. Proc.* **21**:287.

Freney, J. R., 1960, The oxidation of cysteine to sulphate in soil, *Aust. J. Biol. Sci.* **13**:387.

Freney, J. R., 1967, Sulfur-containing organics, in: *Soil Biochemistry*, vol. 1 (A. D. McLaren and G. H. Peterson, eds.), pp. 229–259, Marcel Dekker, New York.

Freney, J. R., and Swaby, R. J., 1975, Sulphur transformations in soils, in: *Sulphur in Australasian Agriculture* (K. D. McLachlan, ed.), pp. 31–39, Sydney University Press, Sydney.

Fried, M., 1948, The absorption of sulfur dioxide by plants as shown by the use of radioactive sulfur, *Soil Sci. Soc. Am. Proc.* **13**:135.

Friend, J. P., 1973, The global sulfur cycle, in: *Chemistry of the Lower Atmosphere* (S. I. Rasool, ed.), pp. 177–201, Plenum Press, New York.

Fuhr, I., Bransford, A. V., and Silver, S. D., 1948, Sorption of fumigant vapors by soil, *Science* **107**:274.

Ghiorse, W. C., and Alexander, M., 1976, Effect of microorganisms on the sorption and fate of sulfur dioxide and nitrogen dioxide in soil, *J. Environ. Qual.* **5**:227.

Gibson, C. R., Hammons, G. A., and Cameron, D. S., 1974, Environmental aspects of El Paso's Burnham I coal gasification complex, in: *Proceedings of Symposium on Environmental Aspects of Fuel Conversion Technology*, pp. 91–100, EPA-65/2-74-118, U.S. Environmental Protection Agency.

Gobert, N., Chaigneau, M., and Savel, J., 1971, Etude des gaz libérés au cours de la coulture en anaérobiose de *Trichomonas vaginalis, C. R. Soc. Biol.* **165**:276.

Gourmelon, C., Boulègue, J., and Michard, G., 1977, Oxydation partielle de l'hydrogène sulfuré en phase aqueuse, *C. R. Acad. Sci., Paris,* Ser. C **284**:269.

Granat, L., Rodhe, H., and Hallberg, R. O., 1976, The global sulphur cycle, in: *Nitrogen, Phosphorus, and Sulphur—Global Cycles* (B. H. Svensson and R. Söderlund, eds.), SCOPE Report 7, Ecological Bulletins 22:89.

Greenwood, D. J., and Lees, H., 1956, Studies on the decomposition of amino acids in soils. I. A preliminary survey of techniques, *Plant Soil* 7:253.

Grey, D. C., and Jensen, M. L., 1972, Bacteriogenic sulfur in air pollution, *Science* 177:1099.

Grice, H. W., Yates, M. L., and David, D. J., 1970, Response characteristics of the Melpar flame photometric detector, *J. Chromatogr. Sci.* 8:90.

Grundon, N. J., 1975, Release of volatile sulphur compounds by plants: development of techniques for studying release by intact plants and oven-drying plant material, Ph.D. Thesis, University of Queensland, Brisbane, Australia.

Gumerman, R. C., 1968, Chemical aspects of H_2S removal in soil, Ph.D. Thesis, University of Washington, Seattle.

Gunkel, W., and Oppenheimer, C. H., 1963, Experiments regarding the sulfide formation in sediments of the Texas Gulf Coast, in: *Symposium on Marine Microbiology* (C. H. Oppenheimer, ed.), pp. 674–684. Charles C. Thomas, Springfield, Ill.

Hales, J. M., Wilkes, J. O., and York, J. L., 1974, Some recent measurements of H_2S oxidation rates and their implications to atmospheric chemistry, *Tellus* 26:277.

Hanst, P. L., Spiller, L. L., Watts, D. M., Spence, J. W., and Miller, M. F., 1975, Infrared measurement of fluorocarbons, carbon tetrachloride, carbonyl sulfide, and other atmospheric trace gases, *J. Air Pollut. Control Assoc.* 25:1220.

Harter, R. D., and McLean, E. O., 1965, The effect of moisture level and incubation time on the chemical equilibria of a Toledo clay loam soil, *Agron. J.* 57:583.

Herbert, R. A., and Shewan, J. M., 1976, Roles played by bacterial and autolytic enzymes in the production of volatile sulphides in spoiling North Sea cod *(Gadus morhua)*, *J. Sci. Food Agric.* 27:89.

Herbert, R. A., Henrie, M. S., Gibson, D. M., and Shewan, J. M., 1971, Bacteria active in the spoilage of certain sea foods, *J. Appl. Bacteriol.* 34:41.

Hicks, B. B., 1976, Transfer of SO_2 and other reactive gases across the air–sea interface, *Tellus* 28:348.

Hidy, G. M., 1973, Removal processes of gaseous and particulate pollutants, in: *Chemistry of the Lower Atmosphere* (S. I. Rasool, ed.), pp. 121–176, Plenum Press, New York.

Hidy, G. M., and Brock, J. R., 1971, An assessment of the global sources of tropospheric aerosols, in: *Proceedings of the Second International Clean Air Congress* (H. M. Englund and W. T. Berry, eds), pp. 1088–1097, Academic Press, New York.

Hill, A. C., 1971, Vegetation: a sink for atmospheric pollutants, *J. Air Pollut. Control Assoc.* 21:341.

Hill, F. B., 1973, Atmospheric sulfur and its links to the biota, in: *Carbon and the Biosphere* (G. M. Woodwell and E. V. Pecan, eds.), pp. 159–181, U.S. Atomic Energy Commission, Washington, D.C.

Hitchcock, D. R., 1975, Dimethyl sulfide emissions to the global atmosphere, *Chemosphere* 3:137.

Hitchcock, D. R., 1976a, Atmospheric sulfates from biological sources, *J. Air Pollut. Control Assoc.* 26:210.

Hitchcock, D. R., 1976b, Microbiological contributions to the atmospheric load of particulate sulfate, in: *Environmental Biogeochemistry* (J. O. Nriagu, ed.), pp. 351–367, Ann Arbor Science Publishers, Ann Arbor, Michigan.

Hitchcock, D. R., 1977a, Biogenic contributions to atmospheric sulfate levels, in: *Proceedings of the Second National Conference on Complete WateReuse: Water's Interface with Energy, Air and Solids* (L. K. Cecil, ed.), pp. 291–310, American Institute of Chemical Engineers, New York.

Hitchcock, D. R., 1977b, Atmospheric sulfate in Monroe County, New York: apparent biogenic contributions, Paper presented at the Third International Symposium on Environmental Biogeochemistry, Wolfenbüttel, F.R.G., March 27–April 3, 1977.

Hitchcock, D. R., Spiller, L. L., and Wilson, W. E., 1977, Biogenic sulfides in the atmosphere in a North Carolina tidal marsh, Paper presented before the Division of Environmental Chemistry, American Chemical Society, New Orleans, Louisiana, March 20–25, 1977.

Hudson, H. J., 1971, The development of the saprophytic fungal flora as leaves senesce and fall, in: *Ecology of Leaf Surface Micro-organisms* (T. F. Preece and C. H. Dickinson, eds.), pp. 447–455, Academic Press, London.

Inoue, H., Iwamoto, R., Fujii, H., and Imai, T., 1955, *Abstr. Annu. Meet. Soc. Sci. Soil Manure*, p. 9 (cited by Takai and Asami, 1962).

Ishida, Y., 1968, Physiological studies on evolution of dimethyl sulfide from unicellular marine algae, *Memoirs of the College of Agriculture, Kyoto University* 94:47.

Iverson, W. P., 1966, Growth of *Desulfovibrio* on the surface of agar media, *Appl. Microbiol.* 14:529.

Iverson, W. P., 1967, Disulfur monoxide: production by *Desulfovibrio, Science* 156:1112.

Jacq, V., and Dommergues, Y., 1970, Influence of light intensity and age of the plant on sulfate reduction in the rhizosphere of maize, *Zentralbl. Bakteriol. Parasitenk.*, Abt. 2 125:661.

Jenkins, D., Medsker, L. L., and Thomas, J. F., 1967, Odorous compounds in natural waters. Some sulfur compounds associated with blue-green algae, *Environ. Sci. Technol.* 9:731.

Jensen, V., 1971, The bacterial flora of beech leaves, in: *Ecology of Leaf Surface Micro-organisms* (T. F. Preece and C. H. Dickinson, eds.), pp. 463–469, Academic Press, London.

Jørgensen, B. B., 1977, Bacterial sulfate reduction within reduced microniches of oxidized marine sediments, *Mar. Biol.* 41:7.

Joshi, M. M., and Hollis, J. P., 1977, Interaction of *Beggiatoa* and rice plant: detoxification of hydrogen sulfide in the rice rhizosphere, *Science* 195:179.

Junge, C. E., 1960, Sulfur in the atmosphere, *J. Geophys. Res.* 65:227.

Junge, C. E., 1963a, Sulfur in the atmosphere, *J. Geophys. Res.* 68:3975.

Junge, C. E., 1963b, *Air Chemistry and Radioactivity*, Academic Press, New York.

Junge, C. E., 1972, The cycle of atmospheric gases—natural and man made, *Quart. J. R. Meteorol. Soc.* 98:711.

Kadota, H., and Ishida, Y., 1968, Effect of salts on enzymatical production of dimethyl sulfide from *Gyrodinium cohnii, Bull. Jap. Soc. Sci. Fish.* 34:512.

Kadota, H., and Ishida, Y., 1972, Production of volatile sulfur compounds by microorganisms, *Annu. Rev. Microbiol.* 26:127.

Kallio, R. E., and Larson, A. D., 1955, Methionine degradation by a species of *Pseudomonas,* in: *Amino Acid Metabolism* (W. D. McElroy and H. B. Glass, eds.), pp. 616–631, Johns Hopkins Press, Baltimore.

Kanivets, V. I., 1970, Reaction of hydrogen, methane, and hydrogen sulfide with the mineral part of the soil, *Sov. Soil Sci.* 2:294.

Kellogg, W. W., Cadle, R. D., Allen, E. R., Lazrus, A. L., and Martell, E. A., 1972, The sulfur cycle, *Science* 175:587.

Kelly, D. P., 1968, Biochemistry of oxidation of inorganic sulphur compounds by microorganisms, *Aust. J. Sci.* 31:165.

Kendler, J., and Donagi, A., 1970, Diffusion of odors from stabilization ponds in arid zones, in: *Developments in Water Quality Research* (H. I. Shuval, ed.), pp. 241–249, Ann Arbor-Humphrey Science Publishers, Ann Arbor, Michigan.

King, J. E., and Coley-Smith, J. R., 1969, Production of volatile alkyl sulphides by microbial degradation of synthetic alliin and alliin-like compounds, in relation to germination of sclerotia of *Sclerotium cepivorum* Berk., *Ann. Appl. Biol.* **64**:303.

Kittrick, J. A., 1976, Control of Zn^{2+} in the soil solution by sphalerite, *Soil Sci. Soc. Am. J.* **40**:314.

Kondo, M., 1923, Über die Bildung des Mercaptans aus 1-Cystin durch Bakterien, *Biochem. Z.* **136**:198.

Kunert, J., 1973, Keratin decomposition by dermatophytes. I. Sulfite production as a possible way of substrate denaturation, *Z. Allg. Mikrobiol.* **13**:489.

Kuznetsov, S. I., 1975, *The Microflora of Lakes and its Geochemical Activity*, University of Texas Press, Austin.

Kuznetsov, S. I., Ivanov, M. V., and Lyalikova, N. N., 1963, *Introduction to Geological Microbiology*, McGraw-Hill, New York.

Laakso, S., Söderling, E., and Nurmikko, V., 1976, Methionine degradation by *Pseudomonas fluorescens* UK1 and its methionine-utilizing mutant, *J. Gen. Microbiol.* **94**:305.

Labarre, C., Chaigneau, M., Bory, J., and Le Moan, G., 1966, Composition des gaz dégagés par *Clostridium sporogenes* et par *Clostridium septicum* cultivés en milieu au thioglycolate, *C. R. Acad. Sci., Paris*, Ser. D **262**:2550.

Labarre, C., Chaigneau, M., and Bory, J., 1972, Composition des gaz dégagés par *Plectridium tetani* cultivé en milieu au thioglycolate, *C. R. Acad. Sci., Paris*, Ser. D **274**:2545.

Last, F. T., and Deighton, F. C., 1965, The non-parasitic microflora on the surfaces of living leaves, *Trans. Br. Mycol. Soc.* **48**:83.

Last, F. T., and Warren, R. C., 1972, Non-parasitic microbes colonizing green leaves: their form and functions, *Endeavour* **31**:143.

LeGall, J., 1974, The sulfur cycle, in: *The Aquatic Environment: Microbial Transformations and Water Management Implications* (L. J. Guarraia and R. K. Ballentine, eds.), pp. 75–85, U.S. Environmental Protection Agency, Washington, EPA 430/G-73-008.

Leinweber, F.-J., and Monty, K. J., 1961, The mode of utilization of cysteine sulfinic acid by bacteria, *Biochem. Biophys. Res. Commun.* **6**:355.

Leinweber, F.-J., and Monty, K. J., 1965, Cysteine biosynthesis in *Neurospora crassa*. I. The metabolism of sulfite, sulfide, and cysteinesulfinic acid, *J. Biol. Chem.* **240**:782.

Levy, H., 1974, Photochemistry of the troposphere, in: *Advances in Photochemistry*, vol. 9 (J. Pitts, Jr., G. Hammond, and K. Gollnick, eds.), pp. 372–524, Wiley, New York.

Lewis, J. A., and Papavizas, G. C., 1970, Evolution of volatile sulfur-containing compounds from decomposition of crucifers in soils, *Soil Biol. Biochem.* **2**:239.

Lewis, J. A., and Papavizas, G. C., 1971, Effect of sulfur-containing volatile compounds and vapors from cabbage decomposition on *Aphanomyces euteiches*, *Phytopathology* **61**:208.

Liss, P. S., and Slater, P. G., 1974, Flux of gases across the air–sea interface, *Nature (London)* **247**:181.

Litchfield, C. D., 1973, Interactions of amino acids and marine bacteria, in: *Estuarine Microbial Ecology* (L. H. Stevenson and R. R. Colwell, eds.), pp. 145–166, University of South Carolina Press, Columbia.

Lovelock, J. E., 1974, CS_2 and the natural sulphur cycle, *Nature (London)* **248**:625.

Lovelock, J. E., Maggs, R. J., and Rasmussen, R. A., 1972, Atmospheric dimethyl sulphide and the natural sulphur cycle, *Nature (London)* **237**:452.

Maloney, T. E., 1963, Research on algal odor, *J. Am. Water Works Assoc.* **55**:481.

Manning, D. J., Chapman, H. R., and Hosking, Z. D., 1976, The production of sulphur compounds in cheddar cheese and their significance in flavour development, *J. Dairy Res.* **43**:313.

Maroulis, P. J., and Bandy, A. R., 1977, Estimate of the contribution of biologically produced dimethyl sulfide to the global sulfur cycle, *Science* **196**:647.

Matheron, R., and Baulaigue, R., 1976, Bactéries fermentatives, sulfato-réductrices et phototrophes sulfureuses en cultures mixtes, *Arch. Microbiol.* **109**:319.

McCready, R. G. L., Laishley, E. J., and Krouse, H. R., 1976, The use of stable sulfur isotope labelling to elucidate sulfur metabolism by *Clostridium pasteurianum, Arch. Microbiol.* **109**:315.

Merkel, J. A., Hazen, T. E., and Miner, J. R., 1969, Identification of gases in a confinement swine building atmosphere, *Trans. A.S.A.E.* **12**:310.

Miller, A., Scanlan, R. A., Lee, J. S., and Libbey, L. M., 1973, Volatile compounds produced in sterile fish muscle *(Sebastes melanops)* by *Pseudomonas putrefaciens, Pseudomonas fluorescens,* and an *Achromobacter* species, *Appl. Microbiol.* **26**:18.

Minárik, E., 1972, SO$_2$-Bildung durch Sulfatreduktion bei verschiedenen Hefearten der Gattung *Saccharomyces, Mitt. Rebe Wein, Obstbau Früchteverwert. (Klosterneuberg)* **22**:245.

Miner, J. R., 1973, *Odors from Livestock Production,* Oregon State University Press, Corvallis.

Miner, J. R., and Hazen, T. E., 1969, Ammonia and amines: components of swine-building odor, *Trans. A.S.A.E.* **12**:772.

Miyamoto, S., Bohn, H. L., Ryan, J., and Yee, M. S., 1974, Effect of sulfuric acid and sulfur dioxide on the aggregate stability of calcareous soils, *Soil Sci.* **118**:299.

Moje, W., Munnecke, D. E., and Richardson, L. T., 1964, Carbonyl sulphide, a volatile fungitoxicant from nabam in soil, *Nature (London)* **202**:831.

Morita, R. Y., 1974, Temperature effects on marine microorganisms, in: *Effect of the Ocean Environment on Microbial Activities* (R. R. Colwell and R. Y. Morita, eds.), pp. 75–79, University Park Press, Baltimore.

Moss, M. R., 1976, Biogeochemical cycles as integrative and spatial models for the study of environmental pollution (the example of the sulphur cycle), *Int. J. Environ. Stud.* **9**:209.

Munnecke, D. E., Domsch, K. H., and Eckert, J. W., 1962, Fungicidal activity of air passed through columns of soil treated with fungicides, *Phytopathology* **52**:1298.

Murakami, F., 1960, The nutritional value of *Allium* plants XXXVI. Decomposition of alliin homologs by microorganism and formation of substance with thiamide-masking activity, *Bitamin (Kyoto)* **20**:126.

Natusch, D. F. S., Klonis, H. B., Axelrod, H. D., Teck, R. J., and Lodge, J. P., Jr., 1972, Sensitive method for measurement of atmospheric hydrogen sulfide, *Anal. Chem.* **44**:2067.

Nicolson, A. J., 1970, Soil sulfur balance studies in the presence and absence of growing plants, *Soil Sci.* **109**:345.

Oaks, D. M., Hartmann, H., and Dimick, K. P., 1964, Analysis of S compounds with electron capture/H flame dual channel gas chromatography, *Anal. Chem.* **36**:1560.

Ochynski, F. W., and Postgate, J. R., 1963, Some biochemical differences between fresh water and salt water strains of sulphate-reducing bacteria, in: *Symposium on Marine Microbiology* (C. H. Oppenheimer, ed.), pp. 426–441, Charles C. Thomas, Springfield.

Ogata, G., and Bower, C. A., 1965, Significance of biological sulfate reduction to soil salinity, *Soil Sci. Soc. Am. Proc.* **29**:23.

Orr, W. L., and Gaines, A. G., 1974, Observations on rate of sulfate reduction and organic matter oxidation in the bottom waters of an estuarine basin: the upper basin of the Pettaquamscutt River (Rhode Island), in: *Advances in Organic Geochemistry 1973* (B. Tissot and F. Bienner, eds), pp. 791–812, Editions Technip, Paris.

Östlund, H. G., and Alexander, J., 1963, Oxidation rate of sulfide in seawater, a preliminary study, *J. Geophys. Res.* **68**:3995.

Parker, B. C., 1970, Life in the sky, *Nat. Hist.* **79**(8):54.

Payrissat, M., and Beilke, S., 1975, Laboratory measurements of the uptake of sulphur dioxide by different European soils, *Atmos. Environ.* **9**:211.

Peck, H. D., Jr., 1962, Comparative metabolism of inorganic sulfur compounds in microorganisms, *Bacteriol. Rev.* **26**:67.

Penkett, S. A., 1972, Oxidation of SO_2 and other atmospheric gases by ozone in aqueous solution, *Nature Phys. Sci.* **240**:105.

Peterson, W. H., 1914, Forms of sulfur in plant materials and their variation with the soil supply, *J. Am. Chem. Soc.* **36**:1290.

Pfennig, N., 1967, Photosynthetic bacteria, *Annu. Rev. Microbiol.* **21**:285.

Ponnamperuma, F. N., 1972, The chemistry of submerged soils, *Adv. Agron.* **24**:29.

Postgate, J. R., 1959, Sulphate reduction by bacteria, *Annu. Rev. Microbiol.* **13**:505.

Powlson, D. S., and Jenkinson, D. S., 1971, Inhibition of nitrification in soil by carbon disulfide from rubber bungs, *Soil Biol. Biochem.* **3**:267.

Pugh, G. J. F., and Buckley, N. G., 1971, The leaf surface as a substrate for colonization by fungi, in: *Ecology of Leaf Surface Micro-organisms* (T. F. Preece and C. H. Dickinson, eds.), pp. 431–445, Academic Press, London.

Quammen, M. L., LaRock, P. A., and Calder, J. A., 1973, Environmental effects of pulp mill wastes, in: *Estuarine Microbial Ecology* (L. H. Stevenson and R. R. Colwell, eds.), pp. 329–344, University of South Carolina Press, Columbia.

Rankine, B. C., and Pocock, K. F., 1969, Influence of yeast strain on binding of sulfur dioxide in wines, and on its formation during fermentation, *J. Sci. Food Agric.* **20**:104.

Rasmussen, K. H., Taheri, M., and Kabel, R. L., 1975, Global emissions and natural processes for removal of gaseous pollutants, *Water Air Soil Pollut.* **4**:33.

Rasmussen, R. A., 1974, Emission of biogenic hydrogen sulfide, *Tellus* **26**:254.

Rasmussen, R. A., and Hutton, R. S., 1972, Utilization of atmospheric organic volatiles as an energy source by microorganisms in the tropics, *Chemosphere* **1**:47.

Rheinheimer, G., 1972, Dissolved gases. Bacteria, fungi and blue-green algae, in: *Marine Ecology*, vol. I, part 3 (O. Kinne, ed.), pp. 1459–1469, Wiley–Interscience, London.

Rheinheimer, G., 1974, *Aquatic Microbiology*, Wiley, London.

Richmond, D. V., 1973, Sulfur compounds, in: *Phytochemistry*, vol. III (L. P. Miller, ed.), pp. 41–73, Van Nostrand Reinhold, New York.

Robbins, R. C., and Kastelic, J., 1961, Fate of tetramethylthiuram disulfide in the digestive tract of the ruminant animal, *J. Agric. Food Chem.* **9**:256.

Roberts, S., and Koehler, F. E., 1965, Sulfur dioxide as a source of sulfur for wheat, *Soil Sci. Soc. Am. Proc.* **29**:696.

Robinson, E., and Robbins, R. C., 1968, *Sources, Abundance, and Fate of Gaseous Atmospheric Pollutants*, Stanford Research Institute Final Report Project PR-6755 (February 1968), pp. 11–48.

Robinson, E., and Robbins, R. C., 1970a, Gaseous atmospheric pollutants from urban and natural sources, in: *Global Effects of Environmental Pollution* (S. F. Singer, ed.), pp. 50–65, Springer-Verlag, New York.

Robinson, E., and Robbins, R. C., 1970b, Gaseous sulfur pollutants from urban and natural sources, *J. Air Pollut. Control Assoc.* **20**:233.

Rodhe, H., 1972, A study of the sulfur budget for the atmosphere over northern Europe, *Tellus* **24**:128.

Ronkainen, P., Denslow, J., and Leppänen, O., 1973, The gas chromatographic analysis of some volatile sulfur compounds, *J. Chromatogr. Sci.* **11**:384.

Roy, A. B., and Trudinger, P. A., 1970, *The Biochemistry of Inorganic Compounds of Sulfur*, Cambridge University Press, Cambridge.

Ruiz-Herrera, J., and Starkey, R. L., 1969, Dissimilation of methionine by fungi, *J. Bacteriol.* **99**:544.

Ruiz-Herrera, J., and Starkey, R. L., 1970, Dissimilation of methionine by *Achromobacter starkeyi, J. Bacteriol.* **104**:1286.

Sachdev, M. S., and Chhabra, P., 1974, Transformation of S^{35}-labelled sulphate in aerobic and flooded soil conditions, *Plant Soil* **41**:335.

Sandalls, F. J., and Penkett, S. A., 1977, Measurements of carbonyl sulphide and carbon disulphide in the atmosphere, *Atmos. Environ.* **11**:197.

Schlegel, H. G., 1974, Production, modification, and consumption of atmospheric trace gases by microorganisms, *Tellus* **26**:11.

Segal, W., and Starkey, R. L., 1969, Microbial decomposition of methionine and identity of the resulting sulfur products, *J. Bacteriol.* **98**:908.

Seim, E. C., 1970, Sulfur dioxide absorption by soil, Ph.D. Thesis, University of Minnesota, St. Paul.

Sharpe, M. E., Law, B. A., and Phillips, B. A., 1976, Coryneform bacteria producing methane thiol. *J. Gen. Microbiol.* **94**:430.

Siegel, L. M., 1975, Biochemistry of the sulfur cycle, in: *Metabolism of Sulfur Compounds,* vol. VII, 3rd ed. (D. Greenberg, ed.), pp. 217–286, Academic Press, New York.

Simán, G., and Jansson, S. L., 1976a, Sulphur exchange between soil and atmosphere, with special attention to sulphur release directly to the atmosphere. 1. Formation of gaseous sulphur compounds in soil, *Swed. J. Agric. Res.* **6**:37.

Simán, G., and Jansson, S. L., 1976b, Sulphur exchange between soil and atmosphere, with special attention to sulphur release directly to the atmosphere. 2. The role of vegetation in sulphur exchange between soil and atmosphere, *Swed. J. Agric. Res.* **6**:135.

Singh, H. B., 1977, Atmospheric halocarbons: evidence in favor of reduced average hydroxyl radical concentration in the troposphere, *Geophys. Res. Lett.* **4**:101.

Smith, K. A., Bremner, J. M., and Tabatabai, M. A., 1973, Sorption of gaseous atmospheric pollutants by soils, *Soil Sci.* **116**:313.

Smith M. S., 1976, Evolution of volatile organic compounds from soils and manure, M.S. Thesis, Cornell University, Ithaca, New York.

Smith M. S., Francis, A. J., and Duxbury, J. M., 1977, Collection and analysis of organic gases from natural ecosystems: application to poultry manure, *Environ. Sci. Technol.* **11**:51.

Soda, K., Novogrodsky, A., and Meister, A., 1964, Enzymatic desulfination of cysteine sulfinic acid, *Biochemistry* **3**:1450.

Somers, E., Richmond, D. V., and Pickard, J. A., 1967, Carbonyl sulphide from the decomposition of captan, *Nature (London)* **215**:214.

Starkey, R. L., 1956, Transformations of sulfur by microorganisms, *Ind. Eng. Chem.* **48**:1429.

Stephens, E. R., 1971, Identification of odors from cattle feedlots, *Calif. Agric.* **25**:10.

Stevens, R. K., Mulik, J. D., O'Keefe, A. E., and Krost, K. J., 1971, Gas chromatography of reactive sulfur gases in air at the parts-per-billion level, *Anal. Chem.* **43**:827.

Stevens, R. K., Baumgardner, R., Paur, R., and McClenny, W. A., 1977, Rural and urban measurements of sulfur dioxide, hydrogen sulfide, oxides of nitrogen, and ammonia, *Abstr. Pap. Am. Chem. Soc.* **173**:58.

Stevenson, L. H., Millwood, C. E., and Hebeler, B. H., 1974, Aerobic heterotrophic bacterial populations in estuarine water and sediments, in: *Effect of the Ocean Environment on Microbial Activities* (R. R. Colwell and R. Y. Morita, eds.), pp. 268–285, University Park Press, Baltimore.

Stoiber, R. E., Leggett, D. C., Kenkins, T. F., Murrmann, R. P., and Rose, W. I., Jr., 1971, Organic compounds in volcanic gas from Santiaguito Volcano, Guatemala, *Bull. Geol. Soc. Am.* **82**:2299.

Stotzky, G., and Schenck, S., 1976, Volatile organic compounds and microorganisms, *CRC Crit. Rev. Microbiol.* **4**:333.

Swaby, R. J., and Fedel, R., 1973, Microbial production of sulphate and sulphide in some Australian soils, *Soil Biol. Biochem.* **5**:773.

Takai, Y., and Asami, T., 1962, Formation of methyl mercaptan in paddy soils I, *Soil Sci. Plant Nutr. (Tokyo)* **8**:40.

Tanner, R. L., and Newman, L., 1976, The analysis of airborne sulfate: a critical review, *J. Air Pollut. Control Assoc.* **26**:737.

Terraglio, F. P., and Manganelli, R. M., 1966, The influence of moisture on the adsorption of atmospheric sulfur dioxide by soil, *Air Water Pollut.* **10**:783.

Toan, T. T., Bassette, R., and Claydon, T. J., 1965, Methyl sulfide production by *Aerobacter aerogenes* in milk, *J. Dairy Sci.* **48**:1174.

Tocher, C. S., Ackman, R. G., and McLachlan, J., 1966, The identification of dimethyl-β-propiothetin in the algae *Syracosphaera carterae* and *Ulva lactuca, Can. J. Biochem.* **44**:519.

Trudinger, P. A., 1975, The biogeochemistry of sulphur, in: *Sulphur in Australasian Agriculture* (K. D. McLachlan, ed.), pp. 11–19, Sydney University Press, Sydney.

Tukey, H. B., Jr., 1971, Leaching of substances from plants, in: *Ecology of Leaf Surface Micro-organisms* (T. F. Preece and C. H. Dickinson, eds.), pp. 67–80, Academic Press, London.

Urone, P., 1976, The primary air pollutants—gaseous. Their occurrence, sources and effects, in: *Air Pollution,* vol. I, 3rd ed. (A. C. Stern, ed.), pp. 23–75, Academic Press, New York.

Vámos, R., 1959, "Brusone" disease of rice in Hungary, *Plant Soil* **11**:65.

Wagner, C., and Stadtman, E. R., 1962, Bacterial fermentation of dimethyl-β-propiothetin, *Arch. Biochem. Biophys.* **98**:331.

Weir, R. G., 1975, The oxidation of elemental sulphur and sulphides in soil, in: *Sulphur in Australasian Agriculture* (K. D. McLachlan, ed.), pp. 40–49, Sydney University Press, Sydney.

White, R. K., 1969, Gas chromatographic analysis of odors from dairy wastes, Ph.D. Thesis, Ohio State University, Columbus.

White, R. K., Taiganides, E. P., and Cole, G. D., 1971, Chromatographic identification of malodors from dairy animal waste, in: *Livestock Waste Management and Pollution Abatement,* pp. 110–113, American Society of Agricultural Engineers, St. Joseph, Mo.

Williams, C. H., 1975, The chemical nature of sulphur compounds in soils, in: *Sulphur in Australasian Agriculture* (K. D. McLachlan, ed.), pp. 21–30, Sydney University Press, Sydney.

Wood, E. J. F., 1965, *Marine Microbial Ecology,* Chapman and Hall, London.

Yee, M. S., Bohn, H. L., and Miyamoto, S., 1975, Sorption of sulfur dioxide by calcareous soils, *Soil Sci. Soc. Am. Proc.* **39**:268.

Young, R. J., Dondero, N. C., Ludington, D. C., and Loehr, R. C., 1971, Poultry waste management and the control of associated odors, in: *Identification and Measurements of Environmental Pollutants* (B. Westley, ed.), pp. 98–104, National Research Council of Canada, Ottawa.

ZoBell, C. E., 1958, Ecology of sulphate-reducing bacteria, *Prod. Mon.* **22**:12.

ZoBell, C. E., 1963, Organic geochemistry of sulfur, in: *Organic Geochemistry* (I. A. Breger, ed.), pp. 543–578, Macmillan, New York.

ZoBell, C. E., 1973, Microbial and environmental transitions in estuaries, in: *Estuarine Microbial Ecology* (L. H. Stevenson and R. R. Colwell, eds.), pp. 9–31, University of South Carolina Press, Columbia.

Microbiological Aspects of Regulating the Carbon Monoxide Content in the Earth's Atmosphere

A. N. NOZHEVNIKOVA AND L. N. YURGANOV

1. Introduction

Carbon monoxide has been known to be a permanent constituent of the earth's atmosphere since 1949, when Migeotte identified absorption lines of CO in the solar spectrum for the first time. Subsequently it was found that the relative content of CO in "clean" atmosphere amounted to about 10^{-7} by volume, which makes the total content of CO in the earth's atmosphere about 5×10^{14} g. Meanwhile, according to recent findings (Seiler, 1974), more than 6×10^{14} g of CO is annually discharged into the atmosphere as a result of human activities.

An examination of the above figures indicates that, in nature, scavenging processes occur so that the growth of anthropogenic production of CO is paralleled by an increase in its consumption, the available experimental data revealing a relative constancy of CO concentration in "clean" atmospheres.

Until recently, very little was known about the natural cycle of CO. The results of recent studies warrant the assumption that the natural turnover of CO is governed by a series of closely interrelated processes, both biogenic and abiogenic. In view of the role of both biological and

A. N. NOZHEVNIKOVA • Institute of Microbiology, U.S.S.R. Academy of Sciences, Moscow, U.S.S.R.   **L. N. YURGANOV** • Institute of Atmospheric Physics, U.S.S.R. Academy of Sciences, Moscow, U.S.S.R.   Copyright for this article is held by the Soviet Union.

nonbiological factors, it is extremely desirable that scientists of various specialities unite their efforts in the study of this problem.

The present chapter deals with the CO cycle from both the geophysical and microbiological points of view. The study considers the distribution of CO in the terrestrial atmosphere and the variations in its concentration. Data are cited concerning the processes of liberation of CO into the atmosphere and of CO consumption, both biogenic and abiogenic. The group of bacteria capable of metabolizing CO (carboxydobacteria) is examined in great detail. The chapter concludes with a consideration of the possible role of microorganisms as regulators of the CO content of the atmosphere and with a comparative evaluation of the biogenic and abiogenic factors in the CO cycle.

2. Carbon Monoxide in the Atmosphere

2.1. Introduction

The distribution of CO in space makes it possible to locate its sources and sinks, as well as the direction and extent of CO diffusion fluxes. Temporal CO variation may shed light on the mechanisms of formation and destruction of CO and the mean residence time of CO in the atmosphere.

2.1.1. Units of Measurement

The distribution of trace gases in the atmosphere may be characterized by several values.

(a) Mixing ratio by volume: the volume of gas contained in a unit volume of air (r), a dimensionless value usually represented in parts per million (ppm or ppmv)

$$r(CO) = p^{CO}/p^{air}$$

where p^{CO} is the partial CO pressure and p^{air} is the air pressure.

(b) Concentration: the mass of gas per unit of air volume (K), usually measured in mg/m^3, or μg/m^3, under normal conditions, $T_N = 273°K$, $p_N = 760$ mm Hg $= 1$ atm.

$K(\text{mg/m}^3) = 1.25\, r(CO)$ (ppm)
$K(\text{mg/m}^3) = 10^{-3}\, K(\text{μg/m}^3)$

It will be noted that for a well-mixed atmosphere the mixing ratio with altitude is constant, whereas the concentration diminishes with altitude, like the air density.

(c) The total amount of gas in the whole atmosphere is measured in grams per square centimeter or in centimeters of the thickness of the gas layer reduced to normal conditions (atm·cm)

$$U_z(\text{atm·cm}) = \int_0^\infty p^{CO}(z)\,dz = \int_0^\infty r(CO)\,(z)\cdot p^{air}\,(z)\;dz \tag{1}$$

$$U_z^*(\text{g/cm}^2) = \frac{\rho_N^{CO}}{p_N} \cdot U_z = 1.25 \cdot 10^{-3} \cdot U_z(\text{atm·cm})$$

where z is the altitude above sea level, and ρ_N^{CO} is the CO density under normal conditions.

It is easy to show that the connection between the total amount of CO U_z and the mean mixing ratio $r(CO)$ [or for a particular case, $r(CO)$ equal to a constant] averaged over a vertical column of atmosphere is represented by the expression

$$U_z(\text{atm·cm}) = 0.8 \; \bar{r}(CO) \; (\text{ppmv}) \tag{2}$$

(d) Production and consumption of CO will be indicated further in the text in metric tons or grams.

2.1.2. Methods of Investigating the CO Content of the Atmosphere

2.1.2a. Integral Methods. The integral methods (as a rule, spectroscopic) make it possible to measure the total amount of the gas in the atmosphere. The abundance of the gas is judged by the extent the whole atmosphere absorbs solar radiation in the spectral intervals containing absorption lines of the gas in question (Frolov, 1976; Yurganov and Dianov-Klokov, 1972; Dianov-Klokov *et al.*, 1975; Shaw 1958). Methods of measuring the total amount of CO from a satellite have been suggested (Kondratyev and Timofeev, 1974).

2.1.2b. Local Methods. Local methods are used for measuring the mixing ratio in a given place. The best known methods are gas chromatography (Swinnerton *et al.*, 1968, 1969) and a method based on the reduction of mercuric oxide to mercury by carbon monoxide with a subsequent recording of the mercury atoms by means of atomic absorption in the ultraviolet line of mercury 253.7 nm (Robbins *et al.*, 1968; Seiler and Junge, 1970).

2.1.3. Transfer of Trace Gases in the Atmosphere

The processes of transferring trace gases in the atmosphere exert no smaller an influence on the distribution of trace gases than do those of their production and consumption; the intensity of the processes of mixing is characterized by the mean mixing time.

The terrestrial atmosphere is in continuous motion. In addition to relatively orderly motion (the general circulation of the atmosphere), there are also disorderly, chaotic motions; that is, turbulence. The minor constituents of the atmosphere are transferred by both the former and latter motions. Several reservoirs may be conditionally distinguished in the terrestrial atmosphere, each of which is characterized by its own mean mixing time. The equatorial plane divides the whole atmosphere into two parts, northern and southern; the upper boundary of the troposphere that divides it from the stratosphere is called the tropopause and is located at an altitude of 10–15 km. The troposphere of a given hemisphere is mixed in the best manner, its mean mixing time amounting to several weeks or a few months. The main mixing mechanism is general circulation; characteristic of the lower troposphere in temperate latitudes, for example, is a west-to-east transfer.

The exchange between the tropospheres of the two hemispheres takes place much more slowly, the mixing time being about 1 year. The time of mixing between the troposphere and stratosphere is about $1\frac{1}{2}$ years. The stratosphere itself is mixed much more slowly than the troposphere, the characteristic time being 1–2 years. The exchange between the stratospheres of the two hemispheres requires still more time, namely, about 5 years.

The above evaluations (Pressman and Warneck, 1970) indicate that the mixing within the tropospheres of the northern and southern hemispheres is a faster process than the exchange between the tropospheres of the hemispheres or between the troposphere and stratosphere. The above-mentioned reservoirs will be considered separately, the principal attention being devoted to the troposphere of the northern hemisphere (cf. Bolin and Rhode, 1973).

It would not be out of place to observe that the process of CO transfer from one reservoir to another will be a sink for the first and a source for the other, although for the atmosphere as a whole this transfer will be neither a source nor a sink.

2.2. Distribution of CO in the Atmosphere

A consideration of the distribution of CO in space makes it possible to distinguish two extreme cases: urban areas with relatively high mixing ratios (from 1 ppm to 100 ppm) and rural areas, where the usual mixing ratios do not exceed 0.5 ppm. In the latter case we shall talk about an "unpolluted" area and a "background" concentration of CO.

2.2.1. Carbon Monoxide in Urban Areas

Carbon monoxide (like SO_2, NO, and NO_2) is one of the most important pollutants of the urban atmosphere, as a result of both its toxicity and the enormous emissions of it during various processes of fuel combustion.

The toxic influence of carbon monoxide on man has been the subject of numerous studies in the USSR. These are summarized in the monographs of Tiunov and Kustov (1969) and Datsenko and Martyniuk (1971). These investigations have served as the basis for establishing maximum permissible concentrations of CO in the air of populated areas, namely, mean daily, maximum single, and maximum permissible concentrations in the air of a working zone of 1, 3, and 20 mg/m³, respectively (Krotov *et al.*, 1975).

Carbon monoxide is regularly measured by local methods in most of the world's large cities (Jaffe, 1973). A review of the extensive findings obtained by Soviet researchers in recent years can be found in Berlyand's study (1972). The results of original studies are published mainly in "Atmospheric Diffusion and Air Pollution," the transactions of the Voyeikov Main Geophysical Observatory.

The mean CO concentration in the air of large cities usually exceeds the maximum permissible values by many fold. For example, the mean annual CO mixing ratio in Sofia, Bulgaria, is 8 ppm, in Zurich, Switzerland, 35 ppm, and in Marseilles, France, 50 ppm (Bezuglaya and Rastorguyeva, 1973).

Within the city limits, the concentration of CO undergoes considerable variations, both spatial and temporal. The maximum CO mixing ratios are usually observed in direct proximity to motor traffic, in some cases (as with tunnels, crossroads, and narrow streets) amounting to 150–200 ppm. At the same time, 20–30 m away from the center of streets the concentration diminishes very greatly. Away from routes of traffic, in the city outskirts, and in parks, the mixing ratio does not, as a rule, exceed the established norms (Zaitsev, 1973).

The temporal variations of CO concentration are normally determined by variations in the intensity of motor traffic and meteorological conditions (wind velocity and distribution of the air temperature with altitude). On working days the morning maximum concentration coincides, as a rule, with the morning rush hours; sometimes there is also a second, afternoon maximum. On nonworking days the diurnal trend of CO is not marked, and the absolute concentrations are much lower (Zaitsev, 1973; Burenin, 1974).

The first investigations of the CO content over a city conducted by an integral method (Lukshin *et al.*, 1976) showed a close connection between the CO content of the whole atmosphere and the meteorological conditions. With a wind of 6–8 m/sec, the total amount of CO over a city hardly differs from the background content; with a wind of very low velocity, the total amount of CO exceeds its background content two- to threefold. An accumulation of CO also favors an inversion of the temperature of the air (an increase in temperature with altitude, which prevents mixing).

In summary it may be said that the main sources of CO for such an atmospheric reservoir as the space over a city are motor transport and industrial emissions. The transfer of CO across the boundaries of this reservoir will be a sink for the city and a source for the rest of the atmosphere.

2.2.2. Distribution of CO in "Clean" Atmospheres

The most interesting peculiarity of geographical distribution is that the southern hemisphere contains much less CO than does the northern hemisphere (Robinson and Robbins, 1972; Seiler, 1974). The latter author estimates the mean mixing ratio in the northern hemisphere at 0.15 ppm and in the southern hemisphere at 0.06 ppm. The above-mentioned measurements made by a local chemical method were confirmed by the integral spectral method (Malkov *et al.*, 1976; Dianov-Klokov *et al.*, 1977), the minimum CO abundance in Antarctica being one-third that of the temperate latitudes of the northern hemisphere (Fig. 1).

Neither the local measurements made by Robinson and Robbins (1972) in the atmosphere close to the surface of the northern part of the Pacific (range of longitudes, 100°E–120°W) nor the integral measurements carried out by Dianov-Klokov *et al.* (1977, 1978) on the territory of the USSR

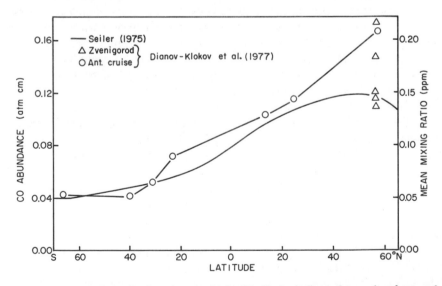

Figure 1. Latitudinal distribution of atmospheric CO. Circles indicate the results of spectral integral measurements during an Antarctic cruise (March–April, 1976) and the triangles correspond to similar measurements in Zvenigorod (near Moscow) in the same period. The curve was calculated using Seiler's (1975) chemical data.

(range of longitudes, 36°E–125°E) showed any essential trend of the CO content. The geographical distribution of CO in temperate latitudes of the northern hemisphere is on the average apparently uniform, although considerable fluctuations of the local concentration are observed (Seiler, 1974).

Measurements of the vertical distribution of CO in the northern hemisphere show, as a rule, a certain mixing-ratio decrease with altitude—from 0.2 ppm at the earth's surface to 0.14–0.16 ppm in the upper troposphere and then a rapid drop to 0.04–0.05 ppm in the lower stratosphere (Seiler, 1974). In the stratosphere, close to altitudes of 40–50 km, the CO mixing ratio apparently does not change (Ehhalt *et al.*, 1975).

A certain increase in the CO mixing ratio with altitude is observed in the tropical latitudes of the southern hemisphere; this increase is most likely a result of invasions of air abounding in CO from the northern hemisphere in the upper layers of the troposphere. In the temperate latitudes of the southern hemisphere, the CO mixing ratios with altitude are constant (Seiler, 1974).

Thus, whereas within the tropospheres of the southern and northern hemispheres the CO is distributed more or less uniformly, considerable differences in its concentration are observed on the boundaries of these reservoirs. The difference in concentration leads to a flux of CO across the boundaries of the reservoirs. Newell *et al.* (1974) estimate the flux across the equator from the northern to the southern troposphere at about 1.6×10^{14} g/year, and the flux through the tropopause into the stratosphere at about 9.3×10^{12} g/year.

The above character of CO transfer between the reservoirs indicates that processes of CO production predominate in the northern troposphere, while processes of CO consumption predominate in the southern troposphere and stratosphere.

2.3. Variations of the CO Content in the Atmosphere

2.3.1. Secular Trend of the CO Content in the Atmosphere

One of the central problems in the matter of atmospheric pollution is the possible influence of anthropogenic emissions on the background concentration of gases, i.e., the secular trend. For example, an increase in the atmospheric carbon dioxide mixing ratio in the surface layer of the atmosphere was discovered in comparisons of the measurements made by similar methods. The increase was from 290 ppm in 1870–1900 to 325 ppm in 1955 (Robinson and Robbins, 1972). According to more recent data (Keeling *et al.*, 1976), the content of CO_2 increased 3.4% from 1959 to 1971.

Robbins *et al.* (1973) analyzed the chemical composition of air contained in the pores of Antarctic and Greenland ice (firn) samples formed 100–2500 years ago. Unfortunately, the attempt to determine the change in the atmospheric concentration of CO failed, the results of the analysis not being amenable to unambiguous interpretation.

Pressman and Warneck (1970) endeavored to compare the measurements of CO content made by a spectral integral method in the beginning of the 1950s with those made by a local method in the surface layer of the atmosphere at the end of the 1960s. Of course, one need hardly comment on the precision of such a comparison, especially taking into account the considerable systematic errors of the spectral method associated with air humidity in particular (Malkov *et al.*, 1976).

Dianov-Klokov *et al.* (1977, 1978) compared the results of integral spectral measurements of CO content in the USA during 1952–1953 (Shaw, 1958) and the USSR during 1970–1975 (Yurganov and Dianov-Klokov, 1972; Dianov-Klokov *et al.*, 1975, 1978). From Fig. 2, which shows the

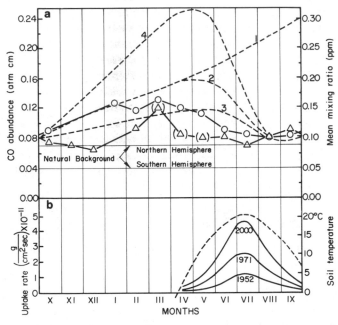

Figure 2. (A) Monthly averaged values of CO abundance (tropospheric mean mixing ratio). Circles: Zvenigorod, USSR, 1971. Triangles: Columbus, Ohio, USA, 1952–1953 (results of single measurements are in parentheses). Dashed curves are theoretical (see text). Curves 1 and 2 correspond to 1971, curve 3 to 1952, and curve 4 to 2000. (B) Dashed curve is a mean climatic seasonal trend of the surface soil temperature for Zvenigorod (right scale). Continuous curves are the temporal dependences of soil absorption rates (left scale) used in the construction of curves 2, 3, and 4 (in g/cm² sec).

monthly averaged CO abundance for two series of measurements, it follows that during June through September both series of data are in good agreement, whereas during October through May, according to our findings, the content of CO is 30–50% higher than in Shaw's data. The seasonal variation in the CO content observed in the 1970s were thus practically absent in the 1950s.

The observed phenomenon (constancy of the summer CO concentration and the increase in winter) suggests the existence of natural processes of CO elimination from the atmosphere, these processes operating during the warm part of the year and being capable of maintaining the background concentration of CO at a relatively constant level that does not depend on the rate of anthropogenic production. During the cold part of the year, this sink is either altogether inactive or less effective, and the CO of anthropogenic origin accumulates in the atmosphere in the course of the winter.

Various sources and sinks of CO, both abiogenic and biogenic, will be dealt with in the following sections.

2.3.2. Regular Temporal CO Variations in the Atmosphere

As was already mentioned (Section 2.2.2), the concentration of CO in the surface layer of the atmosphere in the northern hemisphere fluctuates greatly. For this reason it is difficult to obtain information on the regular seasonal trend of CO from local measurements. The integral method, which averages out the fluctuations of the CO mixing ratio and yields the direct mean CO mixing ratio throughout the troposphere [$\bar{r}$(CO)], is more suitable for this purpose. Figure 2 shows the seasonal dependence of the total amount of CO—U_z [relation between U_z and $\bar{r}$(CO); see Section 2.1.1.] (Dianov-Klokov et al., 1977). The noticeable increase in U_z during October through March is followed in April through June by a decrease in CO content; during July through September the U_z is constant. The maximum amount of CO in March exceeds the summer level by about 50%. If we assume that the increase in the CO content in winter is associated with the action of anthropogenic sources (see Section 2.3.1), then not less than one-third of the atmospheric CO must be of anthropogenic origin in March.

The above results agree qualitatively with those of the local measurements (Stevens et al., 1972). In the winter and spring of 1970–1971, the mean mixing ratios in the surface layer of the air proved higher than in summer (January through May, 0.33 ppm; June and July, 0.25 ppm). A sharp increase in the CO mixing ratio was observed during the second half of October; the authors attribute this to a discharge of CO in the process of the autumnal decay of chlorophyll.

The seasonal variations in the oxygen isotopic composition of atmospheric CO (Stevens *et al.*, 1972) are more significant than those of carbon. The essential differences in the oxygen isotopic composition for atmospheric CO in summer and anthropogenic CO led the authors to conclude that the contribution of motor emissions to the total production of CO in summer does not exceed 10%. In winter the contribution of anthropogenic CO to the total amount in the atmosphere of the northern hemisphere reaches 50%.

Stevens *et al.* (1972) assumed that the observed variations in the isotopic composition are the result solely of the temporal variations in the intensity of sources and the relation between the natural and anthropogenic sources. Moreover, they held that the CO from each source had its own isotopic composition. On the basis of this assumption, the authors distinguished five varieties of atmospheric CO. However, several authors (Seiler, 1975; Zyakun *et al.*, 1976) hold that, in interpreting the observed seasonal variations, it is necessary to take into account the isotopic fractionation during the processes of CO consumption by bacteria. If this consideration is ignored, the division of atmospheric CO into isotopic varieties does not appear quite substantiated. On the other hand, isotope fractionation during the consumption of CO by bacteria in summertime must make atmospheric CO heavier (Zyakun *et al.*, 1976). As a matter of fact, a contrary picture is observed. The conclusion of Stevens *et al.* (1972) that atmospheric CO has different origins in winter and summer can for this reason hardly be doubted.

Lamontagne *et al.* (1971) observed diurnal variations in CO concentration in the layer of atmosphere above the ocean, with a maximum at about noon and a minimum at night. However, several other investigators discovered no regular diurnal variations above either the land (Yurganov and Dianov-Klokov, 1972; Dianov-Klokov *et al.*, 1975; Stevens *et al.*, 1972) or the ocean (Seiler, 1974).

3. Abiogenic Processes in the CO Cycle

3.1. Anthropogenic CO Production

Carbon monoxide is formed in all combustion processes as a result of the incomplete oxidation of carbon. Motor transport is one of the main sources of CO emission. The amount and composition of exhaust gases depend on the type of engine, the quality of fuel used, the technical condition of the motor vehicle, and the speed of its movement. The largest amount of CO is emitted when the engine is idling or is run at low speed (Korenev, 1962). During the combustion of 1 kg of gasoline, the amount of CO emitted may vary between 150 and 800 g (Varshavsky, 1968). According to the norms existing in the USSR, the concentration of CO in exhaust gases of vehicles with a gasoline engine must not exceed 4.5% (Feldman, 1973).

The contribution of motor transport to the total pollution of a city (Burenin and Solomatina, 1975) depends primarily on the character of the industry in the given city. In cities with developed oil-refining and metallurgical industries, the share of motor-transport emissions is about 10–30%. In cities with a less developed industry, the motor-transport emissions are of the same order as the industrial emission and even exceed them. The intensity of CO emission by motor transport alone may, according to Burenin and Solomatina (1975), be considered proportional to the population of the city.

The global anthropogenic production of CO has often been estimated on the basis of data on fuel consumption and data on the amount of CO formed as a result of combustion of a ton of fuel. Global production of CO was estimated for the first time in 1952 as 191×10^6 tons per year (Bates and Witherspoon, 1952). Subsequently, the estimates increased as follows: 1966, 280×10^6 tons (Robinson and Robbins, 1972); 1970, 359×10^6 tons (Jaffe, 1973); and 1971, 640×10^6 tons (Seiler, 1974). It should be noted that the estimates grew not only because of the increase in world fuel consumption but also because of the addition of the formerly unaccounted sources of CO.

The anthropogenic production of CO can be estimated on the basis of the experimental findings of Lukshin *et al.* (1976) dating from 1974. In this study it was found that the total amount of CO over a city in the daytime and in calm weather, with no transfer of CO from the city, increases at a rate of about 0.050 atm·cm/hr (estimate accuracy ±30%), which is equivalent to emissions of about 6.2×10^{-5} g/cm²/hr (per unit of city area). Extrapolation to the world production (with due regard for the area of the city, fuel consumption, and diminution of the CO emission at night and on weekends) yields an estimate of $(7 \pm 2) \times 10^8$ tons per year. A similar estimate, made by measuring the CO concentration in the environs of Munich (Seiler and Zankl, 1975), gave a value of $6–10 \times 10^8$ tons per year. These data indicate that 6.4×10^8 tons per year (Seiler, 1974) is a reasonable estimate of the total emissions of CO.

According to Jaffe (1973), about 40% of the global CO production comes from the United States. About 15% of the total amount of CO is emitted in the southern hemisphere (Seiler, 1974), and the rest is provided mainly by Europe and Japan. Thus, most of the anthropogenic emissions are concentrated in the temperate latitudes of the northern hemisphere (40°N–60°N).

3.2. Natural Sources and Sinks of Atmospheric CO

3.2.1. Liberation of CO by Oceans

The supersaturation of the surface waters of the ocean relative to the overlying air discovered in the Atlantic Ocean (Swinnerton et al., 1969) and later in the Pacific indicates that the oceans are a source of atmospheric CO. The surface layer of the oceans is supersaturated 30-fold relative to the air. The latest estimates of CO production by the oceans are as follows: 1.0×10^8 tons per year [including 0.4×10^8 tons per year for the northern hemisphere and 0.6×10^8 tons per year for the southern hemisphere (Seiler, 1974)]; and 2.2×10^8 tons per year [including 0.9×10^8 tons per year for the northern hemisphere and 1.3×10^8 tons per year for the southern hemisphere (Linnenbom et al., 1973), i.e., 16–37% of the anthropogenic production. However, on the basis of measurements of CO solubility in water, Meadows and Spedding (1974) concluded that the above estimates were too high, and they suggested that the global CO production by the oceans is 10^5 tons per year, which constitutes only about 2.2% of the global anthropogenic CO production.

Several peculiarities of the distribution of CO in seawater and its variations (maximum concentration in the euphotic layer and a decrease in concentration with depth, diurnal variations with a maximum at about noon, and high concentrations in the biologically more active oceanic areas) indicate a biological source of oceanic CO (Seiler and Schmidt, 1973; Linnenbom et al., 1973; see Section 4.1).

3.2.2. Photochemical Reactions in the Atmosphere

The interest in photochemical reactions in the atmosphere with the participation of minor atmospheric constituents, both stable (CO, CH_4, NO, NO_2, H_2CO, H_2, etc.), unstable, and short-lived (OH, HO_2, CH_3, CH_3O_2, and many others), has of late increased. Relatively recently, free radicals were acknowledged to play some role only in the stratosphere and under conditions of smog (Junge, 1963; Leighton, 1961). As is well known, CO is not oxidized by the O_2 of the air at atmospheric temperatures (Bates and Witherspoon, 1952). The possibility of intense photochemical reactions in the unpolluted lower troposphere has also been discussed since the studies carried out by Levy (1971), McConnell et al. (1971), and Weinstock and Niki (1971). A characteristic feature of these processes is that they occur as catalytic chain reactions, viz., the initial substance is continuously regenerated as a result of a series of reactions.

Hydroxyl radicals are, according to the modern view, the "motive power" in the chain of photochemical reactions in the troposphere. The reaction of an excited oxygen atom (which appears in the photolysis of ozone) with a water molecule is the primary source of OH radicals. The OH radicals, furthermore, participate in two reaction chains that are important to CO. The first chain begins with an oxidation of the CO molecule by an OH radical to yield CO_2:

$$CO + OH \longrightarrow CO_2 + H \tag{3}$$

An important role in this chain is played by OH and HO_2 radicals. NO and NO_2 molecules, whose concentration is maintained photochemically, are also important. For this chain Crutzen (1974) suggested the following equation:

$$CO + 2O_2 - hv \longrightarrow CO_2 + O_3 \tag{4}$$

Thus the destruction of CO and the production of O_3 and CO_2 take place in this process.

The second chain reaction begins with the oxidation of methane by hydroxyl:

$$CH_4 + OH \longrightarrow CH_3 + H_2O \tag{5}$$

Ultimately, this leads to the appearance of CO, H_2, and O_3 (Crutzen, 1974):

$$CH_4 + 4O_2 + 2hv_1 + hv_2 \longrightarrow H_2O + CO + H_2 + 2O_3 \tag{6}$$

Such processes are called *smog-type processes*; they were first considered in describing photochemical reactions in polluted urban atmospheres (Leighton, 1961).

Levy (1973) calculated the mean daily production and consumption of CO for latitude 34° in summer in the course of photochemical reactions with the assumption that there is a constant 1.5 ppm mixing ratio for methane throughout the troposphere and obtained a value of 21×10^{-12} g/cm² sec for the production of CO. The consumption of CO depends on the mixing ratio of CO $r(CO)$ and equals $r(CO) \cdot 3.1 \times 10^{-4}$ g/cm² sec. With a mixing ratio of about 0.07 ppm, production equals consumption, i.e., and photochemical equilibrium is reached; with a larger mixing ratio, photochemical processes act as a sink, and this equals:

$$L(CO) = r(CO) \cdot 3.1 \times 10^{-4} - 21 \times 10^{-12} (\text{g/cm}^2 \text{ sec}) \tag{7}$$

If we assume that the photochemical sink is the only sink that compensates for anthropogenic CO production (see Section 2.3.1), the increase of the latter must be accompanied by an increase in $r(CO)$ in summer.

By using the previously mentioned (Section 3.1) estimate of increase in northern hemisphere anthropogenic production for 20 years ($\Delta P \cong 4 \times 10^{14}$ g/year or 5.3×10^{-12} g/cm² sec), it is easy to calculate the corresponding increase in the summer CO mixing ratio; by writing equation 7 for 1952 and for 1971 and substracting one from the other, one gets

$$\Delta n(CO) = \frac{5.3 \times 10^{-12}}{3.1 \times 10^{-4}} = 0.017 \times 10^{-6} = 0.017 \text{ ppm}$$

This figure is about 17% of the mean CO mixing ratio. The conclusion about the relative constancy of the summer CO concentration cited earlier is not in contradiction with this estimate since the increase in CO concentration that follows from the photochemical concepts does not exceed the error of measurement.

More detailed calculations than these were performed by Sze (1977). Based on photochemical considerations, he estimated the increase in mean annual CO concentration during the last 20 years to be about 20–38%. These estimates are in agreement with the observed increase in mean annual CO abundance (from 0.08 to 0.10 atm·cm, i.e., 25%).

With a mixing ratio of 0.16 ppm, according to Levy (1971), consumption exceeds production by 3.9×10^{-11} g/cm² sec. Both production and consumption of CO are proportional to the mean OH concentration. Warneck's recent calculations (1975) specify an OH concentration about 70% lower than that mentioned by Levy (1973): 1.3×10^6 and 4.1×10^6 cm⁻³, respectively), the photochemical sink therefore decreasing 70% and amounting to about 1×10^{-11}g/cm² sec, whereas the mean hemispheric anthropogenic production now constitutes about 0.8×10^{-11}g/cm² sec.

The concentration of OH was recently measured in the troposphere almost simultaneously by several groups of investigators, although different methods were used (Wang et al., 1975; Davis et al., 1976). All the measurements coincide within the limit of errors and give figures of approximately 10^{-6}–10^{-7} cm⁻³, which correspond with the theoretical estimates.

Thus, according to the experimentally confirmed theory, intense photochemical reactions with participation of free radicals must take place throughout the atmosphere. Moreover, in summer these reactions must proceed faster than in winter. Carbon monoxide takes part in these reactions, and its concentration tends toward some equilibrium; however, the numerical estimates for the rate of this process are contradictory. A determination of the quantitative role played by photochemical reactions in the CO cycle requires further investigation of tropospheric hydroxyl, mainly its geographic distribution and temporal variations, as well as measurements of the atmospheric concentrations of the other participants in the photochemical reactions.

In the stratosphere, photochemical consumption is very important; as was already pointed out (Section 2.2.2), the stratosphere provides for consumption of about 9.3×10^{12} g of CO per annum.

3.2.3. Liberation of CO from the Interior of the Earth

Carbon monoxide forms a considerable part of volcanic gases, an average of 1–4% (Sokolov, 1971). According to more recent estimates (Voitov, 1975), the earth annually liberates close to 10^8 tons of carbon-containing gases, and hence the upper limit of CO liberated by the lithosphere may be estimated as 10^7 tons per annum. Degasification of the earth's crust and mantle in the modern CO cycle apparently plays no decisive role. Nevertheless, it should be noted that during the prebiologic period of evolution of the atmosphere, when its composition was determined primarily by the composition of the gases liberated from the interior of the earth, CO probably formed a considerable part of the atmosphere, possibly several percent. It is natural to assume that the decrease in CO concentration to the present level (about 10^{-5}%) results from the activity of the biosphere, although in principle the role of other processes, for example that of tropospheric photochemistry, cannot be excluded.

3.2.4. Other Geophysical Sources

(1) Production of CO by forest fires is estimated at 6×10^7 tons per year (Seiler, 1974).

(2) A considerable supersaturation of CO was discovered in rain water (Swinnerton et al., 1971). On the other hand, Seiler's measurements (1974) showed that samples of rain water analyzed 10 min after they were obtained contained negligibly little CO, whereas several hours later its concentration increased severalfold. The cause of this increase is not known, but from Seiler's results (1974) it follows that the CO transfer through the surface of raindrops cannot serve as a significant source of CO in the atmosphere.

(3) A small amount of CO is formed during thunderstorms (White, 1932). No estimate of the global CO production resulting from this process has been made.

Thus, of the natural abiogenic processes, only tropospheric photochemical reactions appear to play an essential role in the CO cycle. However, quantitative estimates of the rates of photochemical processes are not yet sufficiently reliable. The role of the biosphere in the CO turnover is considered below.

4. Role of Biological Agencies in Forming and Binding CO

The biosphere plays a major role in the formation and metabolism of gases present in the atmosphere. From the Gaya hypothesis, it follows that soon after its emergence on earth, life assumed control over the planet's environment, and this homeostasis has existed ever since (Lovelock and Margulis, 1974).

It is generally recognized that the concentration of the chief components of the atmosphere—O_2, N_2, and CO_2—is determined by the biosphere (Vinogradov, 1964). The content of atmospheric trace gases may also to a large extent be regulated by the biosphere. It has now been established that 80% of the methane present in the atmosphere is of biological origin (Ehhalt, 1974) and that a large part of atmospheric hydrogen is also formed as a result of anaerobic microbiological processes. In this and the next sections, we shall summarize the data on the biological production and binding of carbon monoxide. A reservation should be made at the outset that nearly all of the data are of a qualitative and episodic character, but we shall try, wherever possible, to make quantitative estimates.

4.1. Formation of CO by Living Systems

4.1.1. Production of CO by Animals

In mammals CO is formed as a result of the metabolism of heme-containing compounds such as hemoglobin and myoglobin, and the exhaled air therefore contains a small amount of CO, about 3–3.5 ppm (Sjöstrand, 1970; Goldsmith, 1970). About 75% of the total endogenous CO is formed during the decomposition of erythrocytic hemoglobin upon death of the erythrocytes, and part of the CO is formed as a result of catabolism of hemin compounds in the liver and bone marrow (White, 1970). Carbon monoxide is formed from the carbon of the α-methene bridge of the heme during the breakdown of the tetrapyrrole ring. Moreover, 1 mol of heme gives rise to 1 mol each of CO and linear tetrapyrrole, with the formation of bilirubin (Sjöstrand, 1952; Landaw *et al.*, 1970). The biochemistry of these conversions has been studied in detail. According to Sjöstrand (1970), a person exhales an average of 10–12 ml of CO a day, and hence the total production of CO by the human population amounts to about 15×10^3 tons per annum. Animals may be expected to produce up to $25–30 \times 10^3$ tons per year. Thus, this source is very small.

Some marine invertebrates (Siphonophores) produce considerable amounts of CO. The pneumatophores by means of which *Namonia bijuga* effects vertical migrations, descending to a depth of up to 500 m, contain close to 90% CO (Barham, 1963; Pickwell *et al.*, 1964). The rate of CO production reaches 277 μl/mg of the pneumatophore's tissue per hour at 20°C in the presence of serine as a substrate and 115 μl in its absence; at a temperature of 7°C, which is normal for the depths at which the Siphonophores grow, these rates are 50% lower (Pickwell, 1970). The CO is formed by Siphonophores from serine with the participation of tetrahydrofolate (Wittenberg, 1960; Wittenberg *et al.*, 1962). These invertebrates are very widely distributed; they constitute a large part of the plankton in the warm seas and oceans and may make a considerable contribution to the production of CO by the world's ocean.

4.1.2. Formation of CO by Algae and Cyanobacteria

Carbon monoxide is a natural metabolite and is formed *in vivo* by seaweeds and cyanobacteria. As early as 1917, a short report stated that CO was present in the pneumatocysts of *Nereocystis luetkeana*, the concentration of CO amounting to 5–10% of the total volume of gas (Langdon, 1917). The author assumed that CO was a product of respiration because the gas did not appear if the seaweed autolyzed or was mechanically destroyed or macerated (Langdon and Gailey, 1920). Carbon monoxide was also discovered in the pneumatocysts of other brown algae (Chapman and Tocher, 1966). The mechanism of CO formation in these algae is unknown, but O_2 is necessary, whereas cyanide suppresses CO production. The authors held that, although brown algae are the most active CO producers among plants, they are much less active than the Siphonophores. However, owing to their numbers, these organisms can influence CO production in the oceans.

Small amounts of CO are formed by cyanobacteria (Cyanophyceae) and red algae. In an enclosed illuminated system, *Anacystis nidulans* formed up to 800–900 ppm CO in a few days; in the dark, CO did not appear (Wilks, 1959). Troxler *et al.* (1970) and Troxler and Dokos (1973), who studied the mechanism of CO formation by various cyanobacteria and red algae, showed that CO appeared during the formation of phycocyanobilin and phycocrythrobilin, which are related in structure to bilirubin. The precursor of these pigments is a metalloporphyrin, and the CO originates from its α-methene bridge; thus, the pigments in algae and bilirubin in animals may be formed in similar ways. As a source of CO, these organisms apparently play no significant role.

Nonenzymatic, light-stimulated CO formation by tissues of various algae and plants, including macerated and heated tissues, was observed by Loewus and Delwiche (1963). They also described CO formation from a polyphenol component extracted from *Egregia menzies*. Up to 120×10^{-5} ml CO per liter of water was formed in 30 days in sea and distilled water to which was added a soluble organic fraction of plankton (Wilson *et al.*, 1970).

Thus, CO production in the oceans is effected by Siphonophores, algae, and possibly bacteria, as well as photochemical decomposition of algal remains.

4.1.3. Formation of CO by Higher Plants

Small amounts of CO are formed during the germination of seeds of such plants as cucumbers, turnips, peas, lettuce, and rye in an atmosphere containing about 5% O_2 (Siegel *et al.*, 1962). Moreover, neither chlorophyll nor light was required for the formation of CO. It is possible that in higher plants, CO is a product of the breakdown of cyclic tetrapyrroles during the synthesis of the protoheme prosthetic group of phytochrome, which is structurally similar to phycocyanobilin (Troxler and Dokos, 1973). This source is hardly of any significance, decomposition of chlorophyll of higher plants being of much greater interest. Using standard methods for determining CO in blood, Wilks (1959) found CO in excised leaves of various plants, in dried and ground leaves, and in extracts containing chlorophyll. The CO was formed in the light in the presence of oxygen. The author held that his results indicated a natural CO source, the chlorophyll system. Stevens *et al.* (1972) observed a large burst of CO production during active leaf fall. Determination of the isotopic composition of CO in experiments on small isolated areas led the authors to conclude that CO is formed from fallen leaves, and this source was estimated to be 2–5 × 10^8 tons for the northern hemisphere.

According to Seiler's (1975) preliminary findings, plants produce considerable amounts of CO during their growth; he estimated that this source may yield 0.2–2.0 × 10^8 tons of CO per annum. The process operated only in the light, no CO formation being observed in the dark. The biochemical mechanism of the process is unknown, but Fischer and Seiler (1974) assume that CO formation may be associated with the synthesis and decomposition of pigments.

An increase in CO production was observed in the tissues of higher plants in the presence of quercetin. It is not known whether this compound is a normal CO source (Delwiche, 1970).

It may be stated that, according to the scarce and preliminary data, plants may in the course of photosynthesis contribute significantly to CO production.

4.1.4. Production of CO by Bacteria

Very little is known about the ability of microorganisms, other than some algae, to produce CO. The gas is formed during anaerobic growth of microscopic fungi in media containing flavonoids, quercitrin, and quercetin (Westlake *et al.*, 1961). Moreover, *Aspergillus flavus* converts rutin to protocatechuic acid, phloroglucinic acid, rutinose, and CO (Simpson *et al.*, 1960).

Hemolytic *Streptococcus* and *Bacillus* form CO during anaerobic growth in media containing hemin compounds of animal origin, viz., erythrocytes, hemoglobin, myoglobin, cytochrome *c*, hematin, etc. (Engel *et al.*, 1972). Hemolytic bacteria do not produce CO if the iron in the heme is removed or is replaced by copper, nor do they form CO from copperhemoporphyrin, protoporphyrin, or bilirubin. It is assumed that hemolytic bacteria destroy animal hemin compounds in a way similar to the metabolism of hemin compounds in animal tissues. *Bacillus cereus* and hemolytic *Streptococcus mitis* also formed CO from cobalamin during incubation under aerobic conditions (Engel *et al.*, 1973). These authors maintain that there are a number of common features in the catabolism of oxycobalamin and of hemin compounds.

Some yeasts, enterobacteria, and pseudomonads are capable, during aerobic growth in a complex medium containing glucose, of forming trace amounts of CO (0.4–2.6 ppm or $2–8 \times 10^{-6}$ mol CO/mol glucose) (Junge *et al.*, 1971; Radler *et al.*, 1974). With *Saccharomyces cerevisiae*, it was shown that the CO is formed from glucose, its production rising with an increase in the concentration of oxygen. Marine bacteria such as *Alginomonas* and *Brevibacterium* also produce a small amount of CO while growing in a medium resembling seawater and supplemented with organic substances (Junge *et al.*, 1972). The activity of these organisms is so slight and the quantity of the CO_2 they release is so small that the authors concluded that the role of such organisms in CO turnover is insignificant. The same is apparently also true of hemolytic bacteria.

While investigating some chemical properties of submerged soils, Robinson (1930) observed that when the samples of soils were kept long under water, up to 3% of the gas formed was CO. There was no microbiological control, but the gases were very likely formed as a result of the metabolic activities of soil microorganisms. From short-term experiments in which small amounts of CO were liberated from soil at a temperature above 30°C and also at lower temperatures when the soil was in contact with a CO-free gaseous mixture, Seiler and Junge (1970) and Seiler (1974) concluded that microorganisms were involved. However, these data correspond more to the physicochemical process of desorption. Ingersoll *et al.* (1974) observed the liberation of CO from autoclaved soil at a temperature above 30°C, but they concluded that the evolution resulted from chemical decomposition of soil organic matter. In old studies, liberation of small amounts of CO from manure is stated to occur (Corenwinder, 1865; Löhnis, 1910), but no later data with regard to the process are available.

4.1.5. Conclusion

Carbon monoxide is formed as a result of metabolic reactions in mammals, invertebrates (Siphonophores), algae, and, possibly, higher plants. Fungi and bacteria are also capable of producing CO from certain compounds. Concerning the biochemical mechanisms of CO formation, the following may be noted: in animal tissues, in green and red algae and in cyanobacteria, CO is formed upon conversion of heme-containing compounds into tetrapyrroles with an open chain; destruction of hemin compounds by hemolytic bacteria with the liberation of CO apparently takes place in a similar manner; and fungi form CO during decomposition of flavoroids; siphonophores produce CO from serine with the participation of tetrahydrofolate. The biochemical mechanisms of CO formation by brown algae, higher plants, germinating seeds, and non-hemolytic bacteria are still unknown. Decomposition of chlorophyll with the formation of CO is a moot question.

Thus, the data on biological CO formation are scarce, and the available information is only qualitative. It should be noted that all of the biological sources are of a nonspecific and minor character and, it seems, cannot provide the quantity of CO whose necessary existence follows from the geophysical data (Stevens *et al.*, 1972). Only CO production by higher plants, taking into account their considerable biomass—about 99% of all the biomass (Bazilevich *et al.*, 1971)—can yield a substantial influx of CO to the atmosphere. The situation with the ocean as a source of CO is also fairly clear. It may, nevertheless, be concluded that CO is a normal and fairly widespread metabolite in nature. It is tempting to assume the existence of specific bacteria which produce considerable amounts of CO or to suppose that, under certain conditions, CO is formed as an incompletely oxidized or incompletely reduced product; the latter theoretically seems quite likely as, for example, in fermentation. Regrettably, however, there are as yet no such data.

4.2. Fixation of CO in the Biosphere

4.2.1. Reactions with Hemoproteins

Of the biological reactions in which CO is fixed, those with hemoproteins have been studied in greatest detail. As far back as the middle of the last century, Bernard showed that CO combines with the hemoglobin of the blood and that this reaction occurs concurrently with that of oxygen fixation. An extensive literature is devoted to the mechanism of CO reactions and its influence on man and animals (Coburn, 1970a; Chovin, 1974; Goldsmith and Aronov, 1975). In addition to reacting with hemoglobin, CO forms carbonyl complexes with other hemoproteins, e.g., myoglobin, cytochrome oxidase, cytochromes *o* and P-450, catalase, peroxidase, and other iron-containing enzymes. The strength of the bond, its structure, and the electronic configuration of hemocarbonyl depend on the structure of the hemoprotein. The complexes are, as a rule, destroyed in the light (Cayghey, 1970). Cytochrome oxidase combines with CO in the reduced form and, in the presence of oxygen, catalyzes the oxidation of CO to CO_2 (Tzagoloff and Wharton, 1965). Hemoproteins are found in animals, plants, and microorganisms; all of them are in some measure capable of fixing CO, but the extent of these reactions is so small that, in considering the CO turnover, it is hardly worth attaching serious biogeochemical importance to them.

The fixation of CO by humans, intact animals, and tissue preparations has been reported (Gorbatow and Noro, 1948; Clark, 1950; Fenn and Cobb, 1932a,b; Clark *et al.*, 1950; Breckenridge, 1953), and it is known that the fixation of CO in animal systems is largely conditioned by reactions with hemoglobin and cytochrome oxidase. Very little exogenous CO combines with myoglobin and other hemoproteins in the body of animals (Coburn, 1970b). After a critical examination of a series of reports, including his own, Fenn (1970) arrived at the conclusion that, as a sink for CO, humans and animals play no role at all; moreover, other authors (Luomanmäki and Coburn, 1969) believe that the production of CO by humans and animals exceeds its fixation.

Metabolic reactions in which CO is oxidized to CO_2 have been found in green algae (Chappelle, 1962). *Chlorella vulgaris* and some thermophilic strains of *Scendesmus* assimilated ^{14}CO in the light twice as fast as in the dark, the reaction requiring oxygen in both instances. Chappelle (1962) believed that, in algae, CO is converted to CO_2 by direct oxidation involving a metal-containing enzyme, presumably cytochrome oxidase. No quantitative data are available to make it possible to estimate the role of algae in the process of removing CO; such a role is probably insignificant, especially since processes of formation and not fixation of CO predominate in the ocean.

4.2.2. Fixation of CO by Higher Plants

An increase in the consumption of O_2 in the presence of CO and oxidation of the latter, apparently catalyzed by iron-containing enzymes, especially cytochrome oxidase, was observed in mature leaves (Ducet and Rosenberg, 1962). However, higher plants probably also have other ways of binding CO. It was shown (Krall and Tolbert, 1957; Chappelle and Krall, 1961) that CO is slowly fixed by intact leaves of barley, spinach, and cucumbers, as well as leaf extracts, even when the concentration of CO in the gas phase was 60–90%. The fixation depended on photosynthesis and, on the basis of an analysis of early products of CO fixation, it was concluded that CO combines with serine.

Bidwell and Fraser (1972) showed that, at a low concentration of CO in the gas phase (1–360 ppm), the leaves of different plants are capable of fixing CO in the light and in the dark at a comparable rate, although some plants (coleus and legumes) consumed a large amount of gas while others (tomatoes and ferns) consumed little. In the light, the CO combines with serine and with sucrose, while the process of CO_2 formation operates in the dark. The authors concluded that plants are a powerful agent for eliminating CO from the atmosphere. They estimated the rate of CO fixation at 12–120 $kg/km^2/day$ in summertime; on a global scale, this sink can be estimated to be 3–30 $\times$ 10^8 tons per year, if one considers the mean time for active growth to be 6 months. However, other authors regard these figures as high (Seiler, 1975); moreover, the results of Fischer and Seiler (1974) indicate a contrary process, viz., large-scale formation of CO by higher plants (Fischer and Seiler, 1974; Seiler, 1975). Delwiche (1970) also observed both formation and fixation of CO by higher plants. The question of the role played by the surface vegetation in CO turnover can probably still be considered open.

Thus, according to the data now available, man, animals, higher plants, and algae are not major sinks for atmospheric CO, although they probably play a certain role in maintaining the CO balance in nature.

4.2.3. Absorption of CO by the Soil

The soil with its microorganisms is considered the most important natural agent for eliminating CO from the lower layers of the atmosphere (Jaffe, 1973; Seiler and Junge, 1970; Seiler, 1974, 1975; Stevens et al., 1972). In a number of studies, beginning in the first third of the 20th century, it was shown that CO disappeared from a gas mixture in contact with the soil, sewage, and other natural substrates, under both aerobic and anaerobic conditions. Thus, Wehmer (1926) discovered that garden soil placed in an atmosphere of lighting gas consumed CO. Sterilized soil did not yield such an effect. The author concluded that CO is used by the microorganisms present in the soil. It was shown that the fermented residue from methane tanks causes anaerobic conversion of CO to CH_4 and CO_2 (Fisher et al., 1930, 1931). Bacteria capable of catalyzing these reactions were subsequently isolated (Barker, 1936). From mine gases, CO disappeared faster than could be explained on the basis of gas diffusion (Jones and Scott, 1939a,b). After investigating various soils, especially those containing added organic matter and sewage, the authors concluded that CO and H_2 oxidation results from the activity of microorganisms, which, however, they were unable to isolate.

Extensive studies of the consumption of CO by soils from different areas of the United States under field and laboratory conditions showed that the binding capacities of various soils are within 7.6–115 mg $CO/hr/m^2$ when the soil was brought in contact with air having an initial CO concentration of 0.01% (Inman et al., 1971; Ingersoll et al., 1974). Tropical soils were the most active, and desert soils were the least active. Cultivated soils were always less active than analogous soils from nearby areas with natural vegetation. Samples of soil taken near roads were always more active than analogous soils taken from areas away from roads. The binding activity of the soil increased with an increase in the concentration of CO in the air with which the soil was in contact. The authors therefore concluded that the high activity of roadside soils is conditioned by permanent contact with a high CO level. The rate of consumption of CO by the soil depended on the temperature, and the gas was most intensively absorbed at 30°C. Soil treated with steam or antibiotics did not absorb CO even after 30 days of incubation in an atmosphere containing CO. On introduction of a small amount of active soil into the treated soil, the capacity to absorb CO was gradually restored, the curve reflecting the increase in the binding activity of the soil corresponding to the typical curve of development of a microbial population. As a result, the authors concluded that the activity of a soil is a reflection of the metabolism of the soil microflora and not of the physical or physicochemical absorption to soil colloids. A similar conclusion was reached by German researchers (Seiler and Junge, 1970; Seiler, 1974). On bringing soil in contact with air containing 0.2–0.5 ppm of CO at a temperature of 10–30°C, a decrease in the CO concentration to 0.02 ppm was observed.

The estimates of the binding activity of the soil made by these two groups of researchers differ; thus, Ingersoll et al. (1974) estimate the total activity of the world's soils at 14.2×10^9 tons per year. Seiler's (1974) estimate is about 96% lower, i.e., 4.5×10^8 tons per year. This may be explained by the fact that the German researchers worked with low CO concentrations (0.2–0.5 ppm), which correspond to those in an unpolluted atmosphere. On the other hand, the findings of the group of American researchers reflect the potential capacity of the soil to increase its binding activity with an increase in the concentration of CO in the gas phase.

Thus, of all the above possible biologic agents for removing CO from the atmosphere, only the soil may really be an important sink for CO. Many experiments convincingly show that the active absorption of CO by the soil is determined by the presence of a microflora. The process of binding CO may be either the result of nonspecific action by components of the soil microflora or of a certain specific group of microorganisms for which CO is of fundamental physiological importance. Only in the latter case can the soil be a CO sink of considerable scale and rapid action.

4.2.4. Fixation of CO by Nonspecific Microflora

All microorganisms having hemoproteins and cytochrome oxidase are capable of binding CO to some slight extent. This process can also be catalyzed by microorganisms which have other enzymes whose accessory function is a reaction with CO. Its oxidation by the methane-oxidizing bacteria *Methylomonas albus, Methylosinus trichosporium,* and *Pseudomonas methanica* may serve as an example of such nonspecific fixation of CO (Hubley *et al.*, 1974; Ferenci, 1974). In *P. methanica*, CO is oxidized by a monooxygenase; the process is associated with the oxidation of NADH, while ethanol can serve as a reducing agent (Ferenci *et al.*, 1975). No increase in number of methane-oxidizing bacteria with CO as a substrate has been discovered; moreover, their increase in the presence of CO was suppressed completely with methane as the carbon source and only partly with methanol (Namsarayev, 1974).

Inman and Ingersoll (1971) reported that they had isolated 16 strains of fungi capable of utilizing CO. These fungi belonged to the genera *Penicillium, Aspergillus, Mucor, Haplosporangium,* and *Mortierella.* However, their study is devoid of proof that this process is catalyzed by pure cultures of the fungi.

There are several frequently quoted reports on CO fixation in anaerobic processes. Kluyver and Schnellen (1947) found that *Methanosarcina barkerii* and *M. formicicum* are capable of converting a mixture of CO and H_2 to methane or of forming methane from CO in the absence of H_2 with intermediate formation of CO_2 and H_2. Stephenson (1949) mentioned the capacity of *Clostridium welchii* to bind CO. It was recently shown that growing cultures of *Clostridium pasteurianum* and a cell-free extract from this anaerobe are capable of oxidizing CO to CO_2 at a maximum rate of about 15 nmol/min/mg with flavin nucleotides as electron acceptors (Thauer *et al.*, 1974; Fuchs *et al.*, 1974). The reaction proceeds even at such low concentration of CO as 1 nM, which is equivalent to 1 ppm of CO in the gas phase.

Oxidation of CO to CO_2 by a cell-free extract from anaerobically grown cells of *Desulfovibrio desulfuricans* (*D. vulgaris*) was investigated by Yagi (1958, 1959), who showed that the reaction proceeds in the presence of sulfite as a reducer or in its absence when water serves as the oxidant and that molecular hydrogen is liberated. During normal growth, the organism did not utilize CO.

Hirsch (1968) reported the isolation of *Rhodopseudomonas*, a photosynthesizing bacterium capable of developing under anaerobic conditions in the light using CO as a reductant. He suggested that growth of the microorganisms in an atmosphere containing CO is possible in the light because of photodissociation of the CO and hemoprotein complex. In our recent work (Keppen *et al.*, 1976), this finding was questioned because CO did not ensure the growth of *Rhodopseudomanas sulfidophila, R. palustris,* and *Rhodomicrobium vannielii,* either in the light or in the dark.

There is no reason to assume that CO fixation is an important function for any of the previous organisms or that they can play an important role in binding atmospheric CO. Of much greater interest are microorganisms for which oxidation of CO serves as a source of energy, i.e., CO is used as a substrate of growth. In this case, CO is fixed irreversibly, and in response to the increase in CO in the gas phase, these organisms must increase in number, and the rate of consumption of the substrate rises also. The following section describes this group of microorganisms.

5. Carboxydobacteria

5.1. History of Research

Beijerinck and van Delden (1903) are considered to be the first to have discovered the microorganisms that utilize CO for growth. These authors isolated *Bacillus oligocarbophilus,* which develops as a nonwettable film on the surface of an inorganic salts medium that does not contain an added source of carbon. They held that, to satisfy its energy and biosynthetic requirements, the organism used a volatile substance present in laboratory air. It was concluded by Kaserer (1906) that this substance was CO. Isolation of this organism was subsequently reported several times (Kaserer, 1906; Lantzsch, 1922; Hasemann, 1927), and it was long thought to be the main representative of the aerobic CO-oxidizing bacteria. However, a detailed critical analysis of these investigations carried out by Kistner (1953) led him to the conclusion, with which we are in total agreement (Zavarzin, 1972; Nozhevnikova,1974a), that the results of these studies are unsatisfactory. In the present review there is no need for repetition of the objections, but it should be noted that, after numerous trials, including one carried out by Nozhevnikova (1974a), not one of the strains listed as *Bacillus oligocarbophilus* (synonyms: *Actinomyces oligocarbophilus, Actinobacillus oligocarbophilus,* and *Carboxydomonas oligocarbophilus*) in the collection of microorganisms at the Technical University in Delft (Holland) and in the All-Union Collection of Microorganisms (USSR) has proven capable of growing by consuming CO.

Thus, although the existence of a specific CO-oxidizing microflora has been assumed since Beijerinck's publication, actually, only a few studies may be regarded as proof of their existence. Mention should be made of the work of Kistner, who succeeded for the first time in isolating a pure culture of the bacterium *Hydrogenomonas carboxydovorans*, which aerobically oxidizes CO and uses the energy from the oxidation for its growth (Kistner, 1953, 1954). However, the organism he described was later lost, or it lost its properties (Davis *et al.*, 1970). Japanese researchers reported the isolation from the soil of oil-bearing areas of two unidentified CO-oxidizing bacteria capable of growing without fixed nitrogen (Shinohara and Ooyama, 1972). A series of studies on CO-oxidizing bacteria was recently carried out at the Institute of Microbiology of the USSR Academy of Sciences (Nozhevnikova, 1974a,b).

5.2. Characteristics of the Physiological Group

5.2.1. Isolation of Carboxydobacteria

Bacteria that oxidize CO, or carboxydobacteria, are widely distributed in nature. They are present in various soils, waters, and the slime of water reservoirs. A study of more than 100 samples of soils from various areas of the USSR revealed the presence of a CO-oxidizing microflora in about half of the samples. However, the best source for isolating these organisms is soil of polluted urban areas, mud of water reservoirs polluted by sewage, or the sewage itself.

The standard method of elective culture was used for the isolation of carboxydobacteria. The culture was obtained from a liquid mineral medium that contained no source of added carbon and in an atmosphere containing 30–80% CO and 15–20% oxygen. The procedure for isolating these organisms has often been described (Kistner, 1953; Sanzhiyeva and Zavarzin, 1971; Nozhevnikova and Zavarzin, 1973; Zavarzin and Nozhevnikova, 1977). The existence of an atmosphere with a high CO content is no guarantee against development of saprophytic forms; even after a series of consecutive inoculations into a mineral medium in an atmosphere with CO, inoculation into a peptone or potato agar yielded not less than five to six morphologically different forms of bacteria of which only one or two proved capable of growing in pure culture by oxidizing CO. The capacity of pure cultures to oxidize CO must be checked thoroughly because many microorganisms grow by using volatile organic substances in the air. *Nocardia autotrophica* Z-401 was isolated in a carbon-free mineral medium in an atmosphere of 30% CO and 70% air. After a careful determination, including elimination of traces of organic substances in the air and in the medium, it was found that the organism is on contaminants in air, for autotrophic growth in hydrogen, and for methylotrophic growth in methanol (Kryukov and Nozhevnikova, 1977).

In our opinion, these results help explain the early findings, especially those of Lantzsch (1922).

It should be noted that the method of elective culture ordinarily used for isolating chemoautotrophs does not make it possible to isolate lithotrophic organisms which need organic substances as growth factors. However, under natural conditions, microorganisms have close and complex interrelationships and do not exist as isolated species. On isolation of carboxydobacteria, protocooperation was discovered in two organisms, *Pseudomonas gazotropha* and *Comamonas compransoris*. It was impossible to separate the two-membered culture of these bacteria until a mixture of B vitamins was added to the mineral medium. It turned out that *P. gazotropha* needed an addition of 20 µg/liter of vitamin B_{12} and *C. compransoris*, 250 µg/liter of thiamine. Inoculation of each organism into filtrates of the second organism gave positive results (Nozhevnikova, 1974a,b). The strain of *Hydrogenomonas carboxydovorans* isolated by Kistner (1953, 1954) needed small amounts of peptone or sewage in the medium, and this addition probably supplies needed growth factors.

5.2.2. Morphological and Cultural Characteristics

All the CO-oxidizing bacteria described in this paper are gram-negative, motile rods multiplying by binary fission, forming no spores, possessing no hydrolytic activity, and being incapable of development under anaerobic conditions, but they differ from one another in form and size of cells, number and site of attachment of flagella, pigmentation, appearance of colonies, and character of growth in liquid media. The conditions of cultivation do not noticeably affect their morphology. They are all mesophiles, and the highest rate of growth is observed at 23–28°C. Most of them, with the noteworthy exception of *Achromobacter carboxydus* Z-1171, grow well in an atmosphere of hydrogen, oxygen, and CO_2. All carboxydobacteria grow organotrophically using a number of organic substrates as the sole sources of energy and carbon. Vitamin-dependent strains can grow in media with organic substrates only after the addition of vitamins. The cultures differ considerably in their capacity for utilizing organic compounds. All the organisms utilize organic acids and low-molecular-weight alcohols, and none makes use of formate; true, such information is not available for Kistner's *H. carboxydovorans*. *H. carboxydovorans*, *A. carboxydus*, and *P. carboxydoflava* utilize carbohydrates, and the pseudomonad also uses many amino acids, whereas *P. gazotropha* and *C. compransoris* use only some organic acids and amino acids. The tendency is that the better the growth in a CO atmosphere, the fewer organic compounds are used. When a gradient of organic acids is provided as substrate, the best growth of carboxydobacteria is observed at low concentrations (Zavarzin, 1976). The capacity of some carboxy-dobacteria to utilize methanol is noteworthy. For *P. gazotropha*, methanol is one of the best substrates. This organism was the first example of the capacity for growing by means of three types of nutrition—organotrophy, autotrophy, and methylotrophy.

In determining the taxonomic position of CO-oxidizing bacteria, it was found possible to distinguish them as independent species, ascribing them to genera whose species they resemble in characteristics of organotrophic growth. For the physiological group of bacteria capable of growing on CO as the sole source of energy and carbon, the trivial name of carboxydobacteria was suggested (Nozhevnikova and Zavarzin, 1974). Kistner (1954) placed his isolate in the genus *Hydrogenomonas* in accordance with the then existing classification, although by its common characteristics the organism was a typical pseudomonad. The traits of carboxydobacteria have been described by Nozhevnikova and Zavarzin (1974) and Zavarzin and Nozhevnikova (1977), and their main characteristics are given in Table I.

5.2.3. Growth of Carboxydobacteria under Autotrophic Conditions

A study of the growth of carboxydobacteria under autotrophic conditions has shown that these organisms differ not only in trivial characteristics but also in their relation to the composition of the gas mixture, in particular, the concentration of CO. Upon growth with CO as the only source of carbon and energy, the yield of *Pseudomonas gazotropha* Z-1156 increased with an increase in the CO content in the mixture up to 80%. *H. carboxydovorans* was also resistant to high concentrations of CO (Kistner, 1954). However, for most carboxy-dobacteria high concentrations of CO are toxic. Thus, for *Seliberia carboxydohydrogena* and *Pseudomonas carboxydoflava*, the maximum yield in a stationary culture was observed at 20% CO, and the maximum rate of growth for *Achromobacter carboxydus* was at 40% CO, while *Comamonas compransoris* grew best at 10% CO (Nozhevnikova, 1974a, b; Zavarzin and Nozhevnikova, 1977). It should be noted that in these experiments the cultures grew under stationary conditions without a change in the atmosphere, and the above CO concentrations were actually only those present initially. However, under these conditions when there was a clear deficiency of CO as substrate, satisfactory growth of all carboxydobacteria was observed at CO concentrations of 3–5%. The growth at low concentrations should be studied in systems with a continuous inflow of a gas mixture.

In a mixture of hydrogen, oxygen, and carbon dioxide, carboxy-dobacteria grow much faster than in an atmosphere with CO. Under

Table I. Main Properties of Carboxydobacteria

Organism	Place of isolation	Morphological characteristics	Colonies	Character of growth in liquid media	GC (mol %)	Vitamin requirements	Utilization of substrates		
							Carbohydrates	Organic and amino acids	CH$_3$OH
Hydrogenomonas carboxydovorans, Kistner (1953)	Sewage	Rods, mono-trichous	Colorless, then yellow to brown, 1 mm	Diffuse		0.02% peptone or 1% sewage	Glucose, sucrose, lactose	Lactate	
Seliberia carboxydo-hydrogena Z-1062, Sanzhiyena and Zavarzin (1971)	Mud drum slime	Bacilli, 0.3 × 1-2 μm forming rosettes	Colorless, then brown, 1 mm	Thin film	58.4±0.34		0/10	10/28	
Pseudomonas car-boxydoflava Z-1107, Nozhevnikova and Zavarzin (1974)	Soil	Rods, monotrichous, 0.7 × 1.5-3 μm	Yellow, lobate, up to 3 mm	Diffuse with flakes	63.7±0.20		9/10	17/28	±
Pseudomonas gazo-tropha Z-1156, Nozhevnikova and Zavarzin (1974)	Slime	Oval monotrich-ous cells, 0.7 × 1.1 μm	Microscopic, colorless	Strong mucous film	66.9±0.49	B$_{12}$	0/10	7/28	+
Comamonas compran-soris Z-1155, Nozhev-nikova and Zavarzin (1974)	Slime	Curved bacilli, lophotrichous, 1 × 2.5 μm	Flat, colorless opaque, up to 1.5 mm	Strong loose whitish film	56.0±0.50	B$_1$	0.10	10/28	±
Achromobacter carboxydus Z-1171, Nozhevnikova and Zavarzin (1974)	Soil	Rods, 0.8 × 2 μm, peri-trichous	White 1-2 mm	Diffuse	48.6±0.40		9/10	20/28	–

conditions of cultivation ordinarily used for growing hydrogen bacteria (Schlegel *et al.*, 1961; Saveliyeva and Trykova, 1969) in a gas mixture of H_2, O_2, and CO_2 (7:2:1), *Seliberia carboxydohydrogena* Z-1062 developed as a typical hydrogen bacterium and displayed the change of phases characteristic of organisms using volatile nutrients. In an atmosphere of CO and oxygen (4:1), this organism grows about 80% slower than in an atmosphere with hydrogen (Saveliyeva and Nozhevnikova, 1972). A similar behavior is observed with the other carboxydobacteria.

An addition of CO to the mixture of H_2, O_2, and CO_2 noticeably inhibits the growth of carboxydobacteria but never completely suppresses it; however, these bacteria have different responses to the concentration of CO in the mixture. An addition of CO to hydrogen did not suppress appreciably the growth of *P. gazotropha* Z-1156 and *S. carboxydohydrogena* Z-1062, which grew at a fairly rapid rate even when the mixture contained 20% CO (Nozhevnikova, 1974a,b; Saveliyeva and Nozhevnikova, 1972).

Thus, the physiological group of carboxydobacteria is now represented by bacteria belonging to the genera closely related to *Pseudomonas*. These organisms are related in their properties to hydrogen bacteria. However, not only are the strains of various species of hydrogen bacteria in culture collections incapable of growing on CO, but their growth under hydrogen is suppressed even by a small amount of CO.

5.3. Utilization of CO by Carboxydobacteria

Because of the uniqueness of CO as a substrate, it may be assumed that the metabolism of carboxydobacteria possesses a number of special features. When these organisms grow by the oxidation of hydrogen, their metabolism, as far as is known, does not differ from the autotrophic type of metabolism of the true hydrogen bacteria. The oxidation of molecular hydrogen to water is the energy-yielding reaction, the energy from the oxidation being used for assimilation of carbon dioxide via the Calvin cycle and for biosynthetic purposes. Study of the autotrophic assimilation of carbon dioxide by *Seliberia carboxydohydrogena* Z-1062 grown under hydrogen—including an analysis of the change in the rate of carbon dioxide assimilation in growing cultures, an investigation of the early products of chemosynthesis, a study of the kinetics of appearance of the ^{14}C label from CO_2 in these products, and the discovery of the activity of key enzymes of the reductive pentose phosphate cycle—has shown that the biosynthetic metabolism of this organism closely resembles the anabolism of hydrogen bacteria, *Hydrogenomonas eutropha* Z-1 in particular (Nozhevnikova and Saveliyeva, 1972; Nozhevnikova, 1974a,b).

When growing by utilizing CO as their source of energy, carboxydobacteria oxidize the gas in accordance with the following equation:

$$24\ CO + 11\ O_2 \longrightarrow 23\ CO_2 + 1\ cell\text{-}C \qquad (9)$$

When growing in a CO atmosphere, these organisms accumulate about 1 g of dry substance in the oxidation of 1 mol of CO, whereas hydrogen bacteria accumulate about 5 g of dry biomass during the oxidation of 1 mol of H_2. In the cells of carboxydobacteria, CO and H_2 are apparently oxidized in different ways, or at least the oxidation of CO requires additional enzymatic activities. This assumption is supported by the capacity, discovered in *S. carboxydohydrogena*, for adaptation to CO or H_2 as energy-yielding substrate, which was shown in experiments with neotetrazolium, an artificial electron acceptor, and in experiments on the assimilation of labeled carbon dioxide. Higher rates of oxidation of the substrate and of assimilation of carbon dioxide were observed in an atmosphere corresponding to the composition of the gas mixture in the growth vessel (Nozhevnikova and Saveliyeva, 1972; Nozhevnikova, 1974a,b). Similar results were obtained with the aid of the manometric method of Kistner, who observed adaptation of *Hydrogenomonas carboxydovorans* to H_2, CO, and lactate (Kistner, 1954).

In view of the resistance of carboxydobacteria to high concentrations of CO, which is an inhibitor of cytochrome oxidase, one may assume that growth linked to utilization of CO is associated with some peculiarities in the cytochrome portion of the electron transport chain. However, a study of the cytochromes of various species of carboxydobacteria grown in H_2, CO, and heterotrophically has shown that, regardless of the cultivation conditions, the cells contain the complete cytochrome chain, including cytochromes *b*, *c*, *a*, and *o*. Similarly, quantitative differences in cytochromes were not discovered (Lebedinsky *et al.*, 1976).

Since CO is an inorganic substance, although its level of reduction is close to that of a one-carbon organic compound such as formate, several assumptions may be made as to the type of metabolism of carboxydobacteria using CO. The metabolism may be lithoheterotrophy, lithoautotrophy, or methylotrophy. Elective methods of isolating and maintaining the cultures of carboxydobacteria warrant the exclusion of the lithoheterotrophic type of metabolism for these organisms. *Hydrogenomonas carboxydovorans*, which grows only in the presence of an organic substance, is an exception. Although a requirement for growth factors appears likely, heterotrophic anabolism cannot be excluded for this organism.

Of the two other types of metabolism, autotrophy appears more like for the carboxydobacteria. This is especially indicated by the capacity of most carboxydobacteria to grow well in a mixture of H_2, O_2, and CO_2 and the inability of these organisms to grow with formate as the sole source of carbon and energy. In isotopic experiments with ^{14}CO, $H^{14}CO_3^-$, and $^{14}CO_2$, various species of carboxydobacteria incorporated labeled carbon in acid-stable products of the cells. However, the kinetics of accumulation of the label by the cells differed (Nozhevnikova, 1974a; Zavarzin and Nozhevnikova, 1977). The cells incubated in an atmosphere of unlabeled CO and $H^{14}CO_3^-$ began to assimilate carbon dioxide at once and at a constant rate; moreover, the presence of CO as an energy substrate was necessary. The labeled carbon from ^{14}CO, in the absence of exogenous bicarbonate, was incorporated in the acid-stable products of the cells after a lag period, with the formation of an acid-unstable intermediate(s). The addition of unlabeled carbon dioxide reduced the rate of incorporation of the label from ^{14}CO. The same was observed when the $^{14}CO_2$ formed from the labeled CO was absorbed by NaOH. These findings indicate that the CO-carbon is assimilated by the cells after oxidation of CO to carbon dioxide. A study of the early products of assimilation of ^{14}CO and $^{14}CO_2$ in a CO and O_2 atmosphere (4:1) by *S. carboxydohydrogena* Z-1062 and *A. carboxydus* Z-1171 has shown that the first compounds to be labeled were mono- and diphosphoric esters of sugars and phosphoglyceric acid; this finding indicates that the assimilation of carbon is primarily by the pentose phosphate cycle (Romanova *et al.*, 1977). The activity of ribulose diphosphate carboxylase in the extracts of *S. carboxydohydrogena* grown in CO was similar to that in cells grown with hydrogen and amounted to 0.087 μM $H^{14}CO_3^-$/min/mg of protein. The rather early appearance of the label in asparatic and glutamic acids indicates carboxylation of phosphoenol pyruvate (Nozhevnikova, 1974a,b). Thus, these findings suggest that carbon assimilation in these organisms, when grown in CO, is by the classical scheme of autotrophic carbon dioxide assimilation.

5.4. Conclusion

Thus, in nature there is a group of microorganisms which specifically utilize CO as an energy substrate for their growth. The carboxydobacteria described in this chapter do not, of course, include all members of this group of microorganisms. The elective culture method of isolation has made it possible to isolate mesophilic cultures capable of developing at a fairly fast rate in a mineral medium containing no organic substances or other sources of carbon and in an atmosphere with a high CO content. The following organisms were not considered in this investigation: psychrophilic and thermophilic strains; organisms with a slower rate of growth than those described here; bacteria favored by low concentrations of CO; bacteria needing growth factors, which may be various vitamins, amino acids, and other substances; organisms with a mixotrophic type of nutrition, requiring organic substances for biosynthetic metabolism, etc. The findings nevertheless indicate that carboxydobacteria represent a vast, diverse, and widely distributed group of microorganisms. These organisms are found in soils and water reservoirs. The successful attempts to isolate them from urban samples may indicate that, in response to an increase in the CO content in the environment, these organisms may increase in numbers and thus serve as an increasingly active CO sink. Moreover, carboxydobacteria may influence the composition of atmospheric CO. It was shown (Zyakun *et al.*, 1976) that, in developing in an atmosphere containing CO, carboxydobacteria fractionate the gas, selectively metabolizing and assimilating the lighter isotope, ^{12}C. The residual CO becomes enriched with heavy carbon. It should also be noted that the capacity of the carboxydobacteria to develop at a fast rate in gas mixtures with a high CO content may indicate high local concentrations of CO in their habitat.

6. Comparative Role of Different Factors in the CO Cycle

A number of calculations of the balance of CO in the atmosphere were published during the last decade (Robinson and Robbins, 1972; Seiler, 1974, 1975; Liebl and Seiler, 1975). The estimations for different sources and sinks are summarized in Table II. As is evident from the data presented, attempts to establish CO balances would be arbitrary because of the wide variations in the data cited by different authors. In attempts to elucidate the main features of the CO cycle, it seems more fruitful to consider variations in atmospheric CO content in time and space rather than attempting to give a global balance.

The main feature of the global distribution of CO is its higher content in the northern than in the southern hemisphere. Moreover, the maximum content of CO is observed in the northern hemisphere in winter and spring and the minimum in summer. In the last 20 years, the concentration in winter has tended to increase, while the level in summer has remained constant. During the warm season, the CO content is apparently regulated by intense natural mechanisms of removal.

As for the abiogenic processes, various authors regard photochemical processes in the troposphere as very intense and believe that these processes are an important part of the net CO sink. The seasonal variation in the rate of photochemical elimination of atmospheric CO was estimated by Warneck (1975). His data are in qualitative agreement with seasonal variations in CO concentration observed in our experiments. However, the considerable uncertainties in the quantitative estimates do

Table II. Main Sources and Sinks of Atmospheric Carbon Monoxide
($\times$ 10^6 Tons per Year)

Sources and sinks	Total		Northern hemisphere	
	Min	Max	Min	Max
Sources				
Abiogenic				
Anthropogenic	600[a]	1,000[a]	540	900
Methane oxidation	400[a]	4,000[b]	200	2,000
Biogenic				
Oceans	100[c]	220[d]	40	90
Continental plants	20[e]	200[e]	14	140
Chlorophyll degradation	300[f]	700[f]	200	500
Sinks				
Abiogenic				
Flux across equator	—	—	160[g]	—
Oxidation by OH	600[a]	8,000[b]	400	5,000
Biogenic				
Soil	450[c]	14,000[h]	300	9,000
Continental plants	300[i]	3,000[i]	200	2,000

[a]Seiler (1975).
[b]McConnell et al. (1971).
[c]Seiler (1974).
[d]Linnenbom et al. (1973).
[e]Fisher and Seiler (1974).
[f]Stevens et al. (1972).
[g]Newell et al. (1974). A single value.
[h]Ingersoll et al. (1974).
[i]Bidwell and Fraser (1972).

not as yet warrant the use of these data in interpreting the observed CO variations.

From the data presented in Table II, it is possible to conclude that the major CO sinks are the soil microflora and CO fixation by higher plants. However, the role of vegetation in CO consumption remains obscure. Moreover, there are data (Fischer and Seiler, 1974) suggesting that photosynthetic organisms represent a net source of CO in the atmosphere. This assumption is in good agreement with a set of different biological observations (Wilks, 1959; Delwiche, 1970; Stevens et al., 1972) as well as with the higher CO concentrations in the northern hemisphere, even after subtraction of anthropogenic production. Thus, of the biological sinks, only CO fixation by the soil may ensure the necessary rate. The discovery of a group of microorganisms which specifically utilize CO for growth once more indicates the microbiological nature of the activity in soil. It should be noted that the metabolism of these organisms could make the soil a very large CO reservoir. The results of experiments that showed a gradual increase in the binding activity of the soil with an increase in the CO concentration are in line with proliferation of this specific group of microorganism (Ingersoll et al., 1974). The soil's

CO-oxidizing microflora is apparently more diverse than are the representatives of the group of carboxydobacteria described to date. At any rate, it must be characteristic of the various climatic and soil zones, which is, in part, revealed by the results with the tundra soil that had a fairly high CO-binding activity (Ingersoll *et al.*, 1974).

Considering the possible role played by microbiological processes, the anthropogenic production of CO, fluxes across the borders of the troposphere, and the consumption of CO by the soil, we shall examine the troposphere of the northern hemisphere as a separate, adequately mixed reservoir. We shall consider the anthropogenic production of CO in the northern hemisphere as 2×10^{14} g/year in 1952 (Bates and Witherspoon, 1952), 6×10^{14} g/year in 1971 (Seiler, 1974), and 16×10^{14} g/year in the year 2000 (Sze, 1977). To calculate the fluxes of CO to the southern hemisphere and the stratosphere, we shall use the data of Newell *et al.* (1974). We shall take the temperature dependence of CO absorption by the soil as follows:

$$L(t) = L_{\max} f(t) \qquad (10)$$

Function $f(t) = L(t)/L_{\max}$ will be taken from the laboratory measurements of Ingersoll *et al.* (1974). For our calculations we shall use the mean climatic seasonal trend of the temperature of the soil surface near Moscow (Fig. 2B).

Curve 1 in Fig. 2A represents an increase in the CO content of the northern hemisphere in 1971, taking into account only anthropogenic emissions and the transfer of CO to the southern hemisphere. The curve is in good agreement with experimental data for the period October through March but is not in good agreement for April through September. Curve 2, which is in agreement with all the experimental data, was constructed taking into account CO consumption by soil. The rate of CO consumption, L, presented in Fig. 2B, was calculated using equation 10. $L_{\max}$ was chosen from the best approximation of the experimental data and the constant summer content. Curves 3 and 4 correspond to the years 1952 and 2000.

Thus, the mean rate for the CO sink (in the northern hemisphere in July) required to maintain a constant concentration must be about 1×10^{-11} in 1952, 2×10^{-11} in 1971, and 5×10^{-11} g/cm^2 sec in 2000. The rates of CO absorption by the soil obtained from the experimental data are estimated, in the same units, to be from $0.8–2.5 \times 10^{-11}$ (Seiler, 1974; Liebl and Seiler, 1975) to $6–50 \times 10^{-11}$ g/cm^2 sec (Inman *et al.*, 1971; Ingersoll *et al.*, 1974). It will be noted that the last figures reflect the potential capacity of the soil to fix CO should its concentration in the environment increase. In any case, these data are of the order of magnitude for the necessary CO removal rates. Microbiological processes of CO consumption are consistent with the seasonal variation in CO concentration; if any other process is involved in CO consumption or production, it should exhibit the same seasonal pattern as do microbiological processes. The consumption of CO by soil and its microflora probably accounts for the constancy of the CO concentration in the atmosphere, in spite of significant increases in anthropogenic production in the last 20 years. Furthermore, the seasonal variation in the rate of microbial consumption agrees well with the observed variations in atmospheric CO concentration.

ACKNOWLEDGMENT

We express our gratitude to G. A. Zavarzin, Associate Member of the USSR Academy of Sciences, and Doctor V. I. Dianov-Klokov for their interest and valuable assistance.

References

Barham, E. G., 1963, Siphonophores and deep scattering layer, *Science* **140**:826.
Barker, H.A.,1936,On the biochemistry of the methane fermentation,*Arch.Mikrobiol.* **7**:404.
Bates, D. R., and Witherspoon, A. E., 1952, The photochemistry of some constituents of the earth's atmosphere (CO_2, CO, CH_4, N_2O), *Mon. Not. R. Astron. Soc.* **112**:101.
Bazilevich, N. I., Rodin, L. E., and Rozov, N. N., 1971, Skolko vesit zhivoe veshchestro planety? *Priroda* **1971**(1):46.
Beijerinck, M. W., and van Delden, A., 1903, Über eine farblose Bakterie, deren Kohlenstoff-nahrung aus der atmosphärischen Luft herrührt, *Zentralbl. Bakteriol. Parasitenk.* Abt. 2, **10**:33.
Bernard, C., 1859, Sur la quantité d'oxygène que contient le sang veineux des organes glanduaries, á l'ètat de fonction et à l'état de repos; et sur l'emploi de l'oxyde carbone pour déterminer les proportions d'oxygène du sang, *C. R. Acad. Sci.* **48**:393.
Beryland, M. Y., 1972, Sovremennyie problemy atmosfernoi diffuzii i obespecheniye chistoty atmosfery, *Tr. Vses. Nauchno-Issled. Inst. Meterol.* **3**:164.
Bezuglaya, E. Y., and Rastorguyeva, G. P., 1973, Air pollution in cities of different countries, *Tr. Gl. Geofiz. Obs.* **293**:215.
Bidwell, R. G. S., and Fraser, D. E., 1972, Carbon monoxide uptake and metabolism by leaves, *Can. J. Bot.* **50**:1435.
Bolin, B., and Rhode, H., 1973, A note on the concepts of age distribution and transit time in natural reservoirs, *Tellus* **25**:58.
Breckenridge, B., 1953, Carbon monoxide oxidation by cytochrome oxidase in muscle, *Am. J. Physiol.* **173**:61.
Burenin, N. S., 1974, Nekotoryie rezultaty nablyudenyi za zagryazneniyem vozdukha na avtomagistralyakh, *Tr. Gl. Geofiz. Obs.* **314**:136.
Burenin, N. S., and Solomatina, I. I., 1975, Ob opredelenii vklada vybrosov avtotransporta v zagryazneniye vozdushnogo basseina gorodov, *Tr. Gl. Geof. Obs., Vyp.* **352**:191.
Cayghey, W. S., 1970, Carbon monoxide binding in hemeproteins, *Ann. N.Y. Acad. Sci.* **174**:148.
Chapman, D. J., and Tocher, R. D., 1966, Occurrence and production of carbon monoxide in some brown algae, *Can. J. Bot.* **44**:1438.
Chappelle, E. W., 1962, Carbon monoxide oxidation by algae, *Biochim. Biophys. Acta* **62**:45.
Chappelle, E. W., and Krall, A. R., 1961, Carbon monoxide fixation by cell-free extracts of green plants, *Biochim. Biophys. Acta* **49**:578.
Chovin, P., 1974, Le monoxide de carbon, *Environ. Qual. Safety* **3**:47.

Clark, R. T., 1950, Evidence for conversion of carbon monoxide to carbon dioxide by the intact animal, *Am. J. Physiol.* **162**:560.

Clark, R. T., Stannard, J., and Fenn, W. O., 1950, The burning of CO to CO_2 by isolated tissues as shown by the use of radioactive carbon, *Am. J. Physiol.* **161**:40.

Coburn, R. F., 1970a, Biological effects of carbon monoxide, *Ann. N.Y. Acad. Sci.* **174**:1.

Coburn, R. F., 1970b, The carbon monoxide body stories, *Ann. N.Y. Acad. Sci.* **174**:11.

Corenwinder, M. B., 1865, Les feuilles des plantes exhallentelles de l'oxide de carbon, *C. R. Acad. Sci., Paris* **60**:102.

Crutzen, P. J., 1974, Photochemical reactions initiated by and influencing ozone in unpolluted tropospheric air, *Tellus* **26**:47.

Datsenko, I. I., and Martyniuk, V. Z., 1971, *Intoksikatsiya okisiyu ugleroda i puti yego umensheniya (Intoxication with Carbon Monoxide and Ways of Reducing It)*, izd. Zdozovie, Kiev.

Davis, D. D., Heaps, W., and McGee, T., 1976, Direct measurements of natural tropospheric levels of OH via an aircraft borne tunable dye laser, *Geophys. Res. Lett.* **3**:331.

Davis, D. H., Stanier, R. Y., Doudoroff, M., and Mandel, M., 1970, Taxonomic studies on some gram negative polarly flagellated "hydrogen bacteria" and related species, *Arch. Mikrobiol.* **70**:1.

Delwiche, C. C., 1970, Carbon monoxide production and utilization by higher plants, *Ann. N.Y. Acad. Sci.* **174**:116.

Dianov-Klokov, V. I., Lukshin, V. V., Sklyarenko, I. Y., and Shakula, Y. P., 1975, O variatsiyakh okisi ugleroda vo vsei tolshche zemnoi atmosfery, *Izv. Akad. Nauk SSSR. Fiz. Atmos. Okeana* **11**:320.

Dianov-Klokov, V. I., Fokeeva, Y. V., and Yurganov, L. N., 1977, Atmospheric carbon monoxide abundance above clean and polluted areas, in: *Abstracts of Proceedings of International Symposium on Meteorologic Aspects of Air Pollution, Leningrad, 10–19 March, 1977*, p. 34.

Dianov-Klokov, V. I., Fokeeva, Y. V., and Yurganov, L. N., 1978, Issledovaniye soderzhaniya okisi ugleroda vo vsei tolshche zemnoi atmosfery (obzor), *Izv. Akad. Nauk SSSR Fiz. Atmos. Okeana*, **14**:366.

Ducet, G., and Rosenberg, A. I., 1962, Leaf respiration, *Annu. Rev. Plant Physiol.* **13**:171.

Ehhalt, D. H., 1974, The atmospheric cycle of methane, *Tellus* **26**:58.

Ehhalt, D. H., Heidt, L. E., Lueb, R. H., and Martell, E. A., 1975, Concentrations of CH_4, CO, CO_2, H_2, H_2O and N_2O in the upper stratosphere, *J. Atmosph. Sci.* **32**:163.

Engel, R. R., Matsen, F. M., Chapman, S. S., and Schwartz, S., 1972, Carbon monoxide production from heme compounds by bacteria, *J. Bacteriol.* **112**:1310.

Engel, R. R., Modler, S., Matsen, F. M., and Petryka, Z. I., 1973, Carbon monoxide production from hydroxycobalamin by bacteria, *Biochim. Biophys. Acta.* **313**:150.

Feldman, T. G., 1973, Avtotransport kak istochnik zagryazneniya atmosfernogo vozdukha i ozdorovitelnyie metopriyatiya. *V sb. Gigiyena atmosf. vozdukha, vody i pochvy*, pod red. G. I. Sidorenko (Motor transport as a source of pollution of atmospheric air and sanitation measures, in: *Hygiene of Atmospheric Air, Water and Soil*, G. I. Sidorenko, ed.), VNIIMI, Moscow.

Fenn, W. O., 1970, The burning of CO in tissues, *Ann. N.Y. Acad. Sci.* **174**:64.

Fenn, W. O., and Cobb, D. M., 1932a, The stimulation of muscle respiration by carbon monoxide, *Am. J. Physiol.* **102**:379.

Fenn, W. O., and Cobb, D. M., 1932b, The burning of carbon monoxide by heat and skeletal muscle, *Am. J. Physiol.* **102**:393.

Ferenci, T., 1974, Carbon monoxide stimulated respiration in methane-utilizing bacteria, *FEBS Letters* **41**:94.

Ferenci, T., Strom, T., and Quayle, J. R., 1975, Oxidation of carbon monoxide and methane by *Pseudomonas methanica*, *J. Gen. Microbiol.* **91**:79.

Fischer, F., Lieske, R., and Winzer, K., 1930, Theory and practice of the biological depoisoning of illuminating gas, *Brenn. Chem.* **11**:452.

Fischer, F., Lieske, R., and Winzer, K., 1931, Biologische Gasreaktionen. I. Mitteilung: Die Umsetzungen des Kohlenoxyds, *Biochem. Z.* **236**:247.

Fischer, K., and Seiler, W., 1974, CO-produktion durch höhere Pflanzen. Proceedings of IX Internationale Tagung über die Luftverunreinigung und Forstwirtschaft. 15–18 October, 1974, pp. 61–68, Marianske Lazne, Czechoslovakia.

Frolov, A. D., 1976, Optichesky metod izmereniya soderzhaniya CO v atmosfere, *Tr. Gl. Geofiz. Obs.* **369**:52.

Fuchs, G., Schnitker, U., and Thauer, R. K., 1974, Carbon monoxide oxidation by growing cultures of *Clostridium pasteurianum*, *Eur. J. Biochem.* **49**:111.

Goldsmith, J. R., 1970, Contribution of motor vehicle exhaust industry and cigarette smoking to community carbon monoxide exposures, *Ann. N.Y. Acad. Sci.* **174**:122.

Goldsmith, J. R., and Aronov, W. S., 1975, Carbon monoxide and coronary heart disease: a review, *Environ. Res.* **10**:236.

Gorbatow, O., and Noro, L., 1948, On acclimatization in connection with acute carbon monoxide poisonings, *Acta Physiol. Scand.* **15**:77.

Hasemann, W., 1927, Zersetzung von Leuchtgas und Kohlenoxyd durch Bakterien, *Biochem. Z.* **184**:147.

Hirsch, P., 1968, Photosynthetic bacterium growing under carbon monoxide, *Nature (London)* **217**:555.

Hubley, I. H., Mitton, I. R., and Wilkinson, I. F., 1974, The oxidation of carbon monoxide by methane oxidizing bacteria, *Arch. Microbiol.* **95**:365.

Ingersoll, R. B., Inman, R. E., and Fischer, W. R., 1974, Soil's potential as a sink for atmospheric carbon monoxide, *Tellus* **26**:151.

Inman, R. E., and Ingersoll, R. B., 1971, Uptake of carbon monoxide by soil fungi, *J. Air Pollut. Control Assoc.* **21**:646.

Inman, R. E., Ingersoll, R. B., and Levy, E. A., 1971, Soil: a natural sink for carbon monoxide, *Science* **172**:1229.

Jaffe, L. S., 1970, Sources, characteristics, and fate of atmospheric carbon monoxide, *Ann. N.Y. Acad. Sci.* **174**:76.

Jaffe, L. S., 1973, Carbon monoxide in the biosphere: sources, distribution and concentrations, *J. Geophys. Res.* **78**:5293.

Jones, C. W., and Scott, G. S., 1939a, Oxidation of carbon monoxide and hydrogen by bacteria, *U.S. Bur. Mines Rept. Invest.* **18**:3466.

Jones, C. W., and Scott, G. S., 1939b, Carbon monoxide in underground atmosphere. The role of bacteria in the elimination of carbon monoxide in underground atmosphere, *Ind. Eng. Chem.* **31**:775.

Junge, C. E., 1963, *Air Chemistry and Radioactivity*, Academic Press, New York.

Junge, C. E., Seiler, W., Bock, R., Greese, K. D., and Radler, F., 1971, Uber die CO-Produktion von Mikroorganismen, *Naturwissenschaften* **58**:362.

Junge, C. E., Seiler, W., Schmidt, U., Bock, R., Greese, K. D., Radler, R., and Rüger, H. J., 1972, Kohlenoxyd und Wasserstoffproduktion mariner Mikroorganismen in Nährmedien mit syntatischem Seewasser, *Naturwissenschaften* **59**:514.

Kaserer, H., 1906, Die Oxydation des Wasserstoffs durch Mikroorganismen, *Zentralbl. Bakteriol. Parasitenk.*, Abt. 2, **16**:681.

Keeling, C. D., Bagastow, R. B., Bainbridge, A. E., Ekdahl, C. A., Guenther, P. R., Waterman, L. S., and Chin, J. F. S., 1976, Atmospheric carbon dioxide variations at Mauna-Loa observatory, Hawaii, *Tellus* **28**:538.

Keppen, O. I., Nozhevnikova, A. N., and Gorlenko, V. M., 1976, Temnovoi metabolizm *Rhodopseudomonas sulfidophila*, *Mikrobiologiya* **45**:15.

Kistner, A., 1953, On a bacterium oxidizing carbon monoxide. *Proc. K. Ned. Akad. Wet.*, Ser. C. **56**:443.

Kistner, A., 1954, Conditions determining the oxidation of carbon monoxide and of hydrogen by *Hydrogenomonas carboxydovorans*, *Proc. K. Ned. Akad. Wet.* Ser. C. **57**:186.

Kluyver, A. J., and Schnellen, C., 1947, Fermentation of carbon monoxide by pure cultures of methane bacteria, *Arch. Biochem.* **14**:57.

Kondratyev, K. Ya., Timofeev, Yu. M. (eds.), 1974, *Kosmicheskaya Distantsionnaya Indikatsiya Malykh Gazovykh i Aerozolnykh Komponent Atmosphery (Remote Detection of Atmospheric Trace Gases and Airborn Particulates from Space),* izd. LGU, Leningrad.

Korenev, M. S., 1962, Puti Umensheniya Vrednosti Otrabotavshikh Gazov Avtomobilnykh Dvigatelei. Ser. 12, Avtomobilestroyeniye, Moscow, pp. 1–97.

Krall, A. K., and Tolbert, N. E., 1957, A comparison of the light-dependent metabolism of carbon monoxide by barley leaves with that of formaldehyde, formate and carbon dioxide, *Plant Physiol.* **32**:321.

Kryukov, V. R., and Nozhevnikova, A. N., 1977, Vydeleniye avtotrofnogo aktionomitseta, *Mikrobiologiya* **46**:365.

Kzotov, Y. A. (ed.), 1975, Predelno dopustimyie konsentratsii vrednykh veshchestv v vozdukhe i vode, p. 10, Izd. Khimiya, Leningradskoye otdeleniye.

Lamontagne, R. A., Swinnerton, J. W., and Linnenbom, V. J., 1971, Nonequilibrium of carbon monoxide and methane at the air–sea interface, *J. Geophys. Res.* **76**:5117.

Landaw, S. A., Callagan, E. W., and Schmid, R., 1970, Catabolism of heme in vivo, comparison of the simultaneous production of bilirubin and carbon monoxide, *J. Clin. Invest.* **49**:914.

Langdon, S. E., 1917, Carbon monoxide, occurrence free in kelp (*Nereocystis luetkeana*), *J. Am. Chem. Soc.* **39**:149.

Langdon, S. E., and Gailey, W. R., 1920, Carbon monoxide, a respiration product of *Nereocystis luetkeana, Bot. Gaz.* **70**:230.

Lantzsch, K., 1922, *Actinomyces oligocarbophilus (Bacillus oligocarbophilus* Beij.) sein Formwechsel und seine Physiologie, *Zentrabl. Bakteriol. Parasitenk.*, Abt. 2, **57**:309.

Lebedinsky, A. V., Ivanovsky, R. N., and Nozhevnikova, A. N., 1976, Sostav i soderzhaniye tsitokhromov v kletkakh karboksidobakteryi, *Mikrobiologiya* **45**:176.

Leighton, P. A., 1961, *Photochemistry of Air Pollution,* Academic Press, New York.

Levy, H., 1971, Normal atmosphere: large radical and formaldehyde concentrations predicted, *Science* **173**:141.

Levy, H., II, 1973, Tropospheric budget for methane, carbon monoxide and related species, *J. Geophys. Res.* **78**:5325.

Liebl, K., and Seiler, W., 1976, CO and H_2 destruction at the soil surface, in: *Proceedings of the Symposium on Microbial Production and Utilization of Gases (H_2, CH_4, CO),* 1–6 Sept., 1975, Göttingen, p. 215, E. Goltze KG, Göttingen.

Linnenbom, V. J., Swinnerton, J. W., Lamontagne, R. A., 1973, The ocean as a source for atmospheric carbon monoxide, *J. Geophys. Res.* **78**:5333

Loewus, M. W., and Delwiche, C. C., 1963, Carbon monoxide production by algae, *Plant Physiol.* **38**:371.

Lovelock, J. E., and Margulis, L., 1974, Atmospheric homeostasis by and for the biosphere: the gaia hypothesis, *Tellus* **26**:2.

Löhnis, F., 1910, *Handbuch der landwirtschaftlichen Bakteriologie,* p. 448, Berlin.

Lukshin, V. V., Fokeeva, E. V., and Yurganov, L. N., 1976, Opredeleniye soderzhaniya okisi ugleroda vo vsei tolshche atmosfery nad gorodami, *Izv. Akad. Nauk SSSR, Fiz. Atmos. Okeana* **12**:557.

Luomanmäki, K., and Coburn, R. F., 1969, Effects of metabolism and distribution of carbon monoxide on blood and body stores, *Am. J. Physiol.* **217**:354.

Malkov, I. P., Yurganov, L. N., and Dianov-Klokov, V. I., 1976, Izmereniya obshchego soderzhaniya CO i CH_4 v severnom i yuzhnom polushariyakh (predvaritelnyie rezultaty), *Izv. Akad. Nauk SSSR, Fiz. Atmos. Okeana* **12**:1218.

Meadows, R. W., and Spedding, D. J., 1974, The solubility of very low concentrations of carbon monoxide in aqueous solution, *Tellus* **26**:143.

McConnell, J. C., McElroy, M. B., and Wofsy, S. C., 1971, Natural sources of atmospheric CO, *Nature (London)* **223**:187.

Migeotte, M. V., 1949, The fundamental band of carbon monoxide at 4.7 μ in the solar spectrum, *Phys. Rev.* **75**:1108.

Namsarayev, B. B., 1974, Vzaimootnosheniye mikroorganizmov pri okislenii metana (Interrelation of microorganisms on oxidation of methane), Thesis, Moscow.

Newell, R. E., Boer, G. J., and Kidson, J. W., 1974, An estimate of the interhemispheric transfer of carbon monoxide from tropical general circulation data, *Tellus* **26**:103.

Nozhevnikova, A. N., 1974a, Karboksidobakterii (Bakterii okislyayushchie okis ugleroda), Thesis, Moscow.

Nozhevnikova, A. N., 1974b. Otnosheniye CO-okislyayushchikh bakteryi k okisi ugleroda, *Izv. Akad. Nauk SSSR, Ser. Biol.* **1974**(6):878.

Nozhevnikova, A. N., and Zavarzin, G. A., 1973, Simbioticheskoye okisleniye okisi ugleroda bakteriyami, *Mikrobiologiya* **42**:158.

Nozhevnikova, A. N., and Zavarzin, G. A., 1974, K taksonomii CO-okislyayushchikh gramotritsatelnykh bakteryi, *Izv. Akad. Nauk SSSR, Ser. Biol.* **1974**(3):436.

Nozhevnikova, A. N., and Saveliyeva, N. D., 1972, Avtotrofnaya assimilyatsiya uglekisloty bakterijei okislyayushchei okis ugleroda, *Mikrobiologiya* **41**:939.

Pickwell, G. V., 1970, The physiology of carbon monoxide production by deep-sea coelenterates: causes and consequences, *Ann. New York Acad. Sci.* **174**:102.

Pickwell, G. V., Barham, E. G., and Wilton, I. W., 1964, Carbon monoxide production by a bathypelagic siphonophore, *Science* **144**:860.

Pressman, J., and Warneck, P., 1970, The stratosphere as a chemical sink for carbon monoxide, *J. Atmosph. Sci.* **27**:155.

Radler, F., Greese, K. D., Bock, R., and Seiler, W., 1974, Die Bildung von Spuren von Kohlenmonoxyd durch *Saccharomyces cerevisiae* und andere Mikroorganismen, *Arch. Microbiol.* **100**:243.

Robbins, R. C., Borg, K. M., and Robinson, E., 1968, Carbon monoxide in the atmosphere, *J. Air Pollut. Contr. Assoc.* **18**:106.

Robbins, R. C., Cavanagh, L. A., Salas, L. J., and Robinson, E., 1973, Analysis of ancient atmospheres, *J. Geophys. Res.* **78**:5341.

Robinson, E., and Robbins, R. C., 1972, Emissions, concentrations and fate of gaseous atmospheric pollutants, in: *Air Pollution Control* (W. Strauss, ed.), pp. 1–94, Wiley–Interscience, New York.

Robinson, W. O., 1930, Some chemical phases of submerged soil conditions, *Soil Sci*, **30**:197.

Romanova, A. K., and Nozhevnikova, A. N., 1977, Assimilation of one-carbon compounds by carboxydobacteria, *Proceedings of Second International Symposium on Microbial Growth on C₁-Compounds*, Pushchino, USSR, September 12–16, 1977, pp. 109–110.

Romanova, A. K., Nozhevnikova, A. N., Leontiyev, I. G., and Alekseyeva, S. I.; 1977, Puti assimilyatsii okislov ugleroda u karboksidobaktery *Seliberia carboxydohydrogena* i *Achromobacter carboxydus, Mikrobiologiya* **46**:885.

Sanzhiyeva, E. U., and Zavarzin, G. A., 1971, Bakteriya okisklyayushchaya okis ugleroda, *Dokl. Akad. Nauk SSSR* **196**:956.

Saveliyeva, N. D., and Nozhevnikova, A. N., 1972, Avtotrofny rost *Seliberia carboxydohydrogena* pri okislenii vodoroda i okisi ugleroda, *Mikrobiologiya* **41**:813.

Saveliyeva, N. D., and Trykova, V. V., 1969, Sravnitelnaya kharakteristika rosta razlichnykh vidov vodorodnykh baktery v avtotrofnykh usloviyakh, *Mikrobiologiya* **38**:245.

Schlegel, H. G., Kaltwasser, H., Gottchalk, G., 1961, Ein Submersverfahren zur Kultur wasserstoffoxydierender Bakterien: Wachstumsphysiologische Untersuchungen, *Arch. Mikrobiol.* **38**:209.

Seiler, W., 1974, The cycle of atmospheric CO, *Tellus* **26**:116.

Seiler, W., 1975, The cycle of carbon monoxide in the atmosphere, in: *Proceedings of the International Conference on Environmental Sensing and Assessment*, Sept. 14–19, vol. 1, pp. 1–9, Las Vegas, Nevada.

Seiler, W., and Junge, C., 1970, Carbon monoxide in the atmosphere, *J. Geophys. Res.* **75**:2217.

Seiler, W., and Schmidt, U., 1973. Dissolved nonconservative gases in seawater, in: *The Sea*, vol. 5 (E. D. Goldberg, ed.), pp. 219–243, Wiley–Interscience, New York.

Seiler, W., and Zankl, H., 1976, Man's impact on the atmospheric carbon monoxide cycle, in: *Environmental Biogeochemistry* (J. Nriagu, ed.), pp. 25–37, Ann Arbor Science, Ann Arbor, Michigan.

Shaw, J. H., 1958, The abundance of atmospheric carbon monoxide above Columbus, Ohio, *Astrophys. J.* **128**:428.

Shinohara, T., and Ooyama, I., 1972, N_2 fixation by a CO-oxidizing bacterium, *Rept. Ferment Res. Inst.* **42**:81.

Siegel, S. M., Renwick, G., and Rosen, L. A., 1962, Formation of carbon monoxide during seed germination and seeding growth, *Science* **137**:683.

Simpson, F. J., Talbot, G., and Westlake, D. W. S., 1960, Production of carbon monoxide in the enzymatic degradation of rutin, *Biochem. Biophys. Res. Comm.* **2**:15.

Sjöstrand, T., 1952, The formation of carbon monoxide by in vitro decomposition of haemoglobin in bile pigments, *Acta Physiol. Scand.* **26**:328.

Sjöstrand, T., 1970, Early studies of CO production. *Ann. N.Y. Acad. Sci.* **174**:5.

Sokolov, V. A., 1971, *Geokhimiya prirodnykh gazov (Geochemistry of Natural Gases)*, Izd. Nedra, Moscow.

Stephenson, M., 1949, *Bacterial Metabolism*, Longmans, Green, London.

Stevens, C. M., Krout, L., Walling, D., Venters, A., Ross, L., and Engelkemeir, A., 1972, The isotope composition of atmospheric carbon monoxide, *Earth Planetary Sci. Lett.* **16**:147.

Swinnerton, J. W., Linnenbom, V. J., and Cheek, C. H., 1968, A sensitive gas chromatographic method for determining carbon monoxide in seawater, *Limnol. Oceanogr.* **13**:193.

Swinnerton, J. W., Linnenbom, V. J., and Cheek, C. H., 1969, Distribution of methane and carbon monoxide between the atmosphere and natural waters, *Environ. Sci. Technol.* **3**:193.

Swinnerton, J. W., Lamontagne, R. A., and Linnenbom, V. J., 1971, Carbon monoxide in rainwater, *Science* **172**:943.

Sze, N. D., 1977, Anthropogenic CO emissions: implications for the atmospheric CO-OH-CH_4 cycle, *Science* **195**:673.

Tiunov, L. A., and Kustov, V. V., 1969, *Toksikologiya okisi ugleroda (Toxicology of Carbon Monoxide)*, izd. Medicina, Leningrad.

Thauer, R. K., Fuchs, G., Kaufer, B., and Schnitker, U., 1974, Carbon monoxide oxidation in cell-free extracts of *Clostridium pasteurianum, Eur. J. Biochem.* **45**:343.

Troxler, R. F., and Dokos, I. M., 1973, Formation of carbon monoxide and bile pigment in red and blue-green algae, *Plant Physiol.* **51**:72.

Troxler, R. F., Lester, R., Brown, A., and White, P., 1970, Bile pigment formation in plants, *Science* **167**:192.

Tzagoloff, A., and Wharton, D. C., 1965, Studies on the electron transfer system, LXII. The reaction of cytochrome oxidase with carbon monoxide, *J. Biol. Chem.* **240**:2628.

Varshavsky, I. L., 1968, *Kak obezvredit otrabotavshiye gazy avtomobilya (How to Render Harmless the Exhaust Automobile Gases)*, Transport, Moscow.

Vinogradov, A. P., 1964, *Gazovy rezhim Zemli. Khimiya zemnoi kory (Gas Regimes of the Earth. Chemistry of the Earthcrust)*, T.II, izd. Nauka, Moscow.

Voitov, G. I., 1975, Gazovoye dykhaniye zemli, *Priroda* 1975(3):90.

Wang, C. C., Davis, L. J., Wu, C. H., Japar S., Niki, H., and Weinstock, B., 1975, Hydroxyl radical concentrations measured in ambient air, *Science* 189:797.

Warneck, P., 1975, OH production rates in the troposphere, *Planetary Space Sci.* 23:1507.

Wehmer, C., 1926, Biochemische Zersetzung des Kohlenoxyds, *Chem. Ber.* 59:887.

Weinstock, B., and Niki, H., 1971, Carbon monoxide balance in nature, *Science* 176:290.

Westlake, D. W. S., Roxburgh, J. M., and Talbot, G., 1961, Microbial production of carbon monoxide from flavonoids, *Nature (London)* 189:110.

White, J. J., 1932, Carbon monoxide and its relation to aircraft, *US Naval Med. Bull.* 30:151.

White, P., 1970, Carbon monoxide production and heme catabolism, *Ann. N.Y. Acad. Sci.* 174:23.

Wilks, S. S., 1959, Carbon monoxide in green plants, *Science* 129:964.

Wilson, D. J., Swinnerton, D. W., and Lamontagne, R. A., 1970, Production of carbon monoxide and gaseous hydrocarbons in seawater: relationship to dissolved organic carbon, *Science* 168:1577.

Wittenberg, J. B., 1960, The source of carbon monoxide in the float of the Portuguese man-of-war, *Physalia physalis*, L., *J. Exp. Biol.* 37:698.

Wittenberg, J. B., Horonha, J. M., and Silverman M., 1962, Folic acid derivatives in the gas gland of *Physalia physalis* L., *Biochem. J.* 85:9.

Yagi, T., 1958, Enzymic oxidation of carbon monoxide, *Biochim. Biophys. Acta* 30:194.

Yagi, T., 1959, Enzymatic oxidation of carbon monoxide. II, *J. Biochem. (Tokyo)* 46:949.

Yurganov, L. N., and Dianov-Klokov, V. I., 1972, O sezonnykh variatsiyakh soderzhaniya okisi ugleroda v atmosfere, *Izv. Akad. Nauk SSSR, Ser. Fizika Atmos. Okeana* 8:981.

Zaitsev, A. S., 1973, Struktura polay kontsentratsy okisi ugleroda v gorode, *Tr. Gl. Geofiz. obs.* 293:47.

Zavarzin, G. A., 1972, *Litotrofnyie mikroorganizmy (Lithotrophic Microorganisms)*, Izd-vo "Nauka" (Nauka Publishers), Moscow.

Zavarzin, G. A., 1976, Prinadlezhnost vodorodnykh baktery i karboksidobaktery k mikroflore rasseyaniya, *Mikrobiologiya* 45:20.

Zavarzin, G. A., and Nozhevnikova, A. N., 1976, CO-oxidizing bacteria, in: *Proceedings of the Symposium on Microbial Production and Utilization of Gases (H₂, CH₄, CO)*, 1–6 Sept., 1975, Göttingen, p. 207, E. Goltze KG, Göttingen.

Zavarzin, G. A., and Nozhevnikova, A. N., 1977, Aerobic carboxydobacteria, *Microbial Ecol.* 3:305.

Zyakun, A. M., Bondar, V. A., and Nozhevnikova, A. N., 1976. O fraktsionirovanii stabilnykh izotopov okisi ugleroda bakteriyami, *Doklady Akad. Nauk SSSR* 227:497.

6

Microbial Ecology of the Human Skin

W. C. NOBLE AND D. G. PITCHER

1. The Environment

1.1. Introduction

From a microbial viewpoint the human skin must constitute a poor environment on which to live and reproduce. It is subject to periodic flooding followed by comparatively long periods of drought, it may be exposed to UV light in sunshine, its nutrient exudates may also contain toxic substances, and there may be "pollution" in the form of specific antibacterial compounds, perfumes, or heavy metals derived from other environmental sources. Finally, the physical environment is in a continual state of erosion. Yet almost the whole skin surface is colonized by microorganisms, sometimes in large numbers.

There are of course a number of quite different habitats on the body surface. Indeed the term *microenvironment* used by McLaren and Skujins (1968) in relation to soil microbiology seems apposite here for there is point-to-point variation in the concentration of solutes and gases.

Diversity of habitat led Kligman (1965) to describe the skin as follows:

> The axilla is a tropical rain forest with its lush supplies of sweat and hair; the perineum a veritable swamp draining the cesspool of the anus. Probably no organism can grow on the barren surface of the nail plate, but a deep, dank, humid cave like the external ear canal is a sanctuary for bacterial growth. The scalp is a thick woods seeping with sebum. There are the oily tundras of the face, the comparative deserts of the trunk, and the moist hot recesses of the intertriginous regions.

W. C. NOBLE and D. G. PITCHER • Department of Bacteriology, Institute of Dermatology, Homerton Grove, London E9 6BX, U.K.

1.2. Physical Features

In considering skin as a microbial habitat, it is convenient to deal first with the purely physical features.

The average human has a skin surface area of about 1.75 m² composed of about 10^8 "squames" or flat, pavement-like cells. Individual squames are about $30 \times 30 \times 3/5$ μm (Fig. 1) and are dead cells full of a protein loosely known as "keratin." A surface layer is lost about every four to five days by a continual process of attrition so that more than a million fragments of skin are lost daily. These may lodge in the clothing or become airborne and so dispersed, bearing with them members of the skin flora. These squames are replaced from below. There is a basal cell layer of cuboidal cells, some of which are in active mitosis. When division occurs, one cell remains in the basal layer, but the other is forced outward; as it nears the surface, it gradually loses vitality and fills with keratin flattening to form the typical squame. There are regional differences in the size of squames and also differences relating to sex. Males have smaller squames than do females by a ratio of area of about four to five.

Apart from the individual squames, the skin surface has hollows and folds which, though small to the human eye, are large in relation to the

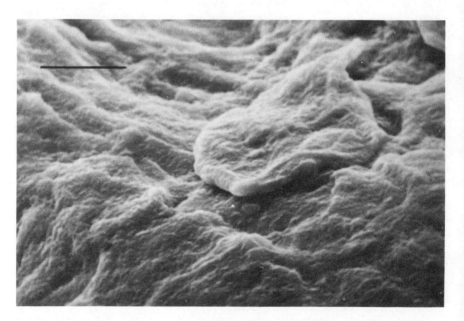

Figure 1. Epithelial cells (squames), which comprise the human skin surface. The bar represents 10 μm.

microbe. The various skin appendages—hair, nails, and eccrine, apocrine, and sebaceous glands—all serve to modify the local skin surface as an environment. It is here that the term *microenvironment* seems most applicable.

1.2.1. Hair

The effect of hair on the microbial environment is largely conjectural. The entire body, except the palms, soles, and terminal phalanges, has hair, which varies in type and density in relation to age, sex, race, and hair color (e.g., blonds have 50% more head hairs than the auburn-haired). We may speculate that hair on the scalp encourages *Pityrosporum* spp. in comparison with the forehead (e.g., McGinley *et al.,* 1975), but a study of the sequential changes in the flora of those with a receding hair line seems not to have been undertaken. It might be difficult to come to firm conclusions since there may be associated changes in sebum secretion and blood flow; Rook (1977) has provided an extensive review of the physiological changes that accompany common baldness. Mongoloid mental patients, who have no axillary hair, appear to have the same experience of the bacterial skin disease erythrasma as do other patients with normally hairy axillae, but there may be other factors that compensate. There are some suggestions that the organisms found in the condition trichomycosis axillaris, in which coryneform organisms colonize the shaft of axillary or even pubic hair, may be different from those found in the unaffected axilla, but this remains to be demonstrated using good taxonomic technique.

1.2.2. pH

In the past much attention has been paid to the so-called acid mantle of the skin, a reference to the fact that skin surface pH is about 5.5. In fact, studies have shown that there is a normal distribution of pH from about 4 to about 8 with a mean of 5.5 on ordinary "glabrous" skin. There are small, though statistically significant, differences between various skin sites and between people. To date it has not been possible to relate pH to the carriage of any specific organisms on the skin (though vaginal pH may be related to the presence of lactobacilli). W. J. Cunliffe and his colleagues (personal communication) believe that intrafollicular pH may govern the production of specific enzymes such as lipase and protease which may be involved in the etiology of common acne. In the disease diaper rash—or diaper dermatitis—bacterial breakdown of urea to ammonia can alter pH to values in excess of 8 with subsequent overgrowth of a pathogenic bacterial or yeast flora, but this is an extreme condition.

The normally acid pH may be the result of the secretion of lactic acid and lactates in the eccrine sweat combined with bacterial production of fatty acids and diffusion of CO_2 through the skin. Skin pH was admirably reviewed by Behrendt and Green (1971).

1.2.3. Temperature

The temperature of glabrous skin is about 33°C in a normal, clothed individual with a circadian rhythm of about 1°C amplitude; the axillae and perineum more closely approach 37°C. Lower temperatures may occur locally and temporarily, for example, on the hands and feet or during hypothermia in the aged, but the effects of this on the microbial flora apparently are not known.

In medical microbiology it is standard practice to incubate all material at human blood temperature, 37°C. However, in cutaneous microbiology it is essential to incubate some samples at lower temperatures; for example, 30°C is necessary for two mycobacterial species. *Mycobacterium ulcerans*, which may perhaps be derived from plant microflora, and *M. marinum*, frequently a fish pathogen, are well documented invaders of human skin and the immediate subcutaneous tissues, yet they do not appear to cause deep infection, and one successful therapeutic measure is to raise the temperature of the skin to 37°C, a temperature at which these organisms fail to grow.

Dermatophyte fungi are also invaders of skin but rarely cause lesions of the deep tissue. The temperature optimum for these organisms lies between 27 and 33°C, though growth will occur at 36°C. We may speculate that the normal skin temperature of 33°C may serve to prevent many plant or soil organisms from colonizing this habitat, since 33°C may be above the normal growth range. Duncan and his colleagues (1969) reported that high temperature as well as high humidity was required to raise the skin flora above its basal level, and Singh (1974) produced good epidemiological evidence that, in India, pyoderma was more frequent when both environmental temperature and humidity were high than when only one of these conditions applied.

1.2.4. Humidity

Humidity has long been the factor felt to be of the greatest importance in influencing the skin flora. Bacteria do not generally grow below about 90% relative humidity, so we must presume that the humidity of the specific site is at about this point, though Christian and Waltho (1962) found *Staphylococcus aureus* to grow at water activities as low as 0.86. However, occluding the skin by the use of plastic, which raises the humidity but also the temperature and probably local concentrations of CO_2 results in a dramatic increase in the flora. Changes are quantitative, but may appear qualitative since gram-negative bacilli may increase disproportionately, resulting in an apparent creation of a bacillary flora *de novo*.

In studies on the effect of temperature and humidity in cattle, Jenkinson and his group (1974) have found changes in pH and in the pattern of protein content of sweat. At 35°C protein nitrogen in sweat amounted to about 1% of digestible dietary protein. Comparable studies in man do not appear to have been undertaken.

1.3. Availability of Nutrients

The chemistry of the skin surface has been much investigated, yet much remains unknown. It is most convenient to consider two classes of chemical substance at the skin surface; the water-soluble—conveniently called "sweat," although this may itself elute substances from the epidermis—and the fat-soluble, lipid substances—conveniently called "sebum," though this too may contain epidermal substances; both may also contain substances of microbial origin.

1.3.1. Sweat

Sweat contains a range of inorganic ions, the principal of which are chloride and sodium, with major contributions from phosphate, sulfate, bromide, fluoride, iodide, potassium, calcium, magnesium, iron, copper, manganese, and zinc. Precise concentrations cannot be given for they vary with temperature and sweat rate, age, sex, and doubtless other factors, but sodium and chloride are frequently about 12 mg/liter each. Among the organic compounds, urea, lactic acid/lactates, and amino acids are the most plentiful with thiourea, creatine, ammonia, uric acid, urocanic acid, and mucoproteins also present. The amino acids total about 1400 mg/liter with serine, alanine, glycine, and citrulline as the most common, though 22 substances in this group have been recorded by various workers. Carbohydrate, quoted as glucose equivalent, may reach 30 mg/liter in normal subjects. Ten vitamins have also been identified.

The general composition of soluble substances at the skin surface has been reviewed by Noble and Somerville (1974). Lee and Baden (1975) have reviewed the chemistry and composition of keratins from a variety of sources; there are of course differences between animal species and also between human epidermis, hair, and nail in the balance of the constituent amino acids. Such differences may underlie the inability of some organisms to attack hair although able to attack stratum corneum.

One group of persons who have a strikingly different sweat composition are those with cystic fibrosis, but no attempt seems to have been made to compare their skin flora with that of normal people. Perhaps too a study of the flora of "rusters," persons whose sweat chloride concentrations is so high that their fingerprints may corrode mild steel, would be of interest. Diabetics may have an increased sugar level in the skin, but changes in the skin flora are minimal (Somerville and Lancaster-Smith, 1973). Erythrasma—the fluorescent, scaling lesions of the axillae, groin, and toewebs caused by *Corynebacterium* species (*Corynebacterium minutissimum*)—is more common in diabetics than in normal individuals, and carriage of the associated diphtheroids is also more frequent. Curiously, *Micrococcus luteus* strains are also more common in persons with diabetes, though these organisms are among the few which do not ferment or oxidize glucose in laboratory tests. There are other apparent anomalies of this nature. Savin (1974) has comprehensively reviewed the evidence for an increased susceptibility to cutaneous infection among diabetics and found contradiction at every stage.

Recent studies by Gloor and his colleagues (1975a,b,c,d, 1976a,b) have revealed differences in the chemistry of the skin surface in patients with various diseases. Such differences in relation to diabetes include increased amino acid content, a *decrease* in reducing substances and an increase in cholesterol, and an increase in free-fatty-acid composition, which they attribute to bacterial action. Similar changes occurred in those who had suffered impetigo contagiosa, a staphylococcal infection of the skin. Clearly the changes may have occurred in response to disease and not have predisposed the person to infection, but these are the first serious steps in seeking chemical changes which may determine the outcome of colonization/infection.

It has proved difficult to study the utilization of sweat substances by skin microorganisms. Smith (1971) showed that cutaneous cocci could use lactic acid as an energy source. Amino acids seem abundant on skin and should not restrict microbial growth. In general, *Micrococcus* species are more versatile at using single amino acids than are *Staphylococcus* species (Halvorson, 1972), and arginine, cystine, and valine are required by some *Staphylococcus aureus* strains but not by others. Emmett and Kloos (1975), studying the amino acid requirements of *Staphylococcus* species from human skin, found that proline, arginine, and valine were the most frequently required amino acids, but ten species of *Staphylococcus* had different requirement patterns.

Farrior and Kloos (1976) examined sulfur amino acid auxotrophs of *Micrococcus* species from human skin and found that only 6% of *M. luteus* isolates required sulfur amino acids but that 92% of *M. sedentarius* isolates had this requirement. Since 90% of humans examined by Kloos and his co-workers carried *M. luteus* but only 13% carried *M. sedentarius,* one can speculate that sulfur amino acids may be limiting for growth on some individuals. Only in certain limited microenvironments would growth of *M. sedentarius* be possible, perhaps where other skin microbes excreted metabolic compounds.

1.3.2. Antibacterial Substances

In her book *The Ecology of Human Skin,* Mary J. Marples (1965) uses the classical ecological terms "coaction" to indicate the effects of one organism on another and "reaction" to indicate changes in the environment brought about by its inhabitants. However, as Marples points out, there is difficulty in applying the classical "woodland" concepts to cutaneous ecology in that the "environment" is itself part of a living organism. Changes in the environment caused by the inhabitants are not brought about on inactive material; the environment may actively seek to control the behavior of its flora and fauna. Such measures include hygiene, in which vast numbers of organisms are removed physically or chemically, and the "disease" response, which may alert the body's cellular and humoral immunity systems.

Potentially antagonistic substances secreted quite normally onto the skin surface are the immunoglobulins. Herrman and Habbig (1976) detected eight proteins in concentrated human sweat; three of these were human serum proteins and one was identified as of epidermal origin. Bacterial protein could not be excluded from the remaining four proteins. On paper electrophoresis, there were three fractions migrating as α, β, and γ globulins, the latter including IgG. Cabau *et al.* (1974) have demonstrated IgG in the sweat glands of children with a variety of diseases; IgA was less frequent, and IgM was found in only a minority but may have pathological significance. IgE has been found in human eccrine sweat by Förström *et al.* (1975). Brodersen and Wirth (1976) found hepatitis B antigen and antibody in human sweat. Thus, sweat may prove a source of infecting virus but also a protective mechanism against reinfection by contact. Earlier reports contain reference to specific antibacterial agents; thus, Neill *et al.* (1931) reported diphtheria antitoxin in sweat. Page and Remington (1967) found IgA, IgG, IgD, and tetanus antibodies in sweat and suggested that these might have some influence on the skin flora. No specific work aimed in this direction seems to have been done, and we are left to speculate. One *Staphylococcus aureus* antigen, known as Protein A, reacts principally with the Fc fragment of IgG but is also known to react with IgA. It seems possible that some nonspecific as well as specific interactions might take place at the skin surface. Other antagonistic substances secreted in normal sweat are likely

Figure 2. Growth of coryneform organisms around an eccrine sweat gland after 4-hr occlusion of the skin. Note the clear zone, free of organisms, which immediately surrounds the gland. The bar represents 20 μm.

simply to be high concentrations of, for example, lactic acid or lactates. These may account for the appearance (Fig. 2) of a clear zone surrounding an eccrine sweat gland taken from skin occluded for about 4 hr with extensive microbial growth surrounding this zone.

Therapeutic substances, especially antibiotics, may be secreted onto the skin surface and profoundly influence the flora. Marples and Kligman (1971) found that therapeutic penicillin did little to alter the skin flora, but presumably it was not excreted. Tetracycline and clindamycin, however, caused a significant drop in numbers of aerobes and anaerobes, while erythromycin had a smaller effect. Qualitative changes also took place. Tetracyclines given by mouth brought about the emergence of a resistant coccal flora and, in the axilla, caused a switch from a predominantly coryneform to a coccal flora. The change occurred over a period of three weeks with an initial delay suggesting that excretion is holocrine via the sebaceous glands rather than via the eccrine sweat. Leyden *et al.* (1973) later reported that prolonged administration of antibiotics such as tetracycline for the treatment of acne would result in major changes in the flora; a gram-negative bacillary flora might be selected. This also occurs in acne itself, resulting in a gram-negative folliculitis more difficult to treat than the original coryneform acne. Valtonen and his colleagues (1976) have recently reported that such long-term, low-dose antibiotic treatment facilitates exchange of R factors among organisms in the gut, but the potential effect of this treatment on genetic exchange among skin organisms has not yet been investigated.

1.3.3. Studies on Sebum

The composition of sebum has been much studied in relation to acne. As with eccrine sweat, it is difficult to produce an average figure for composition, but Greene *et al.* (1970) found the major components of sebum to be triglycerides 57% (w/w), wax esters 26%, squalene 12%, cholesterol esters 3.0%, and cholesterol 1.5%; the data of other workers are in broad agreement. Lipid derived from the epidermis itself may be insignificant in areas rich in sebaceous glands but may make a major contribution to composition where glands are few. A variety of intrinsic and extrinsic factors influence composition of sebum: local temperature changes may affect squalene production, seasonal climatic changes influence triglycerides, increasing age is reflected in decreasing glycerides, and sucrose in the diet may cause an increase in triglycerides. There are also changes in the cholesterol and triglyceride moieties which are related to the menstrual cycle. In any individual the approximate composition of sebum is constant, at least over a period of months, but there are great differences between individuals, especially in the triglyceride–free-fatty-acid ratios, which may not be related to the microbial flora. The complexity of this subject may be judged from the report of Kellum and Strangfeld (1972) that there are 59 fatty acids in the range C_8 to C_{18} alone. It seems unlikely that many, if any, of these free fatty acids are derived from the microbial cells *per se;* there is, however, good evidence that they are products of microbial metabolism and are derived from the skin triglycerides by lipolysis. Morello and Downing (1976), in a paper on the transunsaturated fatty acids of skin, found that the position of unsaturation resembled those in margarine. However, one of their subjects did not eat margarine and bacterial action on sebum was therefore postulated.

In vitro study of lipase is beset with complications for which no direct analogy can be found *in vivo;* protein, peptone, and metabolized sugars are found to inhibit lipase production *in vitro,* and the very substances used to measure lipolysis—the Tweens—may themselves suppress enzyme formation. However, Marples *et al.* (1971) applied antibiotics to the skin in direct ecological experiments, and their results are interesting in comparison with *in vivo* observations. Application of neomycin, which suppresses the coccal flora of the skin, had no effect on the levels of free fatty acids observed, whereas tetracycline, which has its greatest effect on *Propionibacterium* (*Corynebacterium*) *acnes* in the long term, reduced both *P. acnes* and the levels of the fatty acids. However, *in vitro* it is the cocci which prove most versatile in splitting triglycerides and various vegetable oils, although they are less active on the Tweens, which are split by the coryneforms.

The role of sebum in permitting or preventing growth of the cutaneous flora remains controversial to this day. Sebum or some component of sebum is probably an enhancing factor for many coryneforms on the skin, especially the anaerobic species which are found predominantly in the areas of greatest sebaceous gland activity, i.e., the face and thorax. Yet, in the laboratory, few organisms are found to have an absolute requirement for lipid (Somerville, 1973), though many show enhanced growth in its presence. Laboratory studies frequently use Tween 80 (polyoxyethylene sorbitan monooleate) as a source of lipid. The yeasts *Pityrosporum ovale* and *P. orbiculare* appear to have a requirement *in vitro* for lipid and are found in greatest numbers in their yeast form on the face and scalp, yet in the mild disease pityriasis versicolor, a substantial mycelial growth of *P. orbiculare* on the body can occur at any site. This appears so different from the yeast form that it has long been known as *Malazzezia furfur* and was formerly regarded as an entirely distinct organism.

The studies of Matta (1974), Leyden *et al.* (1975) and unpublished studies of W. C. Noble and P. M. White show that *P. acnes* is dependent on the presence of sebum to enable it to grow on the human skin. Fig. 3 shows the change in the numbers of this organism detected by conventional swabbing techniques using soluble swabs on the forehead. Clearly

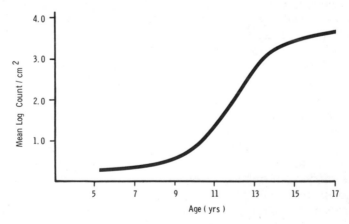

Figure 3. Increase in colonization of the forehead with *Propionibacterium acnes* in relation to age.

P. acnes cannot colonize skin prior to the onset of pubertal changes. Negroid girls, who in this series commenced puberty at an earlier date than caucasoids, also acquired their populations of *P. acnes* earlier. Significant changes in the population of *Pityrosporum* have been shown to occur at puberty (W. C. Noble and G. Midgley, unpublished data), though the technique used is not appropriate to strict quantitative counts. The chemical basis for lipid stimulation seems to be unknown, but it has been suggested that *in vitro* this might be a mechanical effect due to the alteration of surface tension values.

A way in which sebum components may work to modify the action of skin pathogens is seen from the work of Kaplan and Wannamaker (1976). Infection of the throat or deep tissue with *Streptococcus pyogenes* (Lancefield's Group A streptococci) may result in acute glomerulonephritis (AGN) or rheumatic fever as important sequelae. However, infection of the skin rarely if ever results in rheumatic fever, though AGN is well documented and widespread. There is a strong correlation between rheumatic fever and the antibody response to the streptolysin O titer (ASO). Although the ASO titer is raised following acute streptococcal throat infection, it is much less often raised following skin infection for which the anti-DNAase-B is the marker of choice. Kaplan and Wannamaker (1976) have demonstrated that cholesterol (a major component of sebum) suppresses the hemolytic action of SO and the formation of ASO without influencing the anti-DNAase-B. They postulate that the absence of rheumatic fever following skin infection may result from modification by the sebum of the streptococcus antigen output or of the host response to that antigen.

1.4. Location of Skin Flora

The site of growth of skin bacteria has long been a subject for debate. Most microscopic studies have suffered from the necessity to process biopsies before examination, and the scope for the removal or transfer of microbes during any processing in fluid is great. There is general agreement that most aerobic bacteria are situated at or near the surface and that few, if any, are found *within* the epidermis. Transmission electron micrographs have occasionally suggested that some coryneforms may be intracellular, though it is very difficult to see how this might have come about. Studies in our department, using the scanning electron microscope, in which untreated biopsy specimens were dried and then coated with carbon and gold–palladium suggest that at least the aerobic coryneform moiety exists on the skin surface independent of an eccrine or sebaceous gland, but those organisms in the vicinity of eccrine sweat glands receive benefit in the form of increased nutrient or humidity. From the limited studies made so far, it seems possible that the cocci may inhabit the opening of the eccrine sweat glands, for they are found more frequently in the lumen than outside.

The anaerobic coryneforms *P. acnes* complex are assumed to reproduce in the hair follicles or sebaceous gland openings, though they are secreted onto the skin surface and survive in the open. Wolff and Plewig (1976) believe, on the basis of electron microscopic studies, that *Pityrosporum* species inhabit the mouth of the sebaceous follicle; below this are the cocci and at the base the *P. acnes* group.

Roberts (1975) devised a method for sampling from the sebaceous duct which immobilized the surface flora; using this technique, he found that both Micrococcaceae and *P. acnes* could be recovered from the ducts, the latter being more than ten times as common. Roberts demonstrated that substantial recoveries of organisms could be made using his technique even after the surface had been sampled using the scrubbing technique, the best available other technique (Holland *et al.,* 1974).

Transmission electron microscope studies such as those of Montes and Wilborn (1969) confirm findings by other techniques. In these studies microcolonies were reported beneath scales in the superficial layer. Probably most bacteria on the skin surface exist in microcolonies or aggregates which may be quite large. Most studies have used the ratio between surface contact counts (which measure viable units) and the detergent scrub (which is assumed to measure total viable cells). On this basis females are found to have smaller microcolonies of aerobes than are males (see Table I), yet to have larger colonies of anaerobes. The significance of this is not clear.

Table I. Mean Microcolony Size: Pooled Results from Five Males and Five Females [a]

	Forehead	Forearm	Periumbilicus
Females			
Aerobic count	590[b]	124	263
Anaerobic count	2952	224	1063
P. acnes	4725[c]	356	2297[d]
Males			
Aerobic count	2026[b]	163	339
Anaerobic count	1925	202	1074
P. acnes	1924[c]	123	506[d]

[a]Taken from Noble and Somerville (1974). Counts do not differ in *t* test, except as noted.
[b]Male/female ratio, $p < 0.1\%$.
[c]Male/female ratio, $p < 5\%$.
[d]Male/female ratio, $p < 5\%$.

Microcolonies may be envisaged by scanning electron microscopy of biopsies or by gram stains of thin layers of epithelium removed by adhesion to a microscope slide (Marks and Dawber, 1972). Generally it is easier to find microcolonies at areas of infection such as erythrasma or following occlusion of the skin for a few hours. With persistence, however, they may be found on normal, untreated skin (Fig. 4).

The number of organisms per square centimeter of skin varies from site to site, and estimates also vary according to the technique used.

Somerville and Murphy (1973), using the scrubbing technique, found the geometric mean count of anaerobes in a population of 22 normal individuals in a temperate climate to range from $10^4/cm^2$ on the forehead to $10^1/cm^2$ on the calf. Aerobes had a mean of $10^4/cm^2$ to $10^1/cm^2$ on the same sites. Individual differences were more marked, however. On the forehead, anaerobes ranged from less than $7/cm^2$ to $5 \times 10^6/cm^2$, and aerobes showed the same picture. On the calf the range was less than $7/cm^2$ to $10^2/cm^2$ for anaerobes and less than $7/cm^2$ to $10^4/cm^2$ for aerobes. In all scrubbing studies to date, the flora has corresponded to a log-normal distribution (Table II).

Loebl *et al.* (1974), using a biopsy technique in a study relating to infection of burns, considered 10^2 organisms per gram of skin as normal and 10^4 as evidence of infection in cadaver material, Lawrence and Lilly

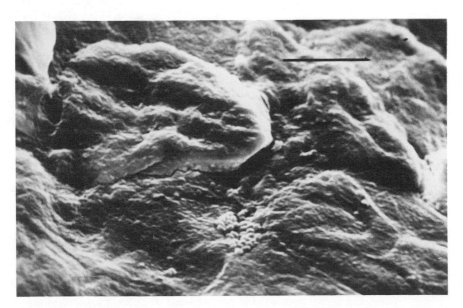

Figure 4. Microcolony of cocci on the surface of the skin. The bar represents 10 μm.

Table II. Geometric Mean Counts per cm² for 22 Normal Healthy Individuals
(11 Male and 11 Female), Derived Using the Scrubbing Technique

	Aerobes, count/cm²	Anaerobes, count/cm²
Forehead	1.1×10^4	4.0×10^4
Subclavicle	7.9×10^2	8.2×10^3
Scapular area	3.1×10^2	4.5×10^3
Deltoid	2.9×10^2	2.5×10^2
Sternum	2.1×10^3	7.2×10^4
Center upper back	1.1×10^3	9.1×10^4
Periumbilicus	1.4×10^3	2.4×10^2
Axilla, lower margin	8.7×10^2	5.2×10
Forearm, anterior	3.8×10^2	4.5×10
Palm of hand	4.4×10^2	2.8×10^2
Thigh		
Upper front	7.7×10^2	7.0×10
Lower front	6.9×10^2	5.6×10
Back	6.4×10^2	2.1×10
Shin	4.8×10^2	3.0×10
Calf	3.1×10^2	1.4×10
Dorsum of foot	3.6×10^2	2.4×10
Sole of foot	1.6×10^4	2.8×10

[a]From Noble and Somerville (1974).

(1972) found 10^4, and Selwyn and Ellis (1972) recorded counts between 10^4 and 10^6 per square centimeter. It seems doubtful whether 1 cm² of full skin thickness biopsy would weigh 1 g, and it is difficult to equate these estimates.

Mustakallio and his colleagues (1967), using full-thickness suction blisters as specimens, found small numbers of organisms (10^2–10^3), commensurate with values obtained by scrubbing, though patients with skin disease had markedly higher counts than did normal individuals. It is clear that this problem has yet to be resolved.

1.5. Microbial Adherence to Skin

There do not appear to be any studies aimed at determining how bacteria adhere to the skin surface. Alkaline solutions remove bacteria from skin more efficiently than neutral or acid solutions, but addition of detergent increases the efficiency. This is the basis for the Williamson and Kligman (1965) prescription for sampling fluid: a phosphate buffer (pH 7.9) with added Triton X-100, a nonionic detergent. It may be relevant that Heckels and his colleagues (1976) have found *Neisseria gonorrhoeae* to have an overall negative charge, which relates to the ability of the cell pili to attach to human cells in tissue culture. Workers in other fields, especially dentistry, have paid much attention to adhesion. Variation has been found in the species able to adhere to tooth surface, and the mechanism suggested is an adsorption of dextran synthetase enzymes on the tooth surface, which binds the small amounts of dextran in the streptococcal cell wall (Mukasa and Slade, 1974). Sugar residues and glycoproteins are also involved in adherence of lactobacilli to chicken crop epithelial cells (Fuller, 1975). Utilization of carbohydrates, especially glycogen by lactobacilli, may influence the ecology of the human vagina (Mardh and Westrom, 1976). Studies of the skin appear to have been carried out only on broiler chickens, where work has been concentrated on acquisition of potential spoilage organisms. Bacterial flagella are found to be important in such systems (Notermans and Kampelmacher, 1975). Principal members of the normal human skin flora do not possess flagella nor have they been described as excreting levan or dextran; the mechanism by which they adhere to skin is not known, and it may be that the surface tension provided by sebum and sweat is sufficient to maintain adherence.

In the only direct study published to date on the adhesion of bacteria to the nasal epithelium, Aly *et al.* (1977) have found differences between staphylococcal species in their ability to adhere to nasal epithelial cells; cells from different individuals also differ in the frequency with which bacteria adhere to them. This study is important in suggesting ways in which the cornified layer could be studied.

1.6. Recovery of Microbes from Skin

To some extent, skin microbiology has suffered from the problems inherent in sampling the human skin surface. Even if biopsy techniques are resorted to (an unpopular measure), the fact that skin microbes exist in microcolonies can result in a yield of 10^6 cells from the first biopsy and zero from a second, adjacent specimen.

The cotton swab is the most popular sampling device but is impossible to use quantitatively. The scrubbing technique devised by Williamson and Kligman (1965) is the best available method but results in some excoriation. More recently, Stringer and Marples (1976) have found equivalent results using an ultrasonic technique that removes far fewer corneocytes. In these tests an area of skin of about 4 cm^2 is sampled, but there are still substantial differences between paired samples. Somerville-Millar and Noble (1974) found the standard deviation in count between pairs of counts on adjacent areas to exceed 10% of the mean. Methods of sampling the flora are discussed fully by Noble and Somerville (1974).

2. Cutaneous Microflora

Skin flora has been divided into three categories on the basis of its apparent constancy. The original concept is that of Price (1938), but Somerville-Millar and Noble (1974) redefined the "resident" flora as those species or biotypes recovered on more than 75% of 25 sampling days over 7-month period. "Transients" were species appearing less than 25% of the time, and the term "nomads" was applied to those present on less than 75% but more than 25% of the occasions. This definition is of course arbitrary and is limited by the sensitivity of the sampling techniques. A proper definition of the resident flora must await techniques for recognizing which microorganisms are reproducing in the specific habitat and which are merely contaminants (transients) not capable of maintaining themselves in the absence of a specific reservoir elsewhere.

As might be expected, some members of the skin flora appear relatively constant and persist over long periods (Somerville-Millar and Noble, 1974). To some extent this is a property of the host (Evans, 1975). Kloos and Musselwhite (1975) studied the predominance and persistence of bacteria on a number of people over one year. Staphylococci and coryneforms were predominant and persisted in the nares, axilla, head, leg, and arms with the addition of *Bacillus* spp. on the head, leg, and arms. The carriage of *Acinetobacter* appeared linked to the climate. In warm humid months, there was a higher carriage rate.

A wide variety of organisms has been isolated from healthy human skin. Bacteria fall into three principle groups—Micrococcaceae, coryneforms, and gram-negative bacilli. Fungi comprise mainly the yeasts *Pityrosporum,* though some people carry dermatophytic fungi in the absence of overt disease. Viruses are not recorded as appearing on normal skin, but few adequate studies have been carried out. It is difficult to define the normal flora of man, but the distribution of the principle components is shown in Table III.

When considering bacteria which are contaminants, a large number

Table III. Principal Components of the Human Skin Flora

Site:	Head	Thorax	Abdomen	Limbs	Groin	Axilla	Toewebs
Aerobic Micrococcaceae	+	+	+	+	+ +	+ +	+ +
Anaerobic Micrococcaceae	?	?	?	?	?	?	+
Aerobic coryneforms	+	+	+	+	+ +	+ +	+ +
Anaerobic coryneforms	+ + +	+ +	−	−	−	−	−
Gram-negative bacilli	−	−	−	−	+ +	+	+
Yeasts (*Pityrosporum*)	+ +	+	−	−	−	−	−

of bacterial genera may be represented on the skin, and standards of hygiene and the occupation of the individual can play a role in this. Day to day contact with the environment must inevitably cause the transfer of microorganisms from a variety of habitats onto the skin. This is particularly true of the hands, which are constantly transferring bacteria from dust, soil, water, and food onto other body sites. A person's occupation may facilitate the transfer to the skin of organisms of a particular genus or even species. For example, a gardener may be constantly contaminated by organisms that live on plants and in the soil; *Bacillus* species are still uncommon but are more frequent in poor hygiene than in good hygiene.

Use of cosmetics may also introduce new bacteria to the skin. It is well known that cosmetics can rapidly become contaminated once opened for use, and a wide variety of organisms have been found in makeup, lotions, and soap (Jarvis *et al.,* 1974).

In some circumstances, there may be reproduction of some of these bacteria on the skin surface, but it is probably the case that if the source of constant contamination is removed for a period, the "resident" flora will once again take over and the contaminants disappear. This type of contamination is most easily described as "transient."

Disease may change our expectation of resident and transient floras. Chin and Davies (1976) have examined the flora of the hands of hemiplegic patients, and they noted a considerable increase in the flora over that of nonparalyzed hands. Only the incidence of *S. aureus* did not increase. Enteric organisms and *Clostridium welchii* were found in abnormal quantities, and the hands had become malodorous. Normal hands are efficient samplers of the environment, and contaminating organisms are to be expected; however, in the immobile hand, it appears that a foreign and permanent flora develops. Chin and Davies (1976) ascribe the phenomenon to an excess sweating associated with many of the diseases of the central nervous system, giving rise to an abnormally humid climate for the hand. It does appear, however, that some breakdown in the antibacterial properties of the skin and its secretions also plays a part. Similar changes have been recorded on the skins of patients with terminal illness (Stratford *et al.,* 1968) or with cancer (McBride *et al.,* 1976).

The relationships of bacteria with the features of particular skin sites that they tend to occupy could be enlightened by the use of ecological "skin flora maps," as developed by Bibel and Lovell (1976), though further progress in the taxonomy of organisms and the chemistry of the skin is necessary to correlate these factors more convincingly.

The composition of the skin flora was extensively reviewed by Noble and Somerville (1974), but since then there have been many advances in our knowledge of the taxonomy of the Micrococcaceae and the coryneforms. These are therefore reviewed below.

2.1. Taxonomy of the Cutaneous Cocci

The family Micrococcaceae as now set out in the eighth edition of *Bergey's Manual* (Baird-Parker, 1974b) comprises three genera of gram-positive, spherical cells. Of these, only *Micrococcus* and *Staphylococcus* are of interest in cutaneous microbiology since they are a dominant part of the skin flora of man. The packet-forming aerobic cocci of skin, previously known as *Sarcina,* are now considered to be *Micrococcus luteus,* and the term *Sarcina* is reserved for a genus of anaerobic packet-forming cocci in the family Peptococcaceae.

Though the anaerobic genus *Sarcina* is not commonly associated with human skin, *Peptococcus,* non-packet-forming, non-chain-forming anaerobic cocci in the same family have been reported to be common on the toewebs of healthy adults. Sanderson (1977) isolated what he described as *Peptococcus magnus* from 13 of 20 subjects.

Infections with bacteria of this genus are frequently cited in the literature, though strains are found in the throat, female urogenital tract, and intestinal tract in normal people. Their pathogenic properties are little known. It may be that closer attention to modern methods of achieving anaerobiosis could reveal a role in the skin flora for these cocci.

2.2. Micrococcaceae

The classification of this family by Baird-Parker (1963) has, with occasional modifications, been of great help to cutaneous ecologists over the years. Its main "tour de force" has been its simplicity.

The accepted laboratory separation of the genera *Micrococcus* and *Staphylococcus* is based on their action on glucose. Oxidation of glucose is carried out by *Micrococcus* and anaerobic fermentation by *Staphylococcus* (Baird-Parker, 1963). The guanine-cytosine content (GC%) of the DNA of the two genera supports this division since the range is 66–75 mol % for *Micrococcus* and 30–40 mol % for *Staphylococcus,* (see, for example, Kocur and Boháček, 1974). Further confirmation of the separation of the two genera has been achieved by peptidoglycan analysis, by which staphylococci are found to have a very high glycine content (Schleifer and Kandler, 1972). *Staphylococcus*-like strains can be isolated in which there is little or no anaerobic acid production from glucose, but these intermediate strains fall clearly into either *Staphylococcus* or *Micrococcus* on the basis of GC%.

However, for the usual laboratory using Baird-Parker's classification, problems of distinguishing *Staphylococcus* and *Micrococcus* still remain. Although the Hugh and Leifson test was standardized by the ICSB committee on the taxonomy of staphylococci and micrococci

(Subcommittee on Taxonomy, 1965), a number of staphylococci give only weak anaerobic growth and may be misclassified as micrococci. In an attempt to remedy the situation, Evans and Kloos (1972) devised a thioglycolate medium containing the redox indicator, methylene blue, to improve anaerobiosis. Results of separating the genera in this way appeared promising since there was a good correlation with the GC% of strains. However, Davis and Hoyling (1974) reported that, in their hands, the test suffered from lack of reproducibility when compared to the Hugh and Leifson test.

Tests of classification of this sort rarely give a complete separation of genera, and it is probably a feature of taxonomy which the operator has to accept while acknowledging the undoubted value of the test from a practical standpoint.

Baird-Parker (1974a) has pointed out that in his experience the majority of strains of *Staphylococcus* from human and animal sources give good anaerobic growth, whereas those giving equivocal results are principally strains derived from airborne dust. Thus, while the problem is of great concern to the "dust ecologist," it has less significance for the skin microbiologist.

Separation of the individually named species of Micrococcaceae has been a controversial topic for some time. *S. aureus,* generally recognized in the laboratory by its production of a carotenoid pigment and its possession of plasma coagulase, is considered to be separable from other species by its ribitol teichoic acid, *S. epidermis* and *S. saprophyticus* possessing glycerol or nonalditol teichoic acids (Schleifer and Kandler, 1972; Schleifer and Kocur, 1973). Studies on DNA and cell walls, though serving to define speciation, are of little value in practical ecological studies, whereas the classification of Baird-Parker (1963, 1965) is simple and rapid enough to differentiate Micrococcaceae where large numbers of isolates are being dealt with.

Baird-Parker (1974a) has revised his classification and delineated six biotypes of *S. aureus,* four of *S. epidermidis,* and four of *S. saprophyticus.* This division appears to have practical value, in that it correlates to some extent with the source of isolation. For example, *S. epidermidis* biotype 1 was the most common type isolated from humans and may be found in systemic infection, while biotype 2 was the most common isolate from porcine skin. Thus, three species of *Staphylococcus* and only three species of *Micrococcus* are now presented in the current edition of *Bergey's Manual* (Baird-Parker, 1974b). Of these only *M. luteus, S. aureus,* and *S. epidermidis* are cited as being common on human skin; others—*M. roseus, M. varians,* and *S. saprophyticus*—are commonly isolated from a variety of environmental habitats but may contaminate the skin.

Schleifer and Kloos with their co-workers have produced alternative taxonomic schemes to Baird-Parker's. Their approach is to attempt to correlate all the known data in order to substantiate a specific epithet. These schemes tend to be more complex than that of Baird-Parker, and a number of variable results are included. The classification of the genus *Micrococcus* by Kocur and Schleifer (1975) is the simplest and relies on 12 characteristics to differentiate 9 species (Table IV). The scheme for *Staphylococcus* uses 21 features to differentiate 11 named staphylococci (Kloos and Schleifer, 1975a,b; Schleifer and Kloos, 1975), and a second scheme to distinguish subspecies of *S. sciuri* sp. nov. from other novobiocin-resistant species (Kloos *et al.,* 1976b).

Coincident with the publication of *Bergey's Manual,* Kloos *et al.* (1974) described two new species of *Micrococcus* from skin, *M. lylae* and *M. kristinae,* and clarified the taxonomic position of earlier named strains, *M. agilis, M. nishinomiyaensis,* and *M. sedentarius.* An identification of these was finally put forward (Kocur and Schleifer, 1975) based on simple physiological and biochemical tests. Schleifer, Kocur, Kloos and their co-workers have the advantage that, in proposing classification schemes, they have the support of differentiation by GC% estimations and peptidoglycan structure. The groundwork for taxonomic peptidoglycan analysis was reviewed by Schleifer and Kandler (1972).

The newness of these publications has not allowed for sufficient experience in their use as an ecological tool to make any critical assessment. It would be useful to attempt to correlate them with Baird-Parker's biotypes using a large collection of strains. A comparison is provided in Table IV.

Subdivision below the level of species or biotype is achieved by phage typing. Phages for *S. aureus* are well characterized, and more recently, phages for *S. epidermidis* and *Micrococcus* spp. have been described (Krynski *et al.,* 1975; Peters and Pulverer, 1975; Pulverer *et al.,* 1975; Talbot and Parisi, 1976).

2.3. *Staphylococcus aureus*

Staphylococcus aureus is the only truly pathogenic species within the Micrococcaceae and has been extensively studied.

Much has been written on the carriage of *S. aureus,* although this is rarely an inhabitant of normal human skin. Studies on *S. aureus* stem principally from its role as a pathogen causing local inflammatory lesions, including surgical wound infection, as well as osteomyelitis, pneumonia, and toxigenic food poisoning. Studies on skin carriage of *S. aureus* are often related to the dispersal of this organism in, for example, operating theaters. It possesses a battery of enzymes and toxins, expression of

Table IV. Comparison of Taxonomic Schemes for Aerobic Cocci[a]

Baird-Parker (1963, 1965)	Bergey's Manual (1974)	Kloos et al.
SI	*Staphylococcus aureus*	*Staphylococcus aureus*
SII/SIV	*S. epidermidis* biotype 1	*S. epidermidis*
		S. hominis
		S. haemolyticus
		S. warneri
SIII	*S. epidermidis* biotype 2	*S. simulans*
SV	*S. epidermidis* biotype 3	*S. epidermidis*
		S. hominis
		S. warneri
		S. capitis
SVI	*S. epidermidis* biotype 4	*S. hominis*
		S. haemolyticus
		S. warneri
		S. capitis
		S. simulans
M1	*S. saprophyticus* biotype 1	*S. saprophyticus*
		S. cohnii
M2	*S. saprophyticus* biotype 2	*S. saprophyticus*
		S. cohnii
M3	*S. saprophyticus* biotype 3	*S. saprophyticus*
		S. cohnii
		S. xylosus
M4	*S. saprophyticus* biotype 4	*S. xylosus*
M5/M6	*Micrococcus varians*	*Micrococcus varians*
M7	*M. luteus*	*M. luteus*
		M. lylae
		M. sedentarius
M8	*M. roseus*	*M. roseus*
—	—	*M. kristinae*
—	—	*M. nishinomiyaensis*

[a]Based on Bibel et al. (1976).

which presumably accounts for the varying pathogenic roles. Hemolysins were recently reviewed by Wiseman (1975).

S. aureus inhabits the anterior nares of about 35% of normal adults, the perineum in about 15%, and the toewebs and axillae of about 5–10%. Variations in levels of carriage between various populations may simply reflect random variation inherent in sampling procedures (Noble et al., 1967), but there appears to be a real difference between carriage in negroids and caucasoids (Noble, 1974), negroids being less often carriers. It may be worth noting that, in normal persons, sites of carriage and reproduction are those which possess apocrine glands. On normal undifferentiated skin, resident, multiplying *S. aureus* is rare, though small numbers can always be found as transients or contaminants. Patients with diseases of the skin, however, are particularly susceptible to colonization with *S. aureus,* and carriage rates may reach 100% on clinically normal, uninvolved skin in patients with atopic eczema (atopic dermatitis). Patients with psoriasis may also become colonized but many do not, despite extensive exposure. The reasons for this are unknown (Noble and Somerville, 1974).

For epidemiological purposes, *S. aureus* is divided into groups on the basis of its susceptibility to phage. It is convenient to recognize four phage groups. Group I contains strains which are frequently the cause of local follicular lesions in populations outside hospitals, though the notorious "phage type 80/81" was responsible for much surgical wound infection. Phage group II contains strains, "type 71" complex, which are involved in impetigo. Group III consists of strains most frequently involved in surgical wound sepsis and also in food poisoning, and group IV is a miscellaneous group of no specific abilities. Group II strains have much in common with animal staphylococci (Galinski, 1975); groups I and III are more closely related to each other than either is to group II. Groups I and III may carry plasmids specifying either of two serologically distinct types of penicillinase, while the penicillinase of group II strains is distinct. Perhaps the most striking finding in relation to skin is the epidermolytic toxin which long appeared to be the prerogative of group II type 71 complex. Non-group-II strains are now known to possess the ability to elaborate this toxin, but the majority of strains which possess the plasmid specifying abundant production fall in group II. This toxin is of great interest in that its action appears confined to the stratum granulosum of the skin, causing the skin to split and peel at that junction. For reviews of this fascinating substance, see Elias *et al.* (1976, 1977).

2.4. The Taxonomy of Skin Coryneforms

The term "coryneform," like "coliform," is a designation of a large, ill-defined group of bacteria. The term is in general use among bacteriologists, unlike the term "diphtheroid," which is used in medical bacteriology for isolates which morphologically resemble *Corynebacterium diphtheriae* (Cowan, 1968).

The coryneforms found on the normal human skin have been conveniently separated into anaerobic and aerobic types. The anaerobic coryneforms are at present placed in the family Propionibacteriaceae, genus *Propionibacterium.* The taxonomic status of the aerobic coryneforms is less clear, though historically they have long been considered as corynebacteria.

2.5. Aerobic Coryneforms

Although several schemes for the classification of the aerobic coryneforms of skin have been proposed (Evans, 1968; Smith, 1969; Marples, 1969; Holt, 1969; Somerville, 1973; Kasprowicz *et al.*, 1974), general agreement on a scheme analogous to the Baird-Parker scheme for cocci has not been reached.

Ecological studies of skin usually distinguish between lipophilic and nonlipophilic isolates or between small- and large-colony isolates. Smith (1969) considered that small-colony strains had a definite lipid requirement. Tests which have been used in classification schemes are those thought to have some correlation with the *in vivo* activity of the bacteria or with some observed ecological distribution; for example, most surveys of the flora have noted the close correlation between nitrate reductase activity and nasal coryneforms.

The phenomenon of coral-red fluorescence under UV light associated with erythrasma and the species *C. minutissimum* proposed by Sarkany *et al.* (1961) has led to porphyrin production being used in the scheme of Somerville (1973). However, Somerville (1972) found that many other named *Corynebacterium* spp. as well as some micrococci are capable of fluorescence.

Classification schemes for aerobic skin coryneforms have not had the benefit of support by GC% or peptidoglycan analysis in the same way as have the Micrococcaceae. However, the distinction between *Corynebacterium* cell walls and those of other coryneforms was revealed by Cummins and Harris (1956, 1958). They showed that *meso*-2,6-diaminopimelic acid (*meso*-DAP) and arabinose are characteristic of *Corynebacterium* of animal origin among the coryneform group. This finding was supported by antigenic relationships within the genus (Cummins, 1962). These major constituents of the cell wall are present in such large amounts (neutral sugars), or are unique to the cell walls of coryneforms (*meso*-DAP), that analysis of whole cell hydrolysates may be carried out with reasonable confidence. Rapid methods of preparing chromatograms to identify cell-wall components have also proved useful among actinomycetes (Boone and Pine, 1968; Lechevalier and Lechevalier, 1970; Berd, 1973; Stanneck and Roberts, 1974; Keddie and Cure, 1977).

Chromatography of whole cell hydrolysates has now been carried out on over 1000 isolates of coryneforms from human skin (Pitcher, 1977). The results showed that about 60% of coryneform skin isolates conform to the genus *Corynebacterium* in possessing *meso*-DAP and arabinose but that substantial numbers of apparently resident coryneforms do not have this pattern (Table V).

Table V. Cell-Wall Composition of Coryneforms from Human Skin[a]

		Percent with cell-wall components				
	Total Strains	meso-DAP Arab	meso-DAP Gal	LL-DAP Arab	DAB + Rha	No DAP or DAB
Possible identity		Coryne- bacterium	Brevibac- terium	?	?	?
Total exposed sites (forehead, ear, forearm, hand, leg)	304	43	25	1	1	30
Intertriginous sites (axilla, perineum, toewebs)	496	67	20	2	2	8

[a]Abbreviations: DAP, diaminopimelic acid; DAB, diaminobutyric acid; arab, arabinose; gal, galactose; rha, rhamnose.

About 20% of the isolates possessed meso-DAP but no arabinose, the major sugar detected being galactose. This combination is found in many coryneforms from dairy products. Sharpe et al. (1976) postulated that their strains of coryneforms from dairy products could be derived from human skin since their high salt tolerance and optimum growth temperature were uncharacteristic of Brevibacterium linens. These strains also lacked the orange pigment associated with the latter organism. A later study compared a number of B. linens strains with other dairy coryneforms isolated by Sharpe and with a number of those from human skin isolated by Pitcher. The cell-wall patterns were extremely similar, as were many other physiological and biochemical features, though B. linens strains were pigmented and had much lower growth temperatures (Sharpe et al., 1977).

We conclude that although members of the genus Corynebacterium constitute the major part of the aerobic coryneform flora of skin, Brevibacterium spp. are also very common. Bacteria named in this genus are listed as species incertae sedis in the eighth edition of Bergey's Manual (Rogosa et al., 1974), but Sharpe and her colleagues (1977) suggest that the genus be reconstructed to include B. linens and coryneforms with similar cell-wall composition.

One of the most interesting observations on this group of organisms is that B. linens together with the dairy and skin strains described by Sharpe et al. (1977) are able to liberate the odoriferous gas, methane thiol from L-methionine. Pitcher (unpublished observations) has not been able to demonstrate this reaction in Corynebacterium sensu stricto in a study of over 400 isolates from skin. This could have implications in the production of a thiol component of human body odor. Our observations indicate that only about half of the isolates having a *B. linens* type of wall can actually produce methane thiol.

A further small but resident group of coryneforms possesses LL-diaminopimelic acid (LL-DAP) and arabinose. This appears to represent a previously undescribed genus, and further studies on the ecology of these strains is required (Pitcher, 1976). Other groups of coryneforms also appear on the skin.

Pitcher (unpublished observations) subjected a representative sample of 70 skin isolates of various cell-wall compositions to numerical taxonomy and compared them with 70 reference strains covering a wide range of genera from human sources and from natural environments. It was apparent from this survey that the majority of coryneforms resident on human skin forms a unique and homogeneous group with little affinity with coryneforms derived from nonanimal sources, despite differences in cell-wall composition among them. It was also noted that skin isolates fell broadly into two groups, between which the most notable difference was in the action on sugars. One group was unable to produce acid from sugars and contained reference strains of *C. pseudodiphtheriticum*.

A few skin strains with cell walls not containing *meso*-DAP were included in the study, and these fell outside the area occupied by the majority, with the exception of strains having LL-DAP and arabinose. This unusual combination has not been reported from any source other than human skin, and its relationship to other coryneform genera remains to be ascertained.

The site distribution for aerobic skin coryneforms gives some clue to their claim as resident skin organisms. Strains without DAP in their walls were found almost exclusively on exposed sites, suggesting that their origin is from the extrahuman environment.

2.6. Anaerobic Coryneforms

Anaerobic coryneforms are possibly the most numerous microorganisms on the skin at the sites where they occur. However, they are more restricted in habitat than most of the aerobic flora (Noble and Somerville, 1974).

More attention has been directed toward this group than toward the aerobic coryneforms, principally because of their apparent involvement in the etiology of acne vulgaris. The anaerobic coryneforms from human skin have traditionally been referred to as *Corynebacterium acnes*, and much controversy has existed as to whether they should be included in this genus. However, on the basis of cell-wall composition and metabolic-end-product studies, they have now been assigned to the genus *Propionibacterium*.

The principal characteristic of the genus lies in the ability to metabolize carbohydrates to a mixture of organic acids, including propionic acid. The only other genus in the family—*Eubacterium*—is unable to produce propionic acid. Eight species are listed in the current edition of *Bergey's Manual* (Moore and Holderman 1974), of which four are species from dairy products. The four remaining species are derived from clinical material and have received considerable attention from taxonomists.

Cummins (1970) examined 43 strains of anaerobic coryneforms and was able to divide them into two groups on the basis of cell-wall analysis. Both groups possessed LL-DAP, glucose, and mannose in their walls, but group I had, in addition, galactose. This work was later extended to cover 80 strains of anaerobic coryneforms (*P. acnes*), 29 named *Propionibacterium* spp. and 8 *Arachnia propionica* strains (Johnson and Cummins, 1972). The *P. acnes* strains were divided into type I and II on the basis of the presence or absence of galactose in the walls. The walls of almost all strains contained LL-DAP in contrast to the *meso*-DAP of most aerobic coryneforms. The authors concluded that the organism previously known as *C. acnes* was rightfully placed in *Propionibacterium*. DNA similarities between classical propionibacteria and *P. acnes* showed rather low homologies despite cell-wall relatedness. As a consequence, three major groups were distinguished in the genus; these three have been provisionally named *P. acnes, P. granulosum,* and *P. avidum.*

Subsequently, Cummins and Johnson (1973) examined strains labeled *Corynebacterium parvum* and found them to be synonymous with *P. acnes*, the species *P. granulosum* pertaining to the group II *C. acnes* described by Voss (1970). Radioimmunoassay of cell-wall antigens of *C. parvum* versus *P. acnes* and *P. avidum* revealed high but not complete cross reactivity (Dawes and McBride, 1975); this supported the classification of Johnson and Cummins (1972).

Confirmation of the synonomy of *C. parvum* and *P. acnes* was given by Azuma *et al.* (1975), who also stressed the considerable differences between these organisms and *C. diphtheriae;* these differences extended to the mechanism of their respective adjuvant activities. The anaerobes enhanced circulating antibody production and delayed hypersensitivity in mice but showed no effect on helper function of carrier-prime T cells. *C. diphtheriae,* on the other hand, showed a potent adjuvant activity toward the latter function.

Considerable interest has been shown of late in the anaerobic coryneforms on two fronts. The first to attract attention was the role of *P. acnes* in the pathogenesis of acne vulgaris. The second was the potent adjuvant activity of *C. parvum*. The former of these has centered around the lipolytic properties of *C. acnes* on sebum (Voss, 1974, Puhvel, *et al.,* 1975). More recently, it has been suggested that hyaluronidase activity by *P. acnes* may also play a part (Holland *et al.,* 1974).

Interest in the adjuvant activity of anaerobic coryneforms has recently proliferated. The effect of *C. parvum* on the lymphoreticular system of animals was summarized by Adlam (1973a) as depression of some T-cell functions, inhibition of graft-versus-host disease, impairment of mixed lymphocyte reactivity, and depression of the phytohemagglutinin response of spleen and blood lymphocytes. Apparently related to these properties is an ability by *C. parvum* to sensitize animals to histamine (Adlam, 1973b).

In view of the apparent synonomy of *C. parvum* and *P. acnes,* an immunological component of acne pathogenesis is not unlikely. Massey *et al.* (1978) have shown that *P. acnes* activates the alternate pathway in complement, causing production of C3b from C3. Dahl and McGibbon (1976) have shown C3 deposition in acne lesions.

However, the etiology of acne is complex; it is well known that predisposing factors arising from the skin itself are coincident with the formation of comedones. The increased sebum secretion may be associated with hormonal changes since acne is prevalent during puberty.

Puhvel *et al.* (1975) have demonstrated that the triglycerides of dissected sebaceous glands were effectively hydrolyzed by *P. acnes.* The discussion of the role of lipolysis goes on. Esterases of various sorts are possessed by a high proportion of the normal resident flora of skin, but it is only because of its situation deep in the sebaceous follicles that *P. acnes* continues to receive attention in this respect (Voss, 1974; Strauss *et al.,* 1976; Leyden, 1976; Leyden and Kligman, 1976; Kellum, 1976).

P. acnes involvement in more serious infections, including systemic disease, has been increasingly reported in recent years, and the etiology of such conditions may also have its origins in the remarkable immunological properties of the organism. Such sources include blood in malignant disease or septicemia (Duborgel, 1974), cerebrospinal fluid (Beeler *et al.,* 1976), bone marrow (Saino *et al.,* 1976), liver abscess (Balfour and Minken, 1971), and rheumatoid arthritis (Bartholomew and Nelson, 1972).

The identification of anaerobic coryneforms from skin as *P. acnes, P. granulosum,* and *P. avidum* is now a lot clearer, though difficulties still exist. Pulverer and Ko (1973) have identified eight biotypes and eleven serotypes of *P. acnes;* though the relationship between these could not be established, the authors recommended a combination of techniques.

Jong and co-workers (1975) recommend that, for routine purposes, the identification of anaerobic coryneforms according to the biotypes of Johnson and Cummins (1972) be retained for the present. Jong *et al.* (1975) have also examined the possibilities of phage typing as an adjunct to chemical studies and have identified 13 suitable bacteriophages which could be used to separate *P. acnes* and *P. granulosum* each into seven phage types. Two strains of *P. avidum* could not be typed with these phages. It was considered essential to continue with biotyping in conjunction with phage typing while stable phages are sought to divide *P. acnes* into further phage types.

2.7. Gram-Negative Bacteria

The only group of gram-negative bacteria which are apparently resident on the skin are members of the genus *Acinetobacter,* which were previously known as the Mima/Herellea group. These bacteria tend to favor the moister regions, such as toewebs, axillae, and groin (Somerville and Noble, 1973).

The current status of the genus *Acinetobacter* in *Bergey's Manual* (Lautrop, 1974) is in the family Neisseriaceae. Only one species is described, *A. calcoaceticus,* and this replaces previous synonyms such as *Mima polymorpha, Acinetobacter anitratus,* and *Herellea vaginocola.* Two types of *A. calcoaceticus* are now described; these differ in their ability to oxidize carbohydrates.

Snell (1973) has indicated that two types of *Acinetobacter* exist—those which are biochemically active (*A. anitritus*) and those which are very inactive (*A. lwoffi*). More recently, Pagel and Seyfried (1976), using numerical taxonomy, have confirmed the existance of two distinct phena within the genus. Phenon I consisted mainly of the least biochemically active strains, which were almost all aquatic isolates. The second phenon was more active and included strains from clinical sources. The authors suggest that the name *Acinetobacter calcoaceticus* be retained for the active strains, which would include skin isolates, and they propose the name *Acinetobacter lwoffi* for the less active strains.

The characteristic of *Acinetobacter* which separates it from other Neisseriaceae is the absence of oxidase activity. Nonfermenting gram-negative bacilli which were oxidase positive and identified as *Alcaligenes, Flavobacterium, Moraxella,* and *Pseudomonas* spp. were found by Juni (1972) and Brooks and Sodeman (1974) to be unable to transform *Acinetobacter* DNA, whereas all oxidase-negative strains tested, whether carbohydrate oxidizers or not (*Mima/Herellea*), were able to transform the *Acinetobacter* DNA sample. This would suggest that the criterion of oxidase reaction is a good one for the genus, which, though it may be divided into two subgroups, is genetically homogeneous.

Members of the genus *Acinetobacter* are ubiquitous in nature, being found in soil, water, and sewage. That they form part of the normal skin flora of man (Noble and Somerville, 1974) is surprising, since other very common gram-negative rods are rather rare on the skin except where obvious enteric contamination (e.g., the perineum) occurs.

2.8. Yeasts—*Pityrosporum*

The only yeasts that appear to form stable populations on normal skin are *Pityrosporum* spp. These have, unlike other yeasts, a requirement for lipid. This requirement can be supplied in the form of myristate or palmitate. *Pityrosporum* spp. are generally confined to the scalp and face, where there is an abundance of lipid material. In these regions, these organisms are present in large numbers, though they are occasionally found on other sites.

The present classification of the genus is unsatisfactory. As Fungi Imperfecti, they are placed in the family Cryptococcaceae (Lodder, 1970). Three species are delineated, two of which are lipophilic and colonize human skin; these are *P. ovale* and *P. orbiculare*. A third species, *P. pachydermatis*, which is usually associated with animals but colonizes diseased human skin, does not require lipid for growth.

The differentiation of *P. ovale* and *P. orbiculare* is at present based on morphology. The latter species can exhibit dimorphism to about the same extent as *Candida* spp. Most workers do not appear to bother with separating the species.

The greatest interest in *Pityrosporum* has for a long time centered around their apparent association with dandruff. However, the recent publication of McGinley *et al.* (1975) and Kligman *et al.* (1976) have clearly shown that suppression of *Pityrosporum* occurred without cessation of dandruff. The principal feature of the condition was the overall increase in the scalp microflora. The authors envisage this as a secondary effect consequent to the increased flow of sebum and high epidermal turnover of the condition, providing an increase in nutrients for the residents.

An interesting observation in association with dandruff was made by McGinley *et al.* (1975) in that the incidence of *P. acnes* on the scalp was inversely related to the *Pityrosporum* count, the cocci remaining unchanged and aerobic coryneforms being rare on the scalp under all conditions.

Pityrosporum made up about 50% of the flora in normal subjects, but a rise to 75% in dandruff and 80% in seborrheic dermatitis mirrored a rise in nucleocytes for these conditions. Corneocytes did not change much. The fall in *P. acnes* was dramatic, from 26% for the normal scalp to 6% and 1% for dandruff and seborrheic dermatitis, respectively.

These conditions were apparently quite different in their flora from other scaly lesions, such as psoriasis, in that the latter provides a climate and increased surface area for an overall increase in aerobic flora. However, it is not known whether psoriasis of the scalp would follow this pattern.

In light of these recent publications, it would appear that pathogenicity in *Pityrosporum* is at least questionable, yet it has been suggested that *P. orbiculare* in its filamentous phase can be invasive, giving rise to the scaling condition known as pityriasis versicolor. In this form the yeast has been referred to as *Malassezia furfur* (Noble and Somerville, 1974).

Pityrosporum folliculitis has been described by Potter *et al.* (1973) on histological evidence of yeast cells in the pilosebaceous follicles. Further identification of the yeasts was not carried out, but if indeed they were *Pityrosporum,* then the erythematous reaction of the skin in the area of the papules may result from mechanisms not unlike those of *P. acnes* association in acne vulgaris. It has been suggested that the primary stimulus for acne is a hormonal one, the finding of bacteria being a consequence of this. Pityriasis versicolor has often been linked to long-term corticosteroid therapy and to Cushing's syndrome, though none of the patients cited with folliculitis could be linked with these states (Potter *et al.,* 1973). It would seem that hormonal influences may play a part in instigating skin complaints in which the bacterial flora is forced into imbalance.

3. Microbes in Their Habitat

3.1. Host Factors

The host factors that govern the number of organisms on the skin are not known. It is well established that there are high-count and low-count persons and that the microbial number tends to remain high or low for a period of years (Noble and Somerville, 1974; Evans, 1975). This is shown graphically in Fig. 5, which is drawn from unpublished data collected by Somerville and Murphy (1973). With one exception, the sole of the foot of male A had skin counts 10–100 times greater than those of male B. Both men were laboratory workers of about the same age and were typical high-count and low-count people. It seems probable that this is genetically determined because the nasal flora of identical twins is more alike than that of nonidentical (binovular) twins of the same sex; further, there is a family susceptibility to carriage of *S. aureus* (Hoeksma and Winkler, 1963; Noble *et al.,* 1967; Aly *et al.,* 1974a).

Males carry between three and five times as many bacteria on their skins as do females. Fig. 6 shows the mean logarithmic count on the skin surface at six sites (shin, thigh, chest, back, forearm, and abdomen) for 38 males and 34 females working in surgical units. It is clear that the distribution in counts for males is about 0.5 logarithm units higher. We may speculate that this reflects the tendency of males to sweat more than

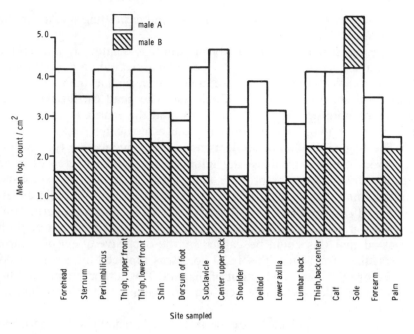

Figure 5. Histogram of aerobic microbial counts on skin in typical high-count and low-count individuals. Note the use of a logarithmic scale.

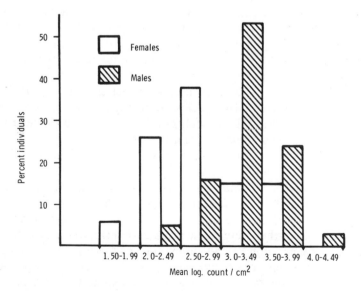

Figure 6. Histogram of aerobic microbial counts per square centimeter of skin in males and females.

females, thereby keeping the skin moister and permitting slightly greater growth rates.

Bearing in mind this diversity in count, it is interesting to look at quantitative counts obtained by the scrubbing method on different areas of the body. With the exception of moist areas, such as the toewebs, the limbs have the lowest counts (Table II). The head carries the largest number of both aerobes and anaerobes, and the thorax is intermediate between these two, both anatomically and ecologically.

The hygiene habits of subjects may also influence the type of flora recovered. Newborn babies and infants have a high degree of gram-negative colonization before the adult pattern of flora is established (Sarkany and Gaylarde, 1967; McAllister et al., 1974). Bollgren and Winberg (1976) found a massive colonization of infants in the periurethral area with gram-negative bacilli and enterococci, and though it might be suggested that this could be caused by the occlusive effect of wearing diapers, they found that this fecal flora was replaced by a normal skin flora before the babies were out of diapers. The authors postulated that before the children were one year old, normal antibacterial mechanisms came into play.

It is well recognized that experimental occlusion exerts a profound effect on the skin flora, and it is most likely that, in everyday life, the wearing of clothing could be described as loose occlusion. It would certainly prevent heat and water loss from a great deal of the body surface, and those organisms which do best in a warm, moist environment would be most successful. Major variation in the flora must therefore occur in different geographic regions (Kloos et al., 1976a).

Though miliaria rubra may be a consequence of occlusion, Henning et al. (1972) could not link any particular bacterial increase to the etiology of the condition. These and other authors (Bibel and LeBrun, 1975; Bibel et al., 1976b) noted that, after occlusion, at least seven days are needed for the flora to return to the preocclusion level and for gram-negative and other intruders to be eliminated from the site.

Dispersal of Skin Flora. There have been a number of studies on dispersal of the skin flora. Two aspects are worth attention here. One is that the dispersal of the human flora takes place principally on squames or skin fragments and is of importance in the operating theater, where microorganisms may cause surgical wound infection; in the manufacturing industry, where they may cause product deterioration, as in cosmetics or pharmaceuticals; or in "clean" areas such as for assembly of electronic components or in spacecraft (Puleo et al., 1977), where either particles or organisms may prove important. The second aspect has centered on the difference in site of principal dispersal. Males disperse more from the thighs and abdomen, while females disperse from the shins. There are other intriguing differences which are yet unexplained. The subject has been reviewed by Noble (1976, 1977a). Fig. 7 shows the way in which squames become detached from the skin surface.

3.2. Microbial Coactions

Coactions between microorganisms on the skin surface are not well documented if one is searching for evidence of coaction *in vivo*. Little enough is known of the stimulation or inhibition of skin microbes *in vitro*. Staphylococci from around the eyes are known frequently to be able to suppress the growth of other gram-positive cocci, *S. aureus* strains of phage group II are active against *Corynebacterium* and *Streptococcus* species, and *"Streptococcus viridans"* can inhibit the growth of some *S. aureus* isolates. These and other *in vitro* coactions are summarized by Noble and Somerville (1974). Bacteriocins are produced by many organisms, and these are active against their own or closely related species. The subject of bacteriocins has been reviewed by Tagg *et al.* (1976). Growth-enhancing substances also occur, and Lilly and Stillwell (1965) proposed the term *probiotic* to stand for substances having the opposite effect of antibiotics. This term is to be preferred to the term *stimulators* since some

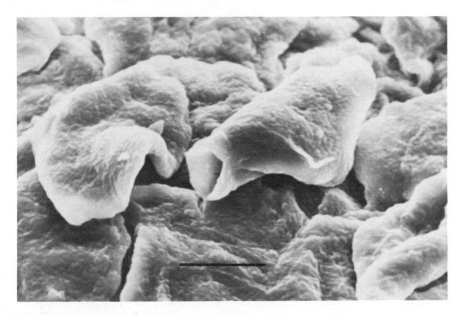

Figure 7. Squames become detached and curl up prior to dispersal from the skin. The bar represents 10 μm.

probiotics may extend the logarithmic phase rather than provide direct stimulation in the form of nutrients or vitamins.

Selwyn and his colleagues have demonstrated the ability of cocci to produce antibiotic-like substances and have shown that, in experimental systems such as the chemostat, antibiotic-producing strains can bring about "ecological" changes by suppression of susceptible organisms (Selwyn *et al.,* 1976; Marsh and Selwyn, 1977a,b; Milyani and Selwyn, 1978). Selwyn (1975) has shown that the presence of these "producer" organisms in lesions of eczema and, to a lesser extent, in psoriasis will protect the host against colonization or infection with exogenous, potentially pathogenic species. In samples from normal skin, however, producer organisms seem to be present in only small numbers. Presumably the antibiotic production does not always have a direct and obvious "use"; if it did, one might expect the producers to be present in greater numbers.

Among the actinomycetes, antibiotic production may be related to cell-wall structure. Kalakoutskii and Agre (1976) suggest that a structurally incorporated antibiotic that was exposed at the cell surface would then function, not as a weapon of attack, but as a defense mechanism against potential attackers; strains possessing such a mechanism would be expected to survive the process of natural selection more successfully. Since fungus mycelium may become heavily colonized in soil by bacteria utilizing fungal exudates, we may appreciate the "need" for production of antibiotics by soil fungi. This does not explain extracellular production of antibiotics, however.

Haavik (1976) has suggested an alternative use for the antibiotic bacitracin. He suggests that the normal function of bacitracin during the growth of its producer strain, *Bacillus licheniformis,* may be to promote the uptake of divalent metal ions. Thus, peptide "antibiotics" may be functional microbial metabolites concerned with cation transport.

Interference between strains of *S. aureus* has been used as a "therapeutic" measure (Aly *et al.,* 1974b). This work resulted from an observation that newborn infants in a nursery who acquired a strain of *S. aureus* from a nurse were not at risk to acquisition of the virulent strains of *S. aureus* then current in hospitals throughout the world. The nurse's strain, designated 502A, was then deliberately applied to the nares of newborn infants and useful reductions in acquisition and sepsis by the newborns was accomplished. Eventually the practice spread to the use of this strain in chronic furunculosis, again with some success. However, strain 502A was not apathogenic, and when the epidemics of *S. aureus* infection ceased, the practice was discontinued. This is one example of direct use of biological control in the ecology of man. Unpublished observations (W. C. Noble) show that, although individuals with impaired immune response may be easy to colonize with antibiotic-producing Micrococcaceae, producer strains do not necessarily oust the pathogenic flora.

The nature of the staphylococcal interference remains rather obscure, but this is not true of the way in which dermatophyte fungi may bring about changes in the associated flora of a lesion. *In vitro,* isolates of *Epidermophyton floccosum, Trichophyton mentagrophytes,* and *T. rubrum* may produce penicillin compounds, including penicillin G; in addition, *Epidermophyton* may produce antibiotics similar to azalomycin F and to fusidic acid, while *Trichophyton* also produces an antibiotic that resembles streptomycin. Only the penicillin moiety seems to be well worked out as yet, but the production of penicillin and other antibiotics has been demonstrated in the fungus lesions of patients (N. Youssef, personal communication; Youssef *et. al.,* 1978), and there is good evidence of a shift toward penicillin resistance in cocci in these lesions (N. Youssef, personal communication; Bibel and Lebrun, 1975). Unpublished observations (D. G. Pitcher and N. Youssef) have shown that the composition of the coryneform flora of dermatophyte lesions may also shift toward the penicillin-resistant species of *Brevibacterium.* The best known example, however, is taken from studies, not of humans, but of hedgehogs, because 92% of skin isolates of *S. aureus* obtained from hedgehogs infected with *T. mentagrophytes* var. *erinacei* were resistant to penicillin. It may be that skin infection is a more chronic lesion in the hedgehog than in humans, allowing more time for the development of a resistant flora (Smith and Marples, 1964).

There is good epidemiological evidence from elsewhere that allergy to penicillin is more common in persons with fungal infection. This may merely be a reflection of the immunological status of the patient, but it is more likely to reflect the *in vivo* production of penicillin, leading to skin hypersensitivity.

3.3. Ecological Genetics

Ford (1975) used the term *ecological genetics* to describe his elegant studies of the distribution of the melanic and normal forms of the moth *Biston betularia,* in which the melanic form appeared to have a selective advantage in heavily industrialized surroundings, for they were less often taken by predators.

There seems little doubt that much the same "ecological genetics" is at work on the human skin among strains of *S. aureus,* though the selective agent(s) have yet to be identified. Noble (1977b) found that strains of *S. aureus* colonizing the skin of patients with skin diseases could persist in an antibiotic-resistant form (resistant to penicillin and/or tetracycline) on the skin surface yet predominate in the nose as a variant that had lost one of these resistant patterns. The mechanism of loss seems clear; penicillinase production and tetracycline resistance are carried on two separate plasmids. *In vitro* studies have shown that strains which appear labile on skin can easily be induced to lose plasmids when exposed to SDS. At present, the ecological factors that maintain a balance of resistant variants on the skin and sensitive variants in the nose are not known, though they may be different for each of the two resistance patterns. Selective factors may include requirements for vitamins and amino acids (J. Naidoo, personal communication).

Acquisition of resistance plasmids by *S. aureus* is phage-mediated. Lacey (1971) has shown that neomycin resistance can be transferred between strains on the skin surface; this transfer may occur naturally under the stimulation of exposure to therapeutic neomycin. Plasmid carriage in *S. aureus* was comprehensively reviewed by Lacey (1975). Naidoo and Noble (unpublished) have demonstrated the transfer of gentamicin plasmids between strains of staphylococci on human skin in the absence of the antibiotic.

3.4. Experimental Ecological Studies

Attempts to study the skin flora in ways which correspond to the habitat are surprisingly recent. Classically, medical microbiologists have used rich media such as blood agar (a nutrient agar containing perhaps 7% whole blood) or various selective or enrichment media. Furthermore, 37°C has been the traditional incubation temperature, except for the dermatophyte fungi, where 26°C or 30°C has been used. Murphy (1975) tried to devise a medium based on the known constituents of the water-soluble fraction of sweat and, although hampered by the insolubility of some substances, produced a defined medium that would support the growth of *Micrococcus luteus* but would support the growth of other skin microbes only if stratum corneum or an aqueous extract of stratum corneum was added. Murphy (1975) established that there was a further, previously uncharacterized substance with certain chemical properties (water solubility, heat stability, and low molecular weight), but she was unable to identify this substance. It could be partially substituted by coenzyme A, but neither coenzyme A nor analogues had the same chromatographic characteristics as the unknown. A variety of substances, including nucleotides, were assayed, but none was found to stimulate growth.

A similar approach to this problem was used by Milayani (1976), whose medium contained a variety of substances that mimicked skin substrates, though some appeared not to have been reported from skin-surface samples. Good growth of all skin microorganisms was claimed for the richest of the media (containing 1.72 g/liter of amino acids; by contrast, Murphy's contained about 2 g/liter), but poorer growth was found on a similar medium containing about half the quantity of amino acid. It was noticed that "diphtheroids" (gram-positive rods that morphologically resemble *C. diphtheriae*) grew with a coccoid morphology on the poorer medium but had a "normal" morphology on the rich medium. At the skin surface there is at least 1.4 g/liter of amino acids but, as evidenced from scanning electron micrographs, diphtheroids have a threadlike, almost mycelial appearance (Noble, 1975).

An alternative approach was that developed by Knight (1973), in which stratum corneum was made to adhere to a sticky tape and sterilized by ethylene oxide gas. When kept in a chamber at high relative humidity and appropriate temperature, fungal spores germinated and produced normal mycelium. Knight has applied this technique to a study of antifungal antibiotics and has determined the conditions of temperature and relative humidity under which fungi will grow (Knight, 1976). Bacteria also grow on stratum corneum as used by Knight.

Experimentally, porcine skin, designed for use in skin grafting to cover burns, has proved a useful tool. Using this sterile, washed material soaked in a synthetic sweat solution, W. C. Noble (unpublished data) has found differences between *M. luteus* and *S. aureus* in ability to grow at relatively low relative humidity. *M. luteus* is less tolerant of 94% relative humidity than is *S. aureus,* though both will reproduce at this level of humidity and both grow equally well at 100% relative humidity. N. Youssef (personel communication) has used this technique to study changes in balance of antibiotic-sensitive and resistant cocci in the presence of dermatophyte fungi. It must be realized, however, that use of skin films *in vitro* is still a step from a study of habitat.

It should be clear that the study of human skin as a habitat for microbes is still in its infancy. Few other habitats are examined and groomed with such care, yet the genera and species of resident microbes are subjects for discovery. We know little of the interaction between skin microbes or their nutrition. Even the interaction between microbe and host, which we call "disease" and which has a massive social and economic cost, is ill understood, and almost any normal skin commensal microbe may function as an opportunist pathogen if the host is stressed or weakened (Savin and Noble, 1977). Skin will long continue to fascinate us as microbiologists and as human beings.

References

Adlam. C., 1973a, *Corynebacterium parvum* and its effects on the lymphoreticular system, *Lab-Lor (Wellcome)* **5**:273.

Adlam. C., 1973b, Studies on histamine sensitization in mice by *Corynebacterium parvum, J. Med. Microbiol.* **6**:527.

Aly, R., Maibach, H. I., Shinefield, H. R., and Mandel, A. D., 1974a, *Staphylococcus aureus* carriage in twins, *Am. J. Dis. Child.* **127**:486.

Aly, R., Maibach, H. I., Shinefield, H. R., Mandel, A., and Strauss, W. G., 1974b, Bacterial interference among strains of *Staphylococcus aureus* in man, *J. Infect. Dis.* **129**:720.

Aly, R., Shinefield, H. I., Strauss, W. G., and Maibach, H. I., 1977, Bacterial adherence to nasal mucosal cells, *Infect. Immun.* **17**:546.

Azuma, I., Sugimura, K., Taniyama, T., Aladin, A. A., and Yamamura, Y., 1975, Chemical and immunological studies on the cell walls of *Propionibacterium acnes* strain C7 and *Corynebacterium parvum* ATCC 11829, *Jpn. J. Microbiol.* **19**:265.

Baird-Parker, A. C., 1963, A classification of micrococci and staphylococci based on physiological and biochemical tests, *J. Gen. Microbiol.* **30**:409.

Baird-Parker, A. C., 1965, The classification of staphylococci and micrococci from worldwide sources, *J. Gen. Microbiol.* **38**:363.

Baird-Parker, A. C., 1974a, The basis for the present classification of staphylococci and micrococci, *Ann. N. Y. Acad. Sci.* **236**:7.

Baird-Parker, A. C., 1974b, Family Micrococcaceae, in: *Bergey's Manual of Determinative Bacteriology* (R. E. Buchanan and N. E. Gibbons, eds.), p. 478, Williams and Wilkins, Baltimore.

Balfour, H. H., and Minken, S. L., 1971, Liver abscess due to *Corynebacterium acnes*. Diphtheroid as pathogen, *Clin. Pediatr.* **10**:55.

Bartholomew, L. E., and Nelson, F. R., 1972, *Corynebacterium acnes* in rheumatoid arthritis. Isolation and antibody studies, *Ann. Rheum. Dis.* **31**:22.

Beeler, B. A., Crowder, J. G., Smith, J. W., and White, A., 1976, *Propionibacterium acnes:* pathogen in central nervous system shunt infection. Report of three cases including immune complex glomerulonephritis, *Am. J. Med.* **61**:935.

Behrendt, H., and Green, M., 1971, *Patterns of Skin pH from Birth through Adolescence,* Charles C. Thomas, Springfield, Ill.

Berd, D., 1973, Laboratory identification of clinically important aerobic actinomycetes, *Appl. Microbiol.* **25**:665.

Bibel, D. J., and LeBrun, J. R., 1975, Effect of experimental dermatophyte infection on cutaneous flora, *J. Invest. Dermatol.* **64**:119.

Bibel, D. J., and Lovell, D. J., 1976, Skin flora maps: A tool in the study of cutaneous ecology, *J. Invest. Dermatol.* **66**:265.

Bibel, D. J., Aly, R., and Maibach, H. I., 1976a, Nomenclature of cutaneous Micrococcaceae: on the brink of chaos, *Int. J. Dermatol.* **15**:670.

Bibel, J. D., Lovell, D. J., and Smiljanic, R. J., 1976b, Effect of occlusion upon population dynamics of skin bacteria, *Br. J. Dermatol.* **95**:607.

Bollgren, J., and Winberg, J., 1976, The periurethral aerobic flora in healthy girls and boys, *Acta Pediatr. Scand.* **65**:74.

Boone, C. J., and Pine, L., 1968, Rapid method for the characterization of actinomycetes by cell wall composition, *Appl. Microbiol.* **16**:279.

Brodersen, M., and Wirth, M., 1976, Detection of HB_s Ag and HB_sAb in sweat, *Acta Hepato-gastroenterol.* **23**:194.

Brooks, K., and Sodeman, T., 1974, Clinical studies on a transformation test for identification of *Acinetobacter* (*Mima* and *Herellea*), *Appl. Microbiol.* **27**:1023.

Cabau, N., Levy, F.-M., and Muller, O., 1974, Les immunoglobulines sudorales chez l'enfant, *Path. Biol.* **22**:883.

Chin, P., and Davis, D. G., 1976, The skin flora of the hemiplegic hand, *J. Hyg. (Camb.)* **77**:93.

Christian, J. H. B., and Waltho, J. A., 1962, The water relations of staphylococci and micrococci, *J. Appl. Bacteriol.* **25**:369.

Cowan, S. T., 1968, *A Dictionary of Microbial Taxonomic Usage,* Oliver and Boyd, Edinburgh.

Cummins, C. S., 1962, Chemical composition and antigenic structure of cell walls of *Corynebacterium, Mycobacterium, Nocardia, Actinomyces* and *Arthrobacter, J. Gen. Microbiol.* **28**:35.

Cummins, C. S., 1970, Cell wall composition in the classification of gram positive anaerobes, *Int. J. System. Bacteriol.* **20**:413.

Cummins, C. S., and Harris, H., 1956, The chemical composition of the cell wall in some gram-positive bacteria and its possible value as a taxonomic character, *J. Gen. Microbiol.* **14**:583.

Cummins, C. S., and Harris, H., 1958, Studies on the cell wall composition and taxonomy of actinomycetales and related groups, *J. Gen. Microbiol.* **18**:173.

Cummins, C. S., and Johnson, J. L., 1973, *Corynebacterium parvum:* a synonym for *Propionibacterium acnes? J. Gen. Microbiol.* **80**:433.

Dahl, M. G. C., and McGibbon, D. H., 1976, Complement in inflammatory acne vulgaris, *Br. Med. J.* **4**:1383.

Davis, G. H. G., and Hoyling, B., 1974, Observations on anaerobic glucose utilization tests in *Staphylococcus Micrococcus, Int. J. System. Bacteriol.* **24**:1.

Dawes, J., and McBride, W. H., 1975, Analysis of anaerobic coryneform cell wall antigens by radioimmunoassay, *Immunochemistry* **12**:855.

Duborgel, S., 1974, Etude de corynebacteries anaerobes isolées à partir de 9000 hemocultures, *Ann. Biol. Clin.* **32**:487.

Duncan, W. C., McBride, M. E., and Knox, J. M., 1969, Bacterial flora: the role of environmental factors, *J. Invest. Dermatol.* **52**:479.

Elias, P. M., Fritsch, P., and Mittermayer, H., 1976, Staphylococcal toxic epidermal necrolysis: species and tissue susceptibility and resistance, *J. Invest. Dermatol.* **66**:80.

Elias, P. M., Fritsch, P., and Epstein, E. H., 1977, Staphylococcal scalded skin syndrome. Clinical features, pathogenesis, and recent microbiological and biochemical developments, *Arch. Dermatol.* **113**:207.

Emmett, M., and Kloos, W. E., 1975, Amino acid requirements of staphylococci isolated from human skin, *Can. J. Microbiol.* **21**:729.

Evans, C. A., 1975, Persistent individual differences in the bacterial flora of the skin of the forehead. Numbers of Propionibacteria, *J. Invest. Dermatol.* **64**:42.

Evans, J. B., and Kloos, W. E., 1972, Use of shake cultures in a semisolid thioglycolate medium for differentiating Staphylococci from Micrococci, *Appl. Microbiol.* **23**:326.

Evans, N. M., 1968, The classification of aerobic diphtheroids from human skin, *Br. J. Dermatol.* **80**:81.

Farrior, J. W., and Kloos, W. E., 1976, Sulfur amino acid auxotrophy in *Micrococcus* species isolated from human skin, *Can. J. Microbiol.* **22**:1680.

Ford, E. B., 1975, *Ecological Genetics,* Chapman and Hall, London.

Förström, L., Goldyne, M. E., and Winkelmann, R. K., 1975, IgE in human eccrine sweat, *J. Invest. Dermatol.* **64**:156.

Fuller, R., 1975, Nature of the determinant responsible for the adhesion of lactobacilli to chicken crop epithelial cells, *J. Gen. Microbiol.* **87**:245.

Galinski, J., 1975, *Staphylococcus aureus* strains of phage group II and their possible relation to animal staphylococci, *Zentralbl. Bakteriol. Hyg. Abt. I, Orig. A* **233**:288.

Gloor, M., Kionke, M., and Friederich, H. C., 1975a, Biochemical and physiological parameters on the skin surface of healthy test persons, *Arch. Dermatol. Res.* **252**:317.

Gloor, M., Kumpel, D., and Friederich, H. C., 1975b, Predisposing factors on the surface of the skin in persons with pityriasis versicolor, *Arch. Dermatol. Res.* **254**:281.

Gloor, M., Marckardt, V., and Friederich, H. C., 1975c, Biochemical and physiological particularities on the skin surface of diabetics, *Arch. Dermatol. Res.* **253**:185.

Gloor, M., Weigel, H. J., and Friederich, H. C., 1975d, Predisposing factors of the skin surface in persons with impetigo contagiosa, *Arch. Dermatol. Res.* **254**:95.

Gloor, M., Baumann, C., and Friederich, H. C., 1976a, Biochemical and physiological parameters on the healthy skin surface of patients with common warts, *Dermatologica* **152**:152.

Gloor, M., Geilhof, A., Ronneberger, G., and Friederich, H. C., 1976b, Biochemical and physiological parameters on the healthy skin surface of persons with candidal intertrigo and of persons with tinea cruris, *Arch. Dermatol. Res.* **257**:203.

Greene, R. S., Downing, D. T., Pochi, P. E., and Strauss, J. S., 1970, Anatomical variation in the amount and composition of human skin surface lipids, *J. Invest. Dermatol.* **54**:240.

Haavik, H., 1976, Possible functions of peptide antibiotics during growth of producer organisms: bacitracin and metal (II) ion transport, *Acta Pathol. Microbiol. Scand. B.* **84**:117.

Halvorson, H., 1972, Utilization of single L amino acids as role source of carbon and nitrogen by bacteria, *Can. J. Microbiol.* **18**:1647.

Heckels, J. E., Blackett, B., Everson, J. S., and Ward, M. E., 1976, The influence of surface charge on the attachment of *Neisseria gonorrhoeae* to human cells, *J. Gen. Microbiol.* **96**:359.

Henning, D. R., Griffin, T. B., and Maibach, H. I., 1972, Studies on changes in skin surface bacteria in induced miliaria and associated hypohidrosis, *Acta Dermatovener. (Stockh.)* **52**:371.

Herrman, W. P., and Habbig, J., 1976, Immunological demonstration of multiple esterases in human eccrine sweat, *Br. J. Dermatol.* **95**:67.

Hoeksma, A., and Winkler, K. C., 1963, The normal flora of the nose in twins, *Acta Leidensia* **32**:124.

Holland, K. T., Roberts, C. D., Cunliffe, W. J., and Williams, M., 1974, A technique for sampling microorganisms from the pilosebaceous ducts, *J. Appl. Bacteriol.* **37**:289.

Holt, R. J., 1969, Studies on the microflora of the normal human skin, Ph.D. Thesis, Council for National Academic Awards, London.

Jarvis, B., Reynolds, A. J., Rhodes, A. C., and Armstrong, M., 1974, A survey of microbiological contamination in cosmetics and toiletries in the UK (1971), *J. Soc. Cos. Chem.* **25**:563.

Jenkinson, D. McE., Mabon, R. M., and Manson, W., 1974, The effect of temperature and humidity on the losses of nitrogenous protein from the skin of Ayrshire cattle, *Res. Vet. Sci.* **17**:75.

Johnson, J. L., and Cummins, C. S., 1972, Cell wall composition and deoxyribonucleic acid similarities among the anaerobic coryneforms: classical propionibacteria, and strains of *Arachnia propionica, J. Bacteriol.* **109**:1037.

Jong, E. C., Ko, H. L., and Pulverer, G., 1975, Studies on bacteriophages of *Propionibacterium acnes, Med. Microbiol. Immunol.* **161**:263.

Juni, E., 1972, Interspecies transformation of *Acinetobacter:* genetic evidence for a ubiquitous genus, *J. Bacteriol.* **112**:917.

Kalakoutskii, L. V., and Agre, N. S., 1976, Comparative aspects of development and differentiation in Actinomycetes, *Bacteriol. Rev.* **40**:469.

Kaplan, E. L., and Wannamaker, L. W., 1976, Suppression of the antistreptolysin O response by cholesterol and by lipid extracts of rabbit skin, *J. Exp. Med.* **144**:754.

Kasprowicz, A., Heczko, P. B., and Kucharczyk, J., 1974, Proba Klasyfikacji Maczogowcow Skornych, *Medycyna Doswiadczalora I Mikrobiollogia* **26**:267.

Keddie, R. M., and Cure, G. L., 1977, The cell wall composition and distribution of free mycolic acids in named strains of coryneform bacteria and in isolates from natural sources, *J. Appl. Bacteriol.* **42**:229.

Kellum, R. E., 1976, Acne vulgaris. Studies in pathogenesis. Suppression of non-specific esterases, *Cutis* **17**:510.

Kellum, R. E., and Strangfeld, K., 1972, Acne vulgaris. Studies in pathogenesis. Fatty acids of human surface triglycerides from patients with and without acne, *J. Invest. Dermatol.* **58**:315.

Kligman, A. M., 1965, The bacteriology of normal skin, in: *Skin Bacteria and Their Role in Infection* (H. I. Maibach and G. Hildick-Smith, eds.), pp. 13–31, McGraw-Hill, New York.

Kligman, A. M., McGinley, K. J., and Leyden, J. J., 1976, The nature of dandruff, *J. Soc. Cos. Chem.* **27**:111.

Kloos, W. E., and Musselwhite, M. S., 1975, Distribution and persistance of *Staphylococcus* and *Micrococcus* species and other aerobic bacteria on human skin, *Appl. Microbiol.* **30**:381.

Kloos, W. E., and Schleifer, K. H., 1975a, Simplified scheme for routine identification of human *Staphylococcus* species, *J. Clin. Microbiol.* **1**:82.

Kloos, W. E., and Schleifer, K. H., 1975b, Isolation and characterization of Staphylococci from human skin. II. Descriptions of four new species: *Staphylococcus warneri, Staphylococcus capitis, Staphylococcus hominis,* and *Staphylococcus simulans, Int. J. System. Bacteriol.* **25**:62.

Kloos, W. E., Tornabene, T. G., and Schleifer, K. H., 1974, Isolation and characterization of micrococci from human skin, including two new species: *Micrococcus lylae* and *Micrococcus kristinae, Int. J. System. Bacteriol.* **24**:97.

Kloos, W. E., Schleifer, K. H., and Noble, W. C., 1976a, Estimation of character parameters in coagulase negative *Staphylococcus* species, in: *Staphylococci and Staphylococcal Disease,* pp. 23–41, *Zentralbl. Bakteriol. Parasitenk. I. k. Abt.* 1, Supplementum 5.

Kloos, W. E., Schleifer, K. H., and Smith, R. F., 1976b, Characterization of *Staphylococcus sciuri* sp. nov. and its subspecies, *Int. J. System. Bacteriol.* **26**:22.

Kocur, M., and Boháček, J., 1974, DNA base composition and the classification of non-pigmented micrococci, *Microbios* **10A**:31.

Kocur, M., and Schleifer, K. H., 1975, Taxonomic status of *Micrococcus agilis,* Ali Cohen 1889, *Int. J. System. Bacteriol.* **25**:294.

Knight, A. G., 1973, Culture of dermatophytes upon stratum corneum, *J. Invest. Dermatol.* **59**:427.

Knight, A. G., 1976, The effect of temperature and humidity on the growth of *Trichophyton mentagrophytes* spores on human stratum corneum *in vitro, Clin. Exp. Dermatol.* **1**:159.

Krynski, S., Becla, E., Sarnet, A., Nowicki, B., and Wielgosz, A., 1975, Usefulness of phage typing for identification of *Staphylococcus epidermidis* and *Micrococcus* strains, *Arch. Immunol. Ther. Exp.* **23**:645.

Lacey, R. W., 1971, High frequency transfer to neomycin resistance between naturally occurring strains of *Staphylococcus aureus, J. Med. Microbiol.* **4**:73.

Lacey, R. W., 1975, Antibiotic resistance plasmids of *Staphylococcus aureus* and their clinical importance, *Bacteriol. Rev.* **39**:1.

Lantrop, H., 1974, Genus *Acinetobacter,* in: *Bergey's Manual of Determinative Bacteriology* (R. E. Buchanan and N. E. Gibbons, eds.), p. 436, Williams and Wilkins, Baltimore.

Lawrence, J. C., and Lilly, H. A., 1972, A quantitative method for investigating the bacteriology of skin: its application to burns, *Br. J. Exp. Pathol.* **53**:550.

Lechevalier, M. P., and Lechevalier, H., 1970, Chemical composition as a criterion in the classification of aerobic actinomycetes, *Int. J. System. Bacteriol.* **20**:435.

Lee, L. D., and Baden, H. P., 1975, Chemistry and composition of the keratins, *Int. J. Dermatol.* **14**:161.

Leyden, J. J., 1976, Antibiotic resistant acne, *Cutis* **17**:593.

Leyden, J. J., and Kligman, A. M., 1976, Acne vulgaris: New concepts in pathogenesis and treatment, *Drugs* **12**:292.

Leyden, J. J., Marples, R. R., Mills, O. H., and Kligman, A. M., 1973, Gram negative folliculitis—A complication of antibiotic therapy in acne vulgaris, *Br. J. Dermatol.* **88**:533.

Leyden, J. J., McGinley, K. J., Mills, O. H., and Kligman, A. M., 1975, Propionibacterium levels in patients with and without acne vulgaris, *J. Invest. Dermatol.* **65**:382.

Lilly, D. M., and Stillwell, R. H., 1965, Probiotics: Growth promoting factors produced by microorganisms, *Science* **147**:747.

Loebl, E. C., Marvin, J. A., Heck, E. L., Curreri, P. W., and Baxter, C. R., 1974, The use of quantitative biopsy cultures in bacteriologic monitoring of burn patients, *J. Surg. Res.* **16**:1.

Lodder, J., 1970, *The Yeasts,* 2nd ed., North-Holland, Amsterdam.

Mardh, P. A., and Westrom, L., 1976, Adherence of bacteria to vaginal epithelial cells, *Infect. Immun.* **13**:661.

Marks, R., and Dawber, R. P. R., 1972, *In situ* microbiology of the stratum corneum, *Arch. Dermatol.* **105**:216.

Marples, M. J., 1965, *The Ecology of the Human Skin,* Charles C. Thomas, Springfield, Ill.

Marples, R. R., 1969, Diphtheroids of normal human skin, *Br. J. Dermatol.,* Suppl. 1, **81**:47.

Marples, R. R., and Kligman, A. M., 1971, Ecological effects of oral antibiotics on the microflora of human skin, *Arch. Dermatol.* **103**:148.

Marples, R. R., Downing, D. T., and Kligman, A. M., 1971, Control of fatty acids in human skin surface lipids by *Corynebacterium acnes, J. Invest. Dermatol.* **56**:127.

Marsh, P. D., and Selwyn, S., 1977a, Studies on antagonism between human skin bacteria, *J. Med. Microbiol.* **10**:161.

Marsh, P. D., and Selwyn, S., 1977b, Continuous culture studies of interactions among skin commensal bacteria, *J. Med. Microbiol.* **10**:261.

Massey, A., Mowbray, J. F., and Noble, W. C., 1978, Complement activation by *Corynebacterium acnes, Br. J. Dermatol.,* in press.

Matta, M., 1974, Carriage of *Corynebacterium acnes* in school children in relation to age and race, *Br. J. Dermatol.* **91**:557.

McAllister, T. A., Givan, J., Black, A., Turner, M. J., Kerr, M. M., and Hutchison, J. H., 1974, The natural history of bacterial colonization of the newborn in a maternity hospital, *Scot. Med. J.* **19**:119.

McBride, M. E., Duncan, W. C., Bodey, G. P., and McBride, C. M., 1976, Microbial skin flora of selected cancer patients and hospital personnel, *J. Clin. Microbiol.* **3**:14.

McGinley, K. J., Leyden, J. J., Marples, R. R., and Kligman, A. M., 1975, Quantitative microbiology of the scalp in non-dandruff, dandruff and seborrhoeic dermatitis, *J. Invest. Dermatol.* **64**:401.

McLaren, A. D. and Skujins, J., 1968, The physical environment of organisms in soil, in: *The Ecology of Soil Bacteria,* pp. 3–24, Liverpool University Press, Liverpool.

Milyani, R. M. M., 1976, Studies on interactions of human skin microorganisms on solid surface, Ph.D. Thesis, University of London.

Milyani, R. M. M., and Selwyn, S., 1978, Quantitative studies on competitive activities of skin bacteria growing on solid media, *J. Med. Microbiol.,* in press.

Montes, L. F., and Wilborn, W. H., 1969, Location of bacterial skin flora, *Br. J. Dermatol.* Suppl. 1 **81**:23.

Moore, W. E. C., and Holdeman, L. V., 1974, Family Propionibacteriaceae, in: *Bergey's*

Manual of Determinative Bacteriology (R. E. Buchanan and N. E. Gibbons, eds.), p. 633, Williams and Wilkins, Baltimore.

Morello, A. M., and Downing, D. T., 1976, Trans-unsaturated fatty acids in human skin surface lipids, *J. Invest. Dermatol.* **67**:270.

Mukasa, H., and Slade, H. D., 1974, Mechanism of adherence of *Streptococcus mutans* to smooth surfaces, *Infect. Immun.* **9**:419.

Murphy, C. T., 1975, Nutrient materials and the growth of bacteria on human skin, *Trans. St. John's Hosp. Dermatol. Soc.* **61**:51.

Mustakallio, K. K., Salo, O. P., Kiistala, R., and Kiistala, U., 1967, Counting the number of aerobic bacteria in full thickness human epidermis separated by suction, *Acta Pathol. Microbiol. Scand.* **69**:477.

Neill, J. M., Gaspari, E. L., Mosley, R. A., and Sugg, J. Y., 1931, Loss of immune substances from the body. III. Diphtheria antitoxin in human sweat, *J. Immunol.* **21**:101.

Noble, W. C., 1974, Carriage of *Staphylococcus aureus* and beta haemolytic streptococci in relation to race, *Acta Dermatovener (Stockh.)* **54**:403.

Noble, W. C., 1975, Skin as a microhabitat, *Postgrad. Med. J.* **51**:151.

Noble, W. C., 1976, Dispersal of skin microorganisms, *Br. J. Dermatol.* **93**:477.

Noble, W. C., 1977a, Dispersal of organisms from human skin, *Cosmetics Toiletries* **92**:38.

Noble, W. C., 1977b, Variation in the prevalence of antibiotic resistance of *Staphylococcus aureus* from human skin and nares, *J. Gen. Microbiol.* **98**:125.

Noble, W. C., and Somerville, D. A., 1974, *Microbiology of Human Skin,* Saunders, London.

Noble, W. C., Valkenburg, H. A., and Wolters, C. H. L., 1967, Carriage of *Staphylococcus aureus* in random samples of a normal population, *J. Hyg. (Camb.)* **65**:567.

Notermans, S., and Kampelmacher, E. H., 1975, Further studies on the attachment of bacteria to skin, *Br. Poultry Sci.* **16**:487.

Page, C. O., and Remington, J. S., 1967, Immunologic studies in normal human sweat, *J. Lab. Clin. Med.* **69**:634.

Pagel, J. E., and Seyfried, P. L., 1976, Numerical taxonomy of aquatic *Acinetobacter* isolates, *J. Gen. Microbiol.* **95**:220.

Peters, G., and Pulverer, G., 1975, Bacteriophages from micrococci, *Zentralbl. Bakteriol. Hyg. Abt. I. Orig. A.* **233**:221.

Pitcher, D. G., 1976, Arabinose with LL-Diaminopimelic acid in the cell wall of an aerobic coryneform organism isolated from human skin, *J. Gen. Microbiol.* **94**:225.

Pitcher, D. G., 1977, Rapid identification of cell wall components as a guide to the classification of aerobic coryneform bacteria from human skin, *J. Med. Microbiol.* **10**:439.

Potter, B. S., Burgoon, C. F., and Johnson, W. C., 1973, *Pityrosporum* folliculitis. Report of 7 cases and review of the *Pityrosporum* organism relative to cutaneous disease, *Arch. Dermatol.* **107**:388.

Price, P. B., 1938, The bacteriology of normal skin. A new quantitative test applied to a study of the bacterial flora and the disinfectant action of mechanical cleansing, *J. Infect. Dis.* **63**:301.

Puhvel, S. M., Reisner, R. M., and Sakamoto, M., 1975, Analysis of lipid composition of isolated human sebaceous gland homogenates after incubation with cutaneous bacteria. Thin layer chromatography, *J. Invest. Dermatol.* **64**:406.

Puleo, J. R., Fields, N. D., Bergstrom, S. L., Oxborrow, G. S., Stabekis, P. D., and Koukol, R. C., 1977, Microbiological profiles of the Viking spacecraft, *Appl. Environ. Microbiol.* **33**:379.

Pulverer, G., and Ko, H. L., 1973, Fermentative and serological studies on *Propionibacterium acnes, Appl. Microbiol.* **25**:222.

Pulverer, G., Pillich, J., and Klein, A., 1975, New bacteriophages of *Staphylococcus epidermidis, J. Infect. Dis.* **132**:524.

Roberts, C. D., 1975, The role of bacteria in acne vulgaris, Ph.D. Thesis, University of Leeds.

Rogosa, M., Cummins, C. S., Lelliott, R. A., and Keddie, R. M., 1974, Coryneform group of bacteria, in: *Bergey's Manual of Determinative Bacteriology,* 8th ed. (R. E. Buchanan and N. E. Gibbons, eds.), Williams and Wilkins, Baltimore.

Rook, A., 1977, Common baldness and alopecia areata, in: *Recent Advances in Dermatology* (A. Rook, ed.), pp. 223–247, Churchill-Livingstone, Edinburgh.

Saino, Y., Eda, J., Nagoya, T., Yoshimura, Y., Yamaguchi, M., and Kobayashi, F., 1976, Anaerobic coryneforms isolated from human bone marrow and skin: chemical, biochemical and serological studies and some of their biological activities, *Jpn. J. Microbiol.* **20**:17.

Sanderson, P. J., 1977, Infection of the foot with *Peptococcus magnus, J. Clin. Pathol.* **30**:266.

Sarkany, I., and Gaylarde, C. C., 1967, Skin flora of the newborn, *Lancet* **1**:589.

Sarkany, I., Taplin, D., and Blank, H., 1961, The etiology and treatment of erythrasma, *J. Invest. Dermatol.* **37**:283.

Savin, J. A., 1974, Bacterial infections in diabetes mellitus, *Br. J. Dermatol.* **91**:481.

Savin, J. A., and Noble, W. C., 1977, Opportunism and skin infections, in: *Recent Advances in Dermatology* (A. Rook, ed.), pp. 1–29, Churchill-Livingstone, Edinburgh.

Schleifer, K.H., and Kandler, O., 1972, Peptidoglycan types of bacterial cell walls and their taxonomic implications, *Bacteriol. Rev.* **36**:407.

Schleifer, K. H., and Kloos, W. E., 1975, Isolation and characterization of Staphylococci from human skin. I. Amended descriptions of *Staphylococcus epidermidis* and *Staphylococcus saprophyticus* and descriptions of three new species: *Staphylococcus cohnii, Staphylococcus haemolyticus* and *Staphylococcus xylosus, Int. J. System. Bacteriol.* **25**:50.

Schleifer, K. H., and Kocur, M., 1973, Classification of Staphylococci based on chemical and biochemical properties, *Arch. Mikrobiol.* **93**:65.

Selwyn, S., 1975, Natural antibiosis among skin bacteria as a primary defence against infection, *Br. J. Dermatol.* **93**:487.

Selwyn, S., and Ellis, H., 1972, Skin bacteria and skin disinfection reconsidered, *Br. Med. J.* **1**:136.

Selwyn, S., Marsh, P. D., and Sethna, T. N., 1976, In vitro and in vivo studies on antibiotics from skin micrococcaceae, in: *Chemotherapy,* vol. 5, p. 391 (J. D. Williams and A. M. Geddes, eds.), Plenum Press, New York and London.

Sharpe, M. E., Law, B. A., and Phillips, B. A., 1976, Coryneform bacteria producing methane-thiol, *J. Gen. Microbiol.* **94**:430.

Sharpe, M. E., Law, B. A., Phillips, B. A., and Pitcher, D. G., 1977, Methane-thiol production by coryneform bacteria: strains from dairy and human skin sources and *Brevibacterium linens, J. Gen. Microbiol.* **101**:345.

Singh, G., 1974, Heat, humidity and pyodermas, *Dermatologica* **147**:342.

Smith, R. F., 1969, Characterisation of human cutaneous lipophilic diptheroids, *J. Gen. Microbiol.* **55**:433.

Smith, R. F., 1971, Lactic acid utilization by the cutaneous *Micrococcaceae, Appl. Microbiol.* **21**:777.

Smith, J. M. B., and Marples, M. J., 1964, A natural reservoir of penicillin resistant strains of *Staphylococcus aureus, Nature (London)* **201**:844.

Snell, J. J. S., 1973, *The Distribution and Identification of Nonfermenting Bacteria,* Public Health Laboratory Service, Monograph 4, London.

Somerville, D. A., 1972, A quantitative study of erythrasma lesions, *Br. J. Dermatol.* **87**:130.

Somerville, D. A., 1973, A taxonomic scheme for aerobic diphtheroids from human skin, *J. Med. Microbiol.* **6**:215.

Somerville, D. A., and Lancaster-Smith, M., 1973, The aerobic cutaneous microflora of diabetic subjects, *Br. J. Dermatol.* **89**:395.

Somerville, D. A., and Murphy, C. T., 1973, Quantitation of *Corynebacterium acnes* on healthy human skin, *J. Invest. Dermatol.* **60**:231.

Somerville, D. A., and Noble, W. C., 1970, A note on the Gram negative bacilli of human skin, *Eur. J. Clin. Biol. Res.* **15**:669.

Somerville-Millar, D. A., and Noble, W. C., 1974, Resident and transient bacteria of the skin, *J. Cutan. Pathol.* **1**:260.

Stanneck, J. L., and Roberts, G. D., 1974, Simplified approach to identification of aerobic actinomycetes by thin-layer chromatography, *Appl. Microbiol.* **28**:226.

Stratford, B., Gallus, A. S., Matthieson, A. H., and Dixson, S., 1968, Alteration of superficial bacterial flora in severely ill patients, *Lancet* **1**:68.

Strauss, J. S., Pochi, P. E., and Downing, D. T., 1976, The role of skin lipids in acne, *Cutis* **17**:485.

Stringer, M. F., and Marples, R. R., 1976, Ultrasonic methods for sampling human skin microorganisms, *Br. J. Dermatol.* **94**:551.

Subcommittee on Taxonomy of Staphylococci and Micrococci, 1965, *Int. Bull. Bacteriol. Nomencl. Taxon.* **15**:107.

Tagg, J. R., Dajani, A. S., and Wannamaker, L. W., 1976, Bacteriocins of gram positive bacteria, *Bacteriol. Rev.* **40**:722.

Talbot, H. W., and Parisi, J. T., 1976, Phage typing of *Staphylococcus epidermidis, J. Clin. Microbiol.* **3**:519.

Valtonen, M. V., Valtonen, V. V., Salo, O. P., and Makela, P. H., 1976, The effect of long term tetracycline treatment for acne vulgaris on the occurrence of R factors in the intestinal flora of man, *Br. J. Dermatol.* **95**:311.

Voss, J. G., 1970, Differentiation of two groups of *Corynebacterium acnes, J. Bacteriol.* **101**:392.

Voss, J. G., 1974, Acne vulgaris and free fatty acids. A review and criticism, *Arch. Dermatol.* **109**:894.

Williamson, P., and Kligman, A. M., 1965, A new method for the quantitative investigation of cutaneous bacteria, *J. Invest. Dermatol.* **45**:498.

Wiseman, G. M., 1975, The hemolysins of *Staphylococcus aureus, Bacteriol. Rev.* **39**:317.

Wolff, H. H., and Plewig, G., 1976, Ultrastruktur der Mikroflora in Follikeln und Komedonen, *Hautarzt* **27**:432.

Youssef, N., Wyborn, C. H. E., Holt, G., Noble, W. C., and Clayton, Y. M., 1978, Antibiotic production by dermatophyte fungi, *J. Gen. Microbiol.* **105**:105.

Index